Immersed tunnel techniques

Immersed tunnel techniques

Proceedings of the conference organized by the Institution of Civil Engineers and held in Manchester on 11—13 April 1989

Thomas Telford, London

Conference organized by the Institution of Civil Engineers

Organizing Committee: C. R. Ford (Chairman), G. W. Davies, J. H. Sargent, M. C. F. Thorn and P. Witham

British Library Cataloguing in Publication Data
Immersed tunnel techniques.
　1. Tunnels & tunnelling
　I. Institution of Civil Engineers
　624.1'93

ISBN 0-7277-1512-7

First published 1990

© The Institution of Civil Engineers, 1989, 1990, unless otherwise stated.

All rights, including translation, reserved. Except for fair copying no part of this publication may be reproduced, stored in a retrieval system, or transmitted in any form or by any means electronic, mechanical, photocopying, recording or otherwise, without the prior written permission of the publisher. Requests should be directed to the Publications Manager, Thomas Telford Ltd, Telford House, 1 Heron Quay, London E14 9XF.

Papers or other contributions and the statements made or opinions expressed therein are published on the understanding that the author of the contribution is solely responsible for the opinions expressed in it and that its publication does not necessarily imply that such statements and/or opinions are or reflect the views or opinions of the ICE Council or ICE committees.

Published for the Institution of Civil Engineers by Thomas Telford Ltd, Telford House, 1 Heron Quay, London E14 9XF.

Printed and bound in Great Britain by Billing and Sons Ltd, Worcester.

Contents

Geotechnical aspects

1. Geotechnical aspects of the Zeeburgertunnel in Amsterdam. J KRUIZINGA and G. L. TAN — 1
2. Construction of immersed tunnels — the land connection. G. A. MUNFAKH and B. SCHMIDT — 13
3. Planning the third harbour tunnel in Boston, Massachusetts, USA. W. C. GRANTZ and A. R. LANCELLOTTI — 25
5. Sedimentation and stability of dredged trenches. M. F. C. THORN — 35
6. Trench siltation: origin, consequences and how to cope with it. J. H. VOLBEDA and G. L. M. van der SCHRIECK — 49

Discussion — 57

Planning

7. Bosporus railroad tunnel crossing design alternatives. V. TANAL, W. C. GRANTZ and L. W. ABRAMSON — 63
8. Immersed steel railway tunnel across the eastern channel of the Great Belt — Denmark. A. GURSOY — 77
9. Planning and design fo the Guangzhou immersed tube road and railway tunnel. AU SHIU KIN, J. D. C. OSORIO, JIN FENG and CHEN SHAO ZHANG — 87
10. The Sydney Harbour tunnel — influence of local conditions on the route and on the marine and land tunnels. N. SAITO and A. M. NEILSON — 97
11. The Sydney Harbour tunnel — evaluation of form construction for an immersed tunnel. M. W. MORRIS, D. G. MORTON and L. GOMES — 111
12. The Conwy tunnel — scheme development and advance works. G. W. DAVIES, G. CRAMP and H. KAMP NIELSON — 125
13. Development of immersed tunnels in Canada and Hong Kong. P. HALL — 145
14. Immersed tunnel techniques in a typical delta country. H. J. C. OUD — 161
15. Privately financed immersed tube tunnel projects. V. L. MOLENAAR — 175
16. Comparative merits of steel and concrete forms of tunnel. D. R. CULVERWELL — 185

Discussion — 199

Design

17. Shell composite construction for shallow draft immersed tube tunnels. M. J. TOMLINSON, A. TOMLINSON, M. Ll. CHAPMAN, A. D. JEFFERSON and H. D. WRIGHT — 209

18. Concrete immersed tunnels: the design process. L. C. F. INGERSLEV — 221

19. A computerized concrete hardening control system and its application in tunnel construction. W. C. HORDEN, E. MAATJES and A. C. J. BERLAGE — 235

20. Electrical and mechanical aspects relating to the civil design of immersed tube tunnels. J. F. L. LOWNDES and C. R. WEEKS — 249

21. Impact of the development of tunnel service systems on the planning and design of the immersed tunnel for the Sydney Harbour crossing. A. G. BENDELIUS and K. R. MEEKS — 263

22. The Conwy tunnel — detailed design. P. A. STONE, R. C. LUNNISS and S. J. SHAH — 277

Discussion — 301

Construction

23. The design and construction of steel immersed tube tunnels: an American technology. N. A. MUNFAH and Y. M. TARHAN — 317

24. New approaches to immersed tunnel construction. F. J. HANSEN and J. F. HANSEN — 331

26. Practical accuracies of construction of immersed tunnels — a case study: The Fort McHenry Tunnel, Baltimore, USA. W. C. GRANTZ and A. R. LANCELLOTTI — 345

27. The Conwy Estuary Tunnel — construction. J. F. McFADZEAN and D. PHILLIPS — 357

28. Fabrication, outfitting and construction techniques proposed for the Storebaelt immersed railway tunnel. A. J. BAST III — 373

29. Eastern Harbour crossing, Hong Kong. Y. MATSUMOTO, D. E. OAKERVEE, A. I. THOMSON and D. G. MORTON — 387

30. The immersed part of the 'Spoortunnel': a railway tunnel in Rotterdam. E. H. M. GROOT and G. M. WOLSINK — 397

Discussion — 413

1. Geotechnical aspects of the Zeeburgertunnel in Amsterdam

Ir J. KRUIZINGA, Delft Geotechnics, Delft, and Ir G. L. TAN, Rijkswaterstaat, Locks, Weirs and Tunnels Division, Utrecht, The Netherlands

SYNOPSIS
In order to design the foundations of the Zeeburgertunnel near Amsterdam, the Ministry of Public Works has carried out an extensive soil investigation programme in cooperation with Delft Geotechnics. Due to the extremely difficult subsoil conditions various pile types and lengths have been used, of which the design is based on CPT's and pile tests.

1.   INTRODUCTION AND GEOMETRIC REQUIREMENTS

The Zeeburgertunnel under the Buiten-IJ is a part of the A10, the Motorway around Amsterdam. Together with the bridge over the IJ-meer and the bridge over the Amsterdam - Rhine canal the tunnel forms the Eastern crossing of Amsterdam, the so-called Zeeburger Crossing. Beside the existing Coentunnel, IJ-tunnel and the Schellingwoude bridges, the Zeeburger Crossing will be the fourth river crossing in and around Amsterdam (Figure 1).
Driving from the South the driver crosses over the bridges at a level of NAP + 14.0 m and is back at ground level on the isle of Zeeburg. At this point he begins to enter the Southern approach of the tunnel. Figure 2 shows the longitudinal section of the tunnel. The deepest level of the tunnel is at NAP - 13.10 m. The maximum gradient in the tunnel is as usual in the Netherlands, 4.5 %. The Northern depressed approach ramp is situated in the water but outside the 350 m wide shipping canal. So the tunnel can be limited to a length of 546 m. The total length including the approaches is 946 m. Northwards, a sand dam in the water connects the approach and the Northern bank. Figure 3 shows the cross section of the tunnel. The tunnel has two tubes, each with three lanes. The width of the lanes are 3.30 m each and the total width of the tunnel is 28.20 m, included the central gallery between both traffic tubes.
The central gallery is divided into two floors, the first floor used as service and emergency gallery and the second floor as cable duct. On the exit sides, the approaches are provided with a crawl lane.

GEOTECHNICAL ASPECTS

The execution of the tunnel started in September 1984 and will be completed in Spring 1990. The tunnel shall be opened for traffic in September 1990 on which the Motorway around Amsterdam is completed.

Figure 1

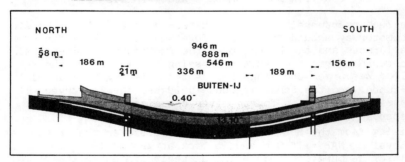

Figure 2   Longitudinal section

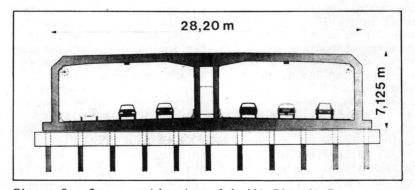

Figure 3   Cross section tunnel built "in situ"

2. SOIL CONDITIONS

General

During the design phase of a tunnel, it is essential to have a thorough knowledge of the subsoil in terms of type of soil, layering, soil properties and groundwater regimes.

Following common practice in the Netherlands, the soil conditions are assessed from a 'tailor made soil investigation programme, comprising mainly Cone Penetration Tests (CPT's) with a few additional borings with undisturbed sampling for laboratory testing.
The programme for the Zeeburgertunnel consisted of 17 CPT's and 4 borings in the axis of the tunnel, 26 CPT's outside the axis and installation of standpipes at the northern and southern approach; the programme was carried out by Delft Geotechnics.

Layering

From this investigation an **important discontunuity** in the normal Amsterdam soil profile has been encountered.

Normal Amsterdam soil profile (figure 4) shows a five-layer system starting with a soft to very soft toplayer consisting of peat, clay and silt (down to NAP - 12 m), follows by a loose to medium dense fine sand layer with interbedded silt layers (down to NAP - 17 m; then a firm sandlayer of 6 m (North) to 10 m (South) thickness is encountered in which high cone resistances are measured, this so-called second Amsterdam sandlayer is the bearing layer for most piled foundations and is overlying a more than 10 m thick, stiff, marine clay layer (Down to NAP - 38 m) with an average cone resistance of 4 MN/m^2. Below this clay layer a firm glacial deposit consisting of middle to coarse sand with occasionally an interbedded boulder clay layer is encounterd. With respect to the free waterlevel of NAP - 0.2 m in the channel, in the sandlayers a waterunderpressure of approximately 2 m has been measured.

The **discontinuity** over approximately 225 m in the axis of the tunnel shows the absence of the normal bearing layer (second Amsterdam sandlayer). Archive information revealed sand dredging activities in the past; the under water excavation was backfilled with a mixture of soft material consisting of clay, silt and sand (sand content increasing with depth) with very low cone resistances (figure 5 shows a picture of a CPT in this area).

Figure 6 shows the geotechnical profile in the axis of the tunnel with the discontinuity.

GEOTECHNICAL ASPECTS

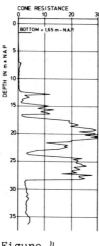

Figure 4

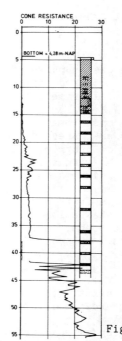

Figure 5

Table 1. Normal Amsterdam soil profile and soil properties

layer	depth m - NAP	soil description	γ_s (kN/m³)	w %	ϕ' [1] °	c' kN/m²	c_u kN/m²	c [2]	k (m/s)
I	down to 3-4	clay and/or mud	12-13	130	20	0.2	2-5	5-7	$2*10^{-9}$
	to 6-7	peat	10-11	300-500	25-30	-	4-6	3-4	$5*10^{-8}$
	to 8.5-9	clay, sandy	17-18		25-26	-	4-6	20-40	$2*10^{-7}$
	to 12-12.5	clay, sometimes overlying 0.5 m peat	13-15		20-23	0.3	6-10	4-7	10^{-8}
II	to 17.5	sand, with silty layers	19-20		30-32	0	0		
III	to 22 (north) to 27 (south)	sand	20		35-37	0	0		
IV	to 38	clay, stiff	16-18	50	30-34	3-10	50÷150	15-20	$2*10^{-9}$
V	deeper than 38	sand, with boulder clay layer	20		> 35	0	0		

Discontinuity:

layer	depth m - NAP	soil description	γ_s (kN/m³)	w %	ϕ' °	c' kN/m²	c_u kN/m²	c	k (m/s)
back fill	to 10	mud and/or clay	12-14	160	20	0-1	0 ÷ 6 to 8	7	$5*10^{-10}$
	to 20-22	clay, slightly sandy	16-17	70	20-22	2-4	8 + 20	20-30	10^{-8} to 10^{-9}
	to 25-27	clay, very sandy	17-19	40	26-28	4	> 30	30-50	$2*10^{-8}$
IV	to 38	as above							
V	> 38	as above							

[1]) determined by Dutch cell test
[2]) Dutch procedure compressibility test

Further detailed CPT-investigations to the extent of the discontinuity revealed that the normal soil profile was encountered again approximately 75 m eastwards of the axis of the tunnel. It also revealed that the transition to the normal soil profile in the axis is rather steep: figure 7 shows a plot of the bottom lines of the discontinuity; as a result of an additional CPT-programme. The soil properties have been collected in table 1.

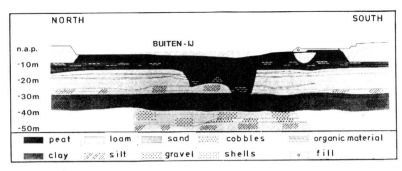

Figure 6

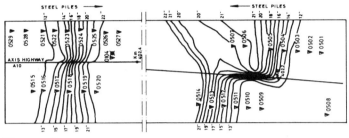

Figure 7

3. FOUNDATION DESIGN CONSIDERATIONS

During the design phase three possibilities are considered:
- immersed tunnel on a spread foundation
- immersed tunnel on a piled foundation
- construction of a tunnel by cut-and-cover (dry docks).

Immersed tunnel on a spread foundation

The soil material of the discontinuity is very compressible and also prone to liquefaction. Total or partly refilling by clean sand would be necessary. This operation with sand dredged from the sea is highly expensive and, moreover, settlements might still occur.

Immersed tunnel on a piled foundation

Normally displacement type of piles are used, preformed (precast concrete piles driven into the ground) or formed in situ. However, the discontinuity required a pile length of over 45 m; therefore steel piles are most suitable. Important phenomena to be expected during driving of large displacement piles are upheaval effects, due to the constant-volume behaviour (initially) of the stiff Eem clay layer. Therefore the spacing of the piles must be sufficient (more than 10 diameters) or very small displacement type of piles have to be used. Like the subway tunnel in Rotterdam, the piles have to be provided with inflatable heads to ensure a proper contact with the immersed tunnel.

GEOTECHNICAL ASPECTS

Construction by cut-and-cover

In this case the tunnel will be build in so called dry docks formed by sheetpiles and under-water-concrete on tension piles. At the discontinuity a special type of piles, like discussed before, have to be used.

Evaluating the three possibilities, it became clear that a location for a dry dock to construct four immersed tunnel sections all together is not available in the direct surroundings. To the East the riverbottom is shallow, which means high dredging costs. To the West an existing dry dock is available but the bottom of shipping locks to be passed (Oranje Sluizen) is too shallow. Construction of a new dry dock using deep well points for dewatering is not permitted due to expected settlements of building in the surrounding area.

Due to the high costs of the first two foundation types it was concluded to build the tunnel by cut-and-cover using dry docks; for reasons of navigation two execution phases are required, allowing for using the sheetpiles twice.

During the tender phase, one of the contractors, Van Hattum & Blankevoort B.V., offered an alternative: building three sections, to be immersed, of 112 m each, using one of the dry docks of the tunnel which has to be provided with a gate, to allow for constructing the sections one by one (figure 8). These sections will be immersed on two rows of piles situated under the outer walls.

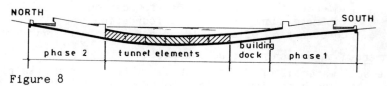

Figure 8

4.   PILE DESIGN RULES

Expressions

The piles to be used for the tunnel are subjected to compression loading in the final stage; however, during construction part of the piles are subjected to tension loading due to uplift forces.
In the Netherlands the ultimate end bearing $Q_{b,ult}$ and the ultimate shaft friction $F_{s,ult}$ of piles are calculated with design rules based on CPT's (in the past test loading results are correlated to CPT's, reference 1).

$$Q_{ult} = Q_{b,ult} + F_{s,ult}$$

$$Q_{ult} = A_b * q_{b,ult} + \sum f_{s,ult} A_{s,i}$$

$$q_{b,ult} = \frac{\frac{I+II}{2} + III}{2} \quad \text{for displacement piles, in}$$
which,

- I = average cone resistance $\bar{q}_c$ from pile base level to a maximum of 3.75 * pile diameter below the pile base
- II = the lowest q_c-value in this area
- III = lowest q_c-values starting from pile base level, taking into account II, and proceeding up to 8 * pile diameter above the pile base level.

The calculation of the shaft friction is based on best loadings under tension by Begemann (reference 2) and correlated to the local frcition values measured by the friction sleeve of the mechanical cone penetration test.

$f_{s_{ult}} = f_{sleeve} * \beta$ ($f_{s_{ult}}$ is limited to 120 kN/m²)

$f_{sleeve} = \frac{1}{50}$ to $\frac{1}{60}$ q_c in n.c. sand (q_c = cone resistance)

$\phantom{f_{sleeve}} = \frac{1}{15}$ to $\frac{1}{30}$ in clay and silt layers

β = pile shape factor

$f_{s_{ult}} = 2$ to 1.7% $q_c * \beta$ (for n.c., fine to medium sand)

$f_{s_{ult}} = a \% q_c$

pile type (reference 2,3)	β	$f_{s_{ult}} = a\, q_c$ a (%)
precast concrete piles		
- with a flat base	0.35	0.6
- with a sharp base	0.4	0.7
steel piles		
- tubular piles	0.35	0.6
- H-shape piles	0.3	0.5
cast-in-situ piles		
- Vibro piles	0.9 - 1	1.5
- MV-piles	0.8 - 0.9	1.4

Reductions

In case of alternating pile loading (tension and compression) a reduction factor has to be applied.

For this project piles will be installed after dredging the trench. This excavation affects the q_c-values; the reduced q_c-values have to be introduced into the pile design expressions:

GEOTECHNICAL ASPECTS

$$q_{c_{corrected}} = q_{c_{original}} * \frac{\sigma i_n}{\sigma i_o} = q_{c,o} * \left(\frac{\sigma_{v_n}' + 2\sigma_{H_n}'}{\sigma_{v_o}' + 2\sigma_{H_o}'} \right)$$

σ_{v_o}' = effective, original vertical stress

σ_{v_n}' = effective, new vertical stress

σ_{H_o}' = effective, original horizontal stress

σ_{H_n}' = effective, new horizontal stress

$\sigma_{H_o}' = K_o \, \sigma_{v_o}' = (1 - \sin\phi') \, \sigma_{v_o}'$ for n.c. soils.

For σ_{H_n}' an assumption has be made: for instance, that part of the original horizontal stress is still present:

$\sigma_{H_n}' = 0.5 \, (K_o \, \sigma_{v_n}' + K_o \, \sigma_{v_o}')$

Pile tests

With respect to deep piling to bridge the discontinuity it was decided to use small displacement piles: steel H-shape piles. To increase the shaft friction, grouting of the shaft in the deep sandlayer was necessary. As there was no experience with such a pile type in a very deep sandlayer (below 40 m) two pile tests have been carried out to determine the pull out capacity and to derive a design expression. The so-called injection piles were installed by Fundamentum near the Southern approach and tested by Delft Geotechnics (see figures 9, 10).
As the soil conditions at the test site (figure 11 shows CPT) differs from the actual site (discontinuity) casings down to NAP - 28 m have been used for both test piles (HE-M-240) to eliminate the friction in this layer. In the deep sand layer grout has been injected along the shaft from NAP - 35 to NAP - 45 m ("short" pile) and from NAP - 35 to NAP - 50 m (long pile); to reduce the friction in the Eem-clay layer along the pile shaft bentonite has been injected down to NAP - 35 m. The piles have been provided with strain transducers at several levels. The test procedure allowed for 5 unloading-loading cycles after each loading stage. The "Short" pile (down to NAP - 45 m) failed at 3600 kN while the test loading for the long pile (down to NAP - 50 m) was stopped at 3900 kN (due to too high steel stresses).

The maximum shaft friction from NAP - 38 to NAP - 45 m and - 50 m varied from 80 kN/m² to approximately 200 kN/m²; these variations are caused by variations in cone resistance.

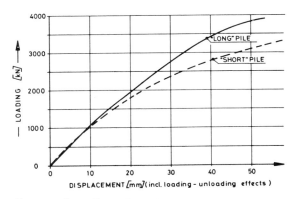

Figure 9   Test loading results

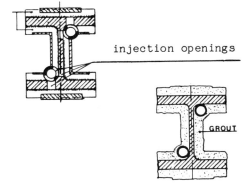

Figure 10   Injection piles

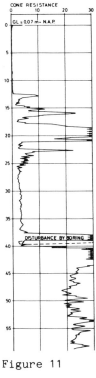

Figure 11
CPT Test pile

After evaluation of the test results, the following design expression has been proposed for the grout injected piles:
$$f_{s_{ult}} = 0.9 \text{ \% } q_c \quad (\beta = 0.5)$$

The allowable pile load has been determined by considering the so-called "point of hesitation" (which is the load giving an additional pile head displacement of 0.2 mm after 4 loading-unloading cylces), a pile toe displacement of 4 mm (= 1.5 % $D_{equiv.}$) and the creep coefficient

$$M_c = \frac{u_2 - u_1}{\log t_2/t_1} = 0.5 \text{ mm}.$$

These criteria resulted into safety factors of 1.45 - 1.79. A safety factor of 1.75 was proposed (corresponding with a pile toe deformation of 4 mm).

GEOTECHNICAL ASPECTS

5. FINAL FOUNDATION DESIGN

During the proces of detailed foundation design the afore mentioned pile design rules were used. This resulted into three different pile systems with varying pile base levels. Figure 12 and 13 show the longitudinal section of the tunnel and dry dock with the different types of pile foundations:

a) <u>The approaches</u>
The depressed approach ramps are founded on precast, pretensioned concrete piles (dimensions 450 x 450 mm^2).
Depending on the stage of execution part of the piles changes from tension piles with a design load of 400 to 600 kN to bearing piles with a design load of 800 kN. The bearing layer of the Southern approach shows a better quality than that of the Northern approach, resulting into pile base levels varying from NAP - 20 to NAP - 28.

b) <u>The dry dock</u>
The dry dock has been partly founded on the same type of tension piles as that of the approaches. In the area of the discontinuity the underwater concrete has been anchored to the grout injected, H-shaped piles. These piles are designed with a load of 1200 kN/pile (tension) and are driven to a depth of NAP - 50 m with a spacing of 3.5 m. This is the biggest pile load in tension in tunnel design in the Netherlands so far.

c) <u>The immersed part of the tunnel</u>
The immersed tunnel sections have been founded on tubular steel piles (diameter 508 mm) with a allowable bearing capacity of 2000 kN/pile and driven to a depth of NAP - 46 m. The spacing is such (5.6 m) that no pile heave of already driven piles is expected. The piles have been provided with a separate head. Contrarily to the inflatable pile caps, a steel tybe with a bottom plate is lowered from the tunnel; the connection is formed by a nylon sleeve. The space between tunnel and tube has been closed by filling the sleeve by cement grout.

During the final design phase the following checks have been carried out:
1) Overall stability
   The whole system of piles and the soil in between has been checked against uplift. The ratio weight of soil + concrete: uplift force is defined as safety factor and should range between 1.2 and 1.5 (depending on whether only the weight or weight and side shearing resistance are taken into consideration.

2) Pile group effect
   The allowable tensile pile load of a single pile has to be reduced in case of a pile group. The reduction depends on the pile grid and spacing. The transition of shear forces from pile to soil results into a reduction of the vertical effective forces and so of the maximum tensile force to be mobilized.

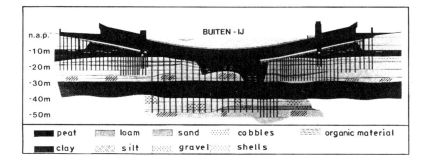

Figure 12   Longitudinal section with different foundations

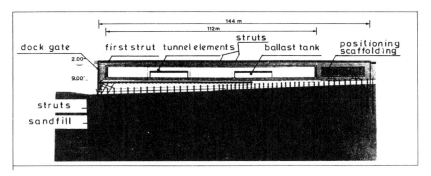

Figure 13   Longitudinal section "building dock"

GEOTECHNICAL ASPECTS

REFERENCES

1. Heijnen, W.J. "Penetration Testing in the Netherlands". ESOPT I, Vol. I p.p. 79-83 (1974).

2. Begemann, H.K.S.Ph. "The Dutch Static Penetration Test with the adhesion jacket cone". LGM-mededelingen XII, XIII 1969.

3. N.N. "Methoden ter voorkoming van opdrijving van constructieve in grondwater". Koninklijk Instituut van Ingenieurs (Sectie Tunneltechniek) 1976.

4. Tan, G.L. "Zeeburgertunnel moet Coentunnel ontlasten". I^2 Bouwkunde en Civiele Techniek no. 9, september 1985.

5. Lohman, A.J.B. and Purmer, C.G. "Uitvoering Zeeburgertunnel gecompliceerder dan voorzien". Cement 1987, nrs. 8, 10, 11

2. Construction of immersed tunnels — the land connection

G. A. MUNFAKH, PhD, and B. SCHMIDT, PhD, Parsons Brinckerhoff Quade and Douglas, Inc., New York and San Francisco

SYNOPSIS. The geotechnical challenges associated with connecting immersed tunnels to ventilation buildings and tunnel approaches on land are studied through four case histories. The land connections were accomplished at the water edge, inshore but close to the water, or on a man-made island. Excavation support systems consisting of interlocking steel H-piles or tied-back soldier pile and lagging walls were used to allow continued operation of busy facilities nearby. Construction dikes and cofferdams, ground improvement, and special dewatering measures were used. In most cases, the land connections were constructed in difficult soils, at highly developed areas, and under strict environmental regulations.

INTRODUCTION
1. Every tunnel has two ends, immersed tunnels, also. The immersed portion has its particular geotechnical concerns -- safe slopes for dredging, disposal of dredged material, safeguarding the excavated trench in the river currents, settlement of the tunnel elements, etc. -- but the most interesting geotechnical challenges usually occur at the land connections where ventilation buildings and approaches are built adjacent to the water. The soils there usually are not very competent, and the proximity of an unlimited source of water often makes it an exciting challenge to keep an open excavation dry.
2. Sometimes the immersed tunnel approaches are built by bored tunnels (e.g., the Detroit Windsor tunnel between the USA and Canada, the Elbe River tunnel in Hamburg, the 63rd Street tunnel in New York,) but usually the land connection is built in an open cut. On occasions, the "land" connection is not even on land but on a man-made island (e.g., several tunnels in Virginia, USA). Often, trenches are excavated into the banks to allow placement of immersed elements beyond the limits of the shoreline. In many cases, the adjacent land is highly developed waterfront property requiring protection during construction.
3. The challenges to the geotechnical engineer are plenty -- construction of stable slopes, support of the excavation, dewatering, protection of adjacent property against ground movements, disposal of excavated (sometimes contaminated)

Immersed tunnel techniques. Thomas Telford, London, 1989.

GEOTECHNICAL ASPECTS

materials, and control of the differential settlement between the finished land structure and the adjacent immersed tunnel element.
4. This paper discusses geotechnical challenges encountered at the land connections of four immersed tunnels designed by Parsons Brinckerhoff.

FORT McHENRY TUNNEL, MARYLAND, USA
5. The eight-lane Fort McHenry Tunnel constructed in Baltimore's congested harbor is the widest vehicular tunnel ever built by the immersed tube method. The immersed tunnel comprises 32 twin binocular steel-lined tube sections placed in a common dredged trench 1620 m (5400 ft.) long and 54 m (180 ft.) wide at the base. The immersed section connects to a ventilation building at each end, with cut-and-cover sections extending east and west for a portal to portal length of 2160 m (7,200 ft.).
6. The subsurface conditions at the site and the geotechnical aspects of the design and construction of the immersed tunnel were described by Sarkar and Munfakh (ref. 1). Generally, the soils at the site consist of both cohesive and granular formations of Cretaceous age overlain by Pleistocene to Recent alluvial deposits and manmade fills. Due to tectonic movements and erosional unloading during the Tertiary and Quartenary ages, joints and shear zones (or fissures) were formed within the overconsolidated clay deposits. The presence of these fissures greatly reduces the shear strength of the intact overconsolidated clay, particularly during and after excavation and with access to free water. Fully softened drained strength parameters of C' = 30 kPa (600 psf) and Φ' = 21 degrees were used in the design.

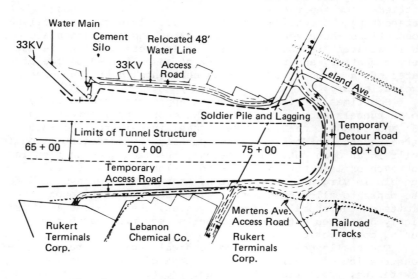

Fig. 1. East end tunnel excavation plan

7. Unlike the west side, where the immersed tunnel section terminated practically at the harbor's edge, the immersed tunnel on the east side, coming under the navigation channel, had to extend about 400 m (1200 ft.) landward requiring a trench 54 m (180 ft.) wide at the bottom and 120 m (400 ft.) wide at the top with the depth of excavation ranging between 19.5 m (65 ft.) and 28.5 m (95 ft.) below the ground surface (Fig. 1). Such a wide excavation through a busy port facility required careful engineering and economic considerations. In the interest of economy, and to minimize disruption of existing port activities, it was decided to excavate the trench slopes as steep as possible and to combine the open cut with an excavation support system (Fig. 2). A tied-back soldier pile and lagging system was installed with a maximum allowable tieback load of 32 metric tons (35 tons) and a maximum allowable tieback spacing of 3 m (10 ft.) vertical and horizontal in most areas and 2.5 m (8 ft.) in critical areas.

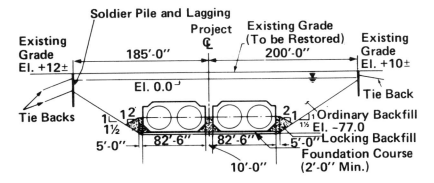

1'= 0.3m

Fig. 2. East end excavation cross-section

8. Straight shaft soil anchors consisting of 25 mm (1-inch) Dywidag bars were installed inside 75 mm (3-inch) diameter holes in sandy soils and 150 mm (6-inch) holes in soft clays. The initial selection of the anchor capacity and estimated length was based on the assumption that the bond section would be in the competent Cretaceous soils, not in the fills. However, because of the variety of soils existing behind the wall, and for the sake of economy, a large number of anchors had to be installed in the variable sand and clay fills. A simple acceptance criterion was, therefore, developed for each production anchor as follows:

  (a) Each anchor was proof loaded to 140% of the design load in increments not to exceed 25% of the proof load, except that the last load increment was not allowed to be less than 15% of the proof load. After reaching the proof load, creep movements were monitored with an independently supported dial gage. Gage readings were recorded at 5 and 15 minutes after application of the final load. The anchor was accepted if the creep movement between 5 and 15 minutes was less than 0.69 mm (0.027 inches).

    (b) Unacceptable anchors were retested at reduced loads. The acceptable design loads of those anchors were calculated by dividing their acceptable proof loads by 1.4.

    (c) If an anchor failed to carry the full proof load, its proposed design load was reduced to 50% of the maximum load carried by the anchor. The anchor was then proof tested as indicated above to confirm the adequacy of the revised capacity.

9. Based on the above criteria, soil anchors were installed in both the natural soils and fills at bond lengths ranging from 4.5 to 9 m (15 to 30 ft.). The anchors performed satisfactorily for the 2 to 3 year duration of the project.

SECOND DOWNTOWN ELIZABETH RIVER TUNNEL, VIRGINIA, USA

10. Connecting the cities of Portsmouth and Berkley in Virginia, this immersed tunnel crosses the Elizabeth River parallel to and about 60 m (200 ft.) distant from an existing immersed tunnel. The 750 m (2500 ft.) immersed tube section, which consists of eight steel-shell prefabricated elements, is connected at both sides to cut and cover tunnels, which exit into walled depressed roadways leading to two interchanges.

11. The subsurface conditions along the tunnel alignment vary in composition from essentially dense sandy material on the Portsmouth side to essentially overconsolidated clayey material on the Berkley side. At the Portsmouth shore, erratic deposits of soft organic clays and silts are believed to have been deposited by meandering channels cut by a creek which has changed its course several times in the general area of the Portsmouth land connection.

12. The undrained shear strength of the clayey soils ranged from $Su = 25 - 40$ kPa (500 - 800 psf) for the soft to medium silty clay to $Su = 47 - 115$ kPa (950 - 2300 psf) for the medium to stiff silty clay. The corresponding drained strength parameters for these soils were $C' = 0$, $\Phi' = 24 - 27$ and $C' = 15$ kPa (300 psf), $\Phi' = 24 - 27$ respectively. The undrained shear strength of the organic soils was $Su = 10 - 20$ kPa (200 -400 psf). Their drain strength parameters were $C' = 0$, $\Phi' = 24 - 27$. A detailed description of the soil stratification and the engineering characteristics of the various soils at the site were provided by Munfakh (ref. 2).

13. Connection of the immersed tunnel to the land cut and cover sections was complicated by the presence of many facilities of the Norfolk Ship Building and Drydock Corp. on the Berkley side, and a culvert, waterfront structures and

other facilities at the Portsmouth side. Construction difficulties were also caused by the close proximity of the old tunnel and its ventilation building to the land connection area.

14. To connect the immersed tunnel elements to the cut and cover sections, construction dikes were placed on top of the end tubes to form water barriers which allow dewatering behind the dikes and connecting the tubes to the cut and cover sections in the dry (Fig. 3). The overall stability of the construction dike and the tunnel backfill material placed beneath the dike were dependent upon maintenance of seepage flow lines away from its downstream (dry) face. This was accomplished by a conventional wellpoint system. Because the highly permeable foundation course beneath the tube units may transmit quantities of water too great to be handled by sumping, the injection of low pressure grout in the foundation course at that location was recommended. However, during construction, the contractor was able to handle all water seepage by heavy pumping.

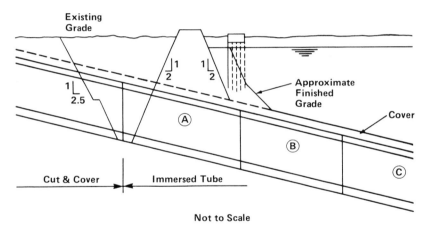

Fig. 3. Construction dike at Portsmouth side

15. In the Portsmouth area, the required extended duration of groundwater drawdown necessitated the use of a groundwater recharge system to protect the untreated timber piles supporting the adjacent ventilation building of the existing tunnel from fungus decay upon exposure to air. The recharge system remained operational for the duration of the dry construction, maintaining the groundwater at the building location above the pile caps. Connection of the immersed tunnel to the cut and cover sections was successfully accomplished at both shores.

GEOTECHNICAL ASPECTS

CHAO PHYA RIVER CROSSING, BANGKOK, THAILAND
16. Soft clays are found throughout Bangkok to depths often exceeding 15 m (50 ft.). These soft clays pose exceptional constraints on construction activities in Bangkok, and the recent river crossing was no exception. Parsons Brinckerhoff in joint venture was commissioned to evaluate bridge and tunnel alternatives for crossing the Chao Phya River, and to design the most feasible and cost effective type of crossing.
17. The subsurface conditions at the site generally consist of about 15 m (50 ft.) of soft silty clay underlain by stiff clays to depths of 22 to 26 m (72 to 85 ft.). Both clay strata contain occasional silt or sand partings. The stiff clay is underlain by dense fine sand of unknown thickness. The soft clays are (nearly) normally consolidated with the upper 1 m (3 ft.) submerged in the river being very recent and underconsolidated. On the north side of the river, the shear strength profile is described by:
$$Su = 2 Z \text{ (kPa)}$$
where Su = undrained shear strength, and Z = depth from ground surface. On the south side, a dry crust was developed resulting in a shear strength profile of
$$Su = 10 + 2 Z \text{ (kPa)}$$
18. An immersed tunnel in soft clay is not without precedence; the Tingstad tunnel in Gothenburg, Sweden, built in the early sixties on a thick layer of soft sensitive clay, had conditions very much similar to those of a tunnel under the Chao Phya. The Bangkok tunnel was encumbered with very limited space for the land connection and heavy river traffic. Other problems included: very flat safe dredging slopes, maintaining the trench free of silt, expensive excavation support system, settlement of adjacent property, and drydock for casting the immersed tube elements.
19. The most feasible immersed tunnel elements were concrete elements cast in a drydock excavated into the river bank. It was difficult to find a suitable location for such a drydock, all suitable land already having been reserved for other uses. Because of the low strength of the clays, slopes of an excavated drydock would have to be laid back at very flat grades, increasing substantially the size of the dock and the volume of the excavated material to be disposed of. A suitable disposal area was equally hard to find; barging the material to sea would be a costly affair.
20. The clay slopes posed another geotechnical challenge. Observations at other sites showed that the clay would crack deeply when exposed to drying (cracks more than one meter deep were observed). Regular rainfalls would fill the cracks with water, resoften the material, and provide an additional hydrostatic force to destabilize the slope. Natural slopes of the river banks were between 3 h/1v and 5h/1v. These existing slopes appeared to be stable indicating possible slopes for excavations.
21. Fig. 4 illustrates the land connection of the immersed tunnel. At the edge of the river, the tunnel alignment is

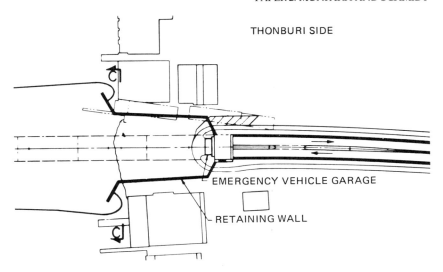

Fig. 4. Land connection plan

bracketed on both side by heavily used buildings leaving no more than 90 m (300 ft.) of available width. A sloped cut at that location would require a total width of at least 180 m (600 ft.). Therefore, the use of an excavation support system was necessary.

22. A vertical steel wall comprised of interlocking H-pile sections (Arbed or similar) and supported by battered compression piles (Fig. 5) was found to be the only economically feasible support system given the extremely limited construction space available and the difficult soil conditions. Approximately 360 m (1200 ft.) of temporary retaining walls were required extending from the portal building cofferdams out into the river (Fig. 4). The distance

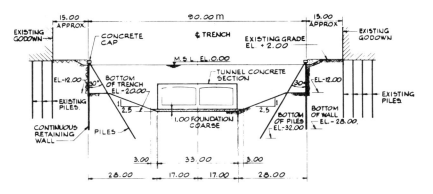

Fig. 5. Excavation cross-section

between the retaining walls was kept to a maximum of 90 m (300 ft.) to allow room for placement of the tube sections while leaving a minimum distance of 15 m (50 ft.) between the walls and the existing structures, sufficient to install the excavation support system.

23. The land connection is accomplished in three stages (Fig. 6). First, a cofferdam and temporary retaining walls would be installed for construction of the tunnel portal and the open approach in the dry. After the open approach and portal structures are built, the retaining walls for the immersed tube placement would be installed -- the raking compression piles would be driven first, followed by the continuous vertical H-pile wall which is connected to the supporting compression piles within a reinforced concrete cap. Once the portal structure is completed, the closure part of its cofferdam facing the river would be removed and the tunnel trench dredged up to the face of the structure (which would include a special joint to which the tube would be attached). After the connection is made and the trench backfilled, the retaining wall may be removed and the steel piles salvaged for reuse.

24. Needless to say, the cable-stayed bridge alternative was selected and its construction was completed in November 1987, a worthy birthday present for King Bhumipol.

SECOND HAMPTON ROADS TUNNEL, VIRGINIA, USA

25. The Second Hampton Roads bridge-tunnel crossing connecting Norfolk with Hampton in Virginia, USA, includes a 2,100 m (7,000 ft.) long two-lane immersed tunnel constructed between two man-made islands, and two trestles connecting the islands with the mainland. The project was completed in 1976, forming a twin to the first bridge-tunnel crossing, which was completed in 1957.

26. The North Island was founded on sands and silty sands presenting no significant settlement or stability problems. At the South Island, however, about 25 m (80 ft.) of soft normally consolidated clays were encountered above the sandy soils. The undrained shear strength of the upper 10 m (35 ft.) of clay was $Su = 22.5 +/- 10$ kPa (450 +/- 200 psf). The 15 m (45 ft.) below had an undrained strength of $Su = 47.5 +/- 15$ kPa (950 +/- 300 psf). The construction of the man-made island here was also complicated by the presence of the existing immersed tunnel and island immediately adjacent. The success of the land connection, therefore, was greatly impacted by the stability and long-term settlement behavior of the man-made island.

27. The existing tunnel island was constructed by hydraulic filling of sand after the soft clays were removed by dredging. This method of construction, however, could not be used for the new tunnel because of the proximity to the existing facility; besides, the huge amount of dredged clays would be difficult to dispose of economically under new environmental regulations. After examination of a number of alternatives

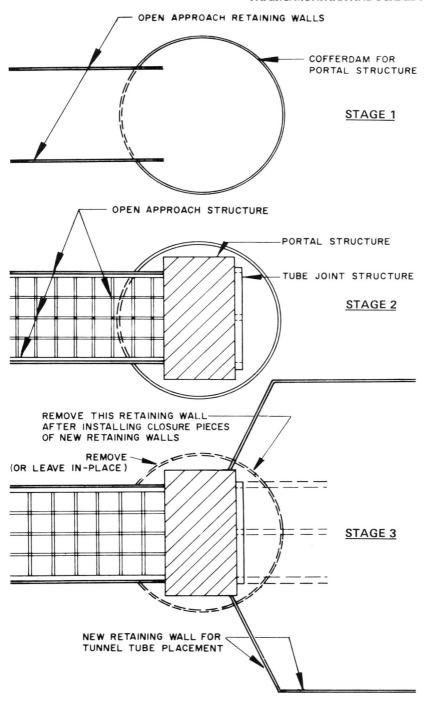

Fig. 6. Land connection staging

(ref. 3), the new island was built using staged construction (Fig. 7) as follows:

- o Place stone dike for containment of sand fill and wave protection.
- o Hydraulically place the first layer of dredged sand.
- o Install sand drains to accelerate consolidation of the clay.
- o Place remainder of the sand in layers while monitoring to ascertain adequate strength gain in the clay before each sand layer is deposited.
- o Build up a surcharge to accelerate consolidation of the clay and minimize residual settlements.
- o After the anticipated long-term settlements are reached, remove the surcharge and dredge a trench into the island to place the end immersed tube; backfill over the tube.
- o Excavate for construction of the ventilation building and open approach and for connection to the immersed tunnel in the dry, using soldier piles and lagging for excavation support.

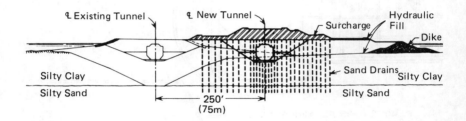

Fig. 7. Cross-section of South Island

28. Because a layer of sandy gravel was placed beneath the tubes as a foundation course, this open-graded material was a conduit for water entering the ventilation building excavation causing construction difficulties. Eventually, this problem was solved by a combination of grouting and heavy pumping.
29. Construction of the island was carefully monitored using piezometers, settlement markers and inclinometers (ref. 4). The horizontal displacements in the soft clay, estimated by finite element analyses and measured by inclinometers, proved to be useful indicators of the overall safety of the slopes. Once the vertical and horizontal soil movements were stabilized, the land connection was completed as planned.

SUMMARY
30. Construction of immersed tunnels is highly dependent on geotechnical considerations. Of their many geotechnical challenges, the most interesting usually occur at the land

connections where ventilation buildings and approaches are built adjacent to the water, in very congested areas, and usually in not very competent soils.

31. The four cases presented in this paper describe a variety of land connections where geotechnical problems were encountered and solved. In every case, construction of the land connection was vital to the successful implementation of the project on schedule and within the constraints imposed by difficult soil, minimum right-of-way, and strict environmental regulations.

REFERENCES
1. SARKAR S.K. and MUNFAKH G.A. Geotechnical aspects of the Fort McHenry Tunnel, design and construction. International Conference on Case Histories in Geotechnical Engineering, St. Louis, 1984.
2. MUNFAKH G.A. Geotechnical aspects of the Second Downtown Elizabeth River Tunnel. Thirteenth Annual Southeastern Geotechnical Engineering Conference, Virginia, 1981.
3. RAFAELI D. Design of the South Island for the Second Hampton Roads crossing. Specialty Conference on the Performance of Earth and Earth-Supported Structures, Purdue, 1972.
4. KUESEL T.R., SCHMIDT B. and RAFAELI D. Settlements and strengthening of soft clay accelerated by sand drains. Highway Research Record, No. 457, 1973.

3. Planning the third harbour tunnel in Boston, Massachusetts, USA

W. C. GRANTZ, PE, and A. R. LANCELLOTTI, PE, Parsons Brinckerhoff Quade and Douglas, Inc., Boston

SYNOPSIS. The planning of Boston's Third Harbour Tunnel project is being influenced by several factors, including commitments to minimize the impact to local businesses, ventilation requirements, available fabrication facilities, and the harbour geology. This paper discusses each of these items and their impact on the planning of the project.
INTRODUCTION
1. The Central Artery (Interstate 93), Third Harbour Tunnel (Interstate 90) Project in Boston, Massachusetts, U.S.A. is a $4.4 Billion (U.S.) project involving major improvements to the highway facilities that pass through Boston. These improvements can be divided into three basic areas (as they are referred to on the project - see Figure 1). Interstate Highway I-93 is the "Central Artery" which runs north and south through the financial center of the city, a large portion of which is currently carried on an elevated structure. This section is to be reconstructed in wider form underground while maintaining traffic on the overhead structure. The overhead structure will later be demolished. The second area involves an extension of Interstate Highway I-90 (Massachusetts Turnpike) through South Boston, then crossing the harbour by immersed tube tunnel and emerging in East Boston at Logan International Airport. Finally the third area involves a major interchange connecting these two highways. The Project also involves other complicated interchanges, numerous ramps and local street connections, and high occupancy vehicle facilities. The Project consists of:

(a) 19.2 km of unidirectional tunnels

(b) 22.4 km of unidirectional bridges

(c) 27.2 km of unidirectional at-grade roads

(d) 11.2 km of unidirectional depressed sections (open boat sections)

2. In total, the Project consists of over 160 lane km of new roadways.

GEOTECHNICAL ASPECTS

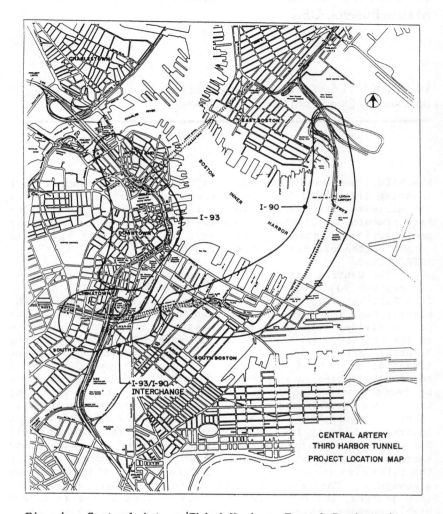

Fig. 1 - Central Artery/Third Harbour Tunnel Project Areas.

3. As noted above, large portions of this Project, apart from the Third Harbour Tunnel, are planned to be in tunnel. Most of this tunnel work will involve cut-and-cover methods although short sections of immersed tunnel are being considered for the crossing of the Fort Point Channel.

4. The Third Harbour Tunnel, as its name implies, is the third vehicular tunnel to cross the harbour. Currently the Sumner and Callahan Tunnels carry traffic across the north end of the harbour between downtown Boston proper and East Boston. They are very heavily loaded and carry close to the 100,000 vehicles per day. The construction of this third link is long overdue.

5. This paper will focus only on the planning of the harbour crossing portion of this major project. Currently this work is in the early stages of the preliminary design phase. Considerable work has been done leading to this point, including studies to identify the best alignment, type of tunnel element construction, interior clearances, ventilation concepts, location of ventilation buildings and other key elements needed to allow the design to go forward. This paper is intended only as a "snapshot" of where the design presently stands in concept; it may well change in some important aspects before going to construction.

PLANNING CONSIDERATIONS

6. Figure 2 shows the current alignment chosen for the Third Harbour Tunnel and I-90 as it traverses South Boston. This Figure indicates the principle constraints in setting the alignments through these areas.

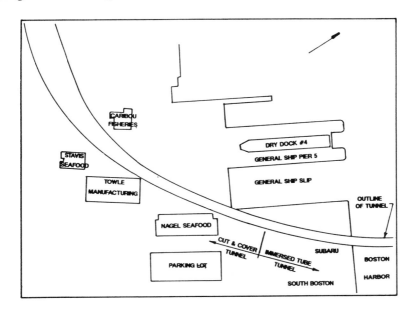

Fig. 2 - Constraints on tunnel alignment in South Boston.

South Boston

7. In South Boston the alignment must provide for a widened area for a toll plaza for the westbound traffic. The principle constraints on the alignment are to avoid the fishing industries scattered in the area and a busy shipyard doing ship repair contracts for the United States Navy. Originally it had been expected to put the alignment through the existing slip adjacent to Pier 5 which is currently used to outfit ships after dry dock work has been completed. Because of the requirement to excavate hard rock in this slip by drilling and blasting, studies by a

GEOTECHNICAL ASPECTS

Marine Consultant showed that this work could impact the ship repair operation. This was because it was likely that the Navy would not allow its ships to be worked on in dry dock, or at Pier 5, while blasting was being done less than 30 m away. The alignment was therefore shifted through the existing land fill to the south. This had certain benefits including:

(a) the profile could rise sooner and thus reduce the amount of rock excavation

(b) in combination with constraints in East Boston which set the alignment there, the amount of tunnel curvature could be reduced; and

(c) improve the roadway gradients in the approach to the toll plaza.

East Boston

8. In East Boston the alignment constraints were largely due to existing airport facilities. Operations and future development of Terminal A were one factor. The alignment must be constructed between an existing boundary access road and a major taxiway. Currently, the westbound lanes are planned to be in cut-and-cover tunnel and the eastbound lanes are to surface in open boat sections. Another factor is the planned development of the southwest corner of Bird Island Flats (the name of the section of the airport that forms the northwest shoreline) with a hotel and conference center. This planning has not been entirely settled and could be subject to further change, particularly with respect to the amount of covered approach tunnels and the ultimate location of the ventilation building.

NAVIGATIONAL CONSTRAINTS

9. The navigation channel width and depth were established early in the planning. The U.S. Army Corps of Engineers and the Massachusetts Port Authority (Massport) were consulted and it was determined that the current plan is to extend the 12.2 m deep channel by an additional 183 m to a total of 366 m. No additional depth is anticipated. The reason for this is the clearance over the existing two vehicular tunnels and Blue Line transit tunnel. The present 9.1 m depth of anchorage area east of the channel must also be respected. On the west side of the channel, the channel depth must be extended to the bulkhead line to permit deep draft ships to dock along the filled area which may be developed as a future port facility by the City of Boston. (See Figure 3.)

10. The tunnel top of structure profile will be set so that it provides a minimum of 2.1 m of clearance below the navigational requirements described above. This includes 1.5 m of protective cover and a 0.6 m allowance for over-dredging.

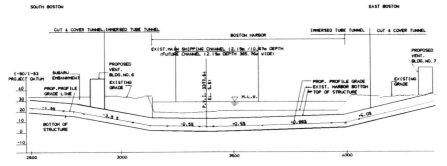

Fig. 3 - Tunnel profile.

GEOTECHNICAL FINDINGS

11. Geologically the Boston Area is characterized by the effects of glaciation. Typically four basic materials are found: Deposited fill materials, Boston "Blue Clay", hard "glacial till", and weathered to sound argillaceous rock. The bedrock surface ranges widely in elevation and is generally below the level of the structures of the Project as a whole.

12. Bottom Conditions: As luck would have it, a pilot boring programme across the harbour indicated that for almost three quarters of the crossing hard argillite will have to be removed underwater. This material will necessarily have to be removed by drilling and blasting. The borings provided additional bad news in that where rock is not a factor, soft clays are. The eastern end of the tunnel alignment traverses a deep area of "weight-of-rod" material. Further, the rock profile varies significantly with high peaks and low valleys separated by short distances.

13. Seismic sub-bottom reflection and refraction studies were used to attempt to extrapolate the rock-soil interface contours beyond the limits of the boring information. Unfortunately the results were inconclusive and did not correlate well with the boring data. This was attributed to the soft clays and hard glacial till materials masking the reflection from the rock-soil interface. The major finding from the seismic studies was that the interface depth appears to vary widely.

14. The rock interface was characterized consistently in the borings as being very fresh rock, not weathered; as if the glacier had scraped the surface clean.

15. Consideration was given to the feasibility of a bored tunnel instead of an immersed tunnel or of a bored tunnel in the sound rock and an immersed tunnel in the soft ground. These alternatives were eventually discarded because of the risks associated with mixed face tunnelling such large diameter bores

underwater and the impact that a bored tunnel would have on the already tight tunnel approaches in South Boston and East Boston.

16. **Dredging and Disposal Methods**: There will be basically three categories of dredging materials involved in this immersed tunnel project. The first will be a surface layer of harbour mud which is considered to be generally contaminated with heavy metals. This material will be disposed of into a sealed containment site. The second material will be clays and tills which qualify for bottom dumping in a designated spoil site at sea. Finally, the most difficult material to dredge, the hard argillite rock which may either be disposed of at the spoil site or stockpiled for backfill or bank protection.

17. Rock excavation will necessarily be done primarily using barges equipped with rows of drill towers. Rock excavation depth will reach as much as 6.1 m in some areas and require at least two benches. Fortunately, the tunnel site is pretty well protected from heavy swells or waves, and drilling and blasting should not be affected unduly by weather conditions. Removal will be done using clamshell buckets discharging into hopper barges.

18. Soil excavation will be segregated depending on whether it is contaminated or not. Also the disposal site is too far from the Project to consider cutter head suction/pipeline operation. Either clamshell or hopper dredge operation is the likely means for removing this material.

19. **Anticipated Settlements**: The condition of having the immersed tunnel transition from very firm support to soft ground is a matter of concern especially considering that the Boston area is a significant seismic zone (zone 3). Very preliminary studies of relative settlements have been made which appear to indicate that this could be a severe problem. Further subsurface borings and laboratory tests are required before this matter can be fully resolved.

20. **Seismic Joints**: The need for seismic joints which allow for uniaxial or even tri-axial movement between tunnel elements or tunnel elements and terminating structures such as cut-and-cover tunnels or ventilation buildings are currently under investigation.

SELECTION OF IMMERSED ELEMENT CONFIGURATION

21. **Ventilation Requirements**: The total length of the Third Harbour Tunnel between ventilation buildings is only 1.2 km. This permits various options of ventilation to be considered. **Fully Transverse**; where air is introduced through ports along the tunnel at the roadway level and extracted at similarly spaced ports in the ceiling. **Semi-transverse**; where supply air is again introduced through ports along the tunnel but returned to single locations either at the portals or at the ventilation buildings through the tunnel roadway. **Longitudinal**; where air is moved through the tunnel by fans mounted on the tunnel walls. The latter scheme, while widely used in European Tunnels has not been accepted as yet in the United States for use in major tunnels.

22. __Tunnel Cross Section Selection__: A major factor in picking the tunnel section configuration was the ventilation system and the number of ventilation buildings to be used for the harbour crossing. If a "Full Circular Binocular" section were to be used, it was possible to provide full transverse ventilation using only a single ventilation building located on Bird Island Flats in East Boston. If two smaller ventilation buildings were used, one in South Boston and one at the same location in East Boston, full transverse ventilation can be achieved using a "Flat-Bottom Binocular" section or a full binocular section. The problem with the full circular binocular section is that it requires some US $7-10 million for additional rock excavation. Studies made of all these alternatives have indicated that a full transverse ventilation system using a flat-bottom section and two ventilation buildings provide the best lowest capital cost but higher operating costs. Further, the full circular section provided for better operational features such as more uniform air distribution. Therefore, despite the higher initial capital cost, the full circular section was selected (see Figure 4 and 5).

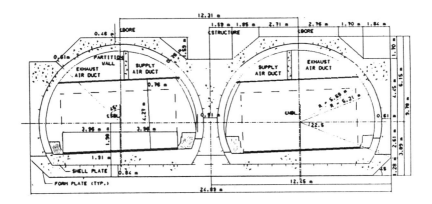

Fig. 4 - Flat bottom binocular tunnel section.

23. __Concrete Box versus Steel Shell Elements__: In the United States, one of the most difficult problems to solve in the use of Concrete Tube Elements is the environmental aspects of having to excavate a basin in which to construct the elements. If the elements can be fabricated in an existing dry dock facility, they may well become competitive. Such a condition occurs in the Boston area. South of Boston in Quincy, Massachusetts, there exists a massive shipyard facility recently shut down due to lack of sufficient shipbuilding business to keep it alive. This shipyard easily provides enough dry dock space to build several concrete elements at a time. It was therefore important to study the possibility of using concrete box elements as an

GEOTECHNICAL ASPECTS

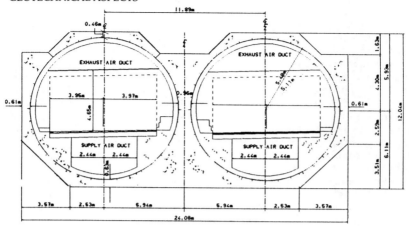

Fig. 5 - Full circular binocular tunnel section.

alternative to steel shell units. It also had the advantage to the area that local labor would be used.

24. Unfortunately there were several factors working against this idea. Currently the facility is owned by the Massachusetts Water Resources Administration which has plans for its use in conjunction with a multi-billion $US harbour clean-up project. Local shipbuilders have formed an association to try and reactivate at least a portion of the plant. But all of these conditions leave the future viability of the yard rather uncertain. Further, it is considered unwise to make the Third Harbour Tunnel Project contingent on this viability. The effect would be to greatly limit bidding flexibility which is contrary to Federal bidding requirements.

25. The result of this was the recommendation to use steel shell elements. Cost estimates indicated higher cost for concrete box elements when the cost of the construction basin was included. Also the concrete box element tunnel was estimated to take some six months longer to complete even without considering the potential for difficulties with environmental and permitting approvals.

26. <u>Internal Arrangement and Clearances</u>: The cross section of the element is set by two basic factors: Space for ventilation and roadway space.

27. The selected section provides full transverse ventilation from two ventilation buildings. The fresh air is to be supplied to the roadway space from the fresh air located below the roadway. This space will connect to separate flues in the sides of the patrol walk which will introduce the fresh air near vehicle exhaust level. The exhaust duct will be located above the ceiling and will collect vitiated air through transverse ports of appropriate size and spacing. The roadway space requirements are to provide two 3.7 m lanes with additional one-foot setbacks for a total of 7.9 m. Vertically a 4.4 m clearance is provided for vehicles which includes 150 mm for

later pavement overlay. Additionally, 760 mm is provided as space for signs. The total vertical clearance is therefore 5.2 m.

28. Although still somewhat unresolved, for the basis of preliminary design, provision is being made for a tunnel patrol walk on the inside lane side of the tunnel (left hand traffic) and a safety walkway along the outside (right) lane of traffic.

TUNNEL CONFIGURATION

29. Based on the foregoing the overall configuration of the tunnel is being developed. It is expected that 12 elements will be required; each approximately 100 metres long. The portion on the South Boston side which lies in the existing land fill will be constructed by dredging a trench up to the location of the ventilation building, then placing elements in this trench, and backfilling. On the East Boston side, elements will also be carried inshore as far as possible given certain lateral limitations on the width of excavation and the final location of the ventilation building.

30. It is anticipated that of the 12 tubes, four will be tangent, two will be horizontally mitred, one will be vertically mitred and six will be horizontally and vertically mitred. Miters are introduced in a straight tube to produce a configuration that approximates a curve. This is accomplished by fabricating the normally rectangular shell modules into trapezoidal shapes and welding several trapezoidal modules together.

SCHEDULE

31. A schedule analysis has shown that the Third Harbour Tunnel and its appurtenant facilities will take approximately 36 months to complete. Scheduled completion of the I-90 connection including the Third Harbour Tunnel is in the last quarter of 1994.

CONCLUSION

32. The Third Harbour Tunnel crossing of Boston Harbour will be a vital link in the Massachusetts transportation system when it is completed. Funding by the Federal Highway Administration has been assured by act of Congress in the fall of 1987. The Project, like all tunnel projects, has several interesting and different engineering challenges.

ACKNOWLEDGEMENTS

33. The preliminary design of the Third Harbour Tunnel Project is being carried out by the joint venture Project Management Team of Bechtel/Parsons Brinckerhoff for the Massachusetts Department of Public Works. Thanks are given to the MDPW for permission to publish this paper. Mr. Walter C. Grantz is the Area Project Engineer for the Third Harbour Tunnel Area of the CA/THT/I-93/I-90 Project. Mr. Lancellotti is the Senior Structural Engineer in charge of the Third Harbour Tunnel Preliminary Design.

5. Sedimentation and stability of dredged trenches

M. F. C. THORN, MA, MS, Hydraulics Research, Wallingford

SYNOPSIS. The interaction between a trench dredged across a tidal channel and the tidal flow and sediment transport regime is discussed in the light of the most recent research developments. The fundamental differences of approach required in sandy and muddy estuaries are highlighted and sound methods for estimating sedimentation and stability of dredged trenches in both types of estuary are suggested.

INTRODUCTION

1. Immersed tunnels commonly require a substantial trench to be dredged across the waterway to receive the tunnel units. In estuaries where sediment is in motion during the tidal cycle, reliable predictions of the rate at which sedimentation will occur in the dredged trench are needed to aid design and planning of construction, particularly in respect of maintenance of the trench between completion of dredging and completion of positioning of the tunnel units. The physical processes of sedimentation in a dredged trench, and the stability of the underwater side-slopes of the trench, under the action of tidal currents and wave activity are complex and depend very much on the type of sediment involved.

2. This paper attempts to summarise present knowledge and practical methods of approach to the problem, which is still the subject of fundamental research. The detailed theory and experimental results are to be found in the references quoted.

GENERAL PRINCIPLES

3. A trench of substantial dimensions, such as is required for an immersed tunnel, dredged across an estuary causes a localised change in flow. If the axis of the trench is at, or close to an angle of 90° to the flow direction, the effect is simply to reduce the water velocity across the trench. In principle, the reduction in water velocity also reduces the sediment transporting capacity of the flow, so that a proportion of the sediment being carried in suspension in the water is deposited. However, the reponse time in which the surplus sediment can actually be transferred from the water

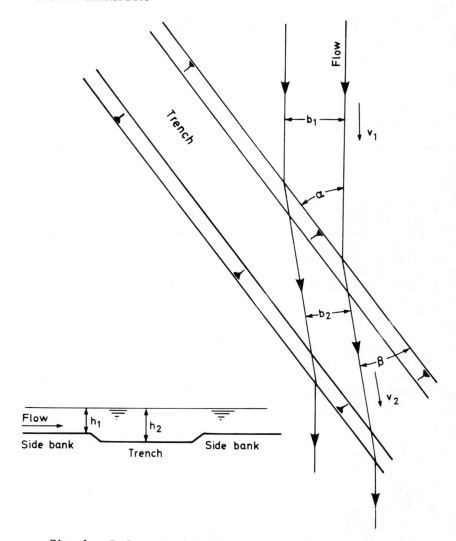

Fig. 1. Refraction of flow across a skewed trench

to the bed is finite, and generally much greater than the time taken for the water to cross the trench and resume its previous, higher velocity. Therefore the actual rate of sedimentation in a trench of finite width is considerably less than the potential rate of sedimentation in a "trench" of infinite width, and this must be taken into account in practical calculations.

4. If the flow, expanding into a deeper trench, loses sediment to the bed, then as its velocity increases at the

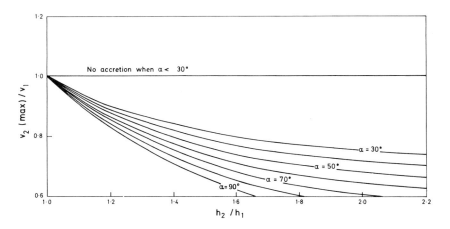

Fig. 2. Changes in velocity in a skewed trench

opposite side of the trench its sediment transporting capacity is restored and the deficiency is made up by eroding sediment from the bed. Thus there is a tendency for the edges of the trench to be infilled by deposition at its "up-flow" side, and eroded by re-suspension at its "down-flow" side. In most tidal estuarine situations the effect is reversed as the tides reverse, so that the axis of the trench remains static and it simply loses depth. But if there are large imbalances between flood and ebb flows, or if these principles are applied to uni-directional flow situations, the trench will tend to migrate in the dominant flow direction, and control of migration by dredging is likely to be a greater problem than maintenance of channel depths.

5. The effect of the trench on the flow becomes more complex when the axis of the trench is skewed to the predominant flow direction. Laboratory experiments have demonstrated that the flow refracts as it crosses the trench, as shown in Fig. 1. The flow streamlines narrow, and so the reduction in velocity is less than would be assumed, as shown in Fig. 2. However, the flux of sediment in suspension in the water is correspondingly compressed into a narrower flow width so the difference between sand actually in suspension and the transport capacity of sediment per unit width of flow must increase in inverse proportion. These interacting effects must all be taken into account when predicting how much of the sediment in suspension will find itself on the bed of the trench in the time that it takes the flow to cross it. The theoretical treatment of these processes and its application to engineering problems is presented by Lean in ref. 1, together with examples of results from the laboratory experiments on skewed channel alignments. The refracting

effects of a skewed channel have subsequently been seen in a general computer model representation of tidal flow crossed by an oblique dredged trench in the Conwy estuary.

6. Although it is obviously important to understand and to reproduce correctly the effect of a dredged trench on flow, prediction of sedimentation inherently depends on accurate understanding and measurement of the relevent sediment properties. Non-cohesive (sandy) sediments have quite different characteristics from cohesive (muddy) sediments and are discussed separately.

SANDY SEDIMENTS

7. There are a number of well-established formulae that relate the transport of sand grains to water velocity through the agency of the bed shear stress (τ_b) generated at the bed by the flow of water over it. The bed shear stress is calculated from the flow velocity using well-established hydro-dynamic theory, and so the <u>potential</u> load of sand transported by the flow can be estimated if the grain size distribution of the bed sediments is known. Recent development of this approach by incorporating sand transport laws into a two-dimensional computer flow model of an estuary have demonstrated that the general characteristics of the time-varying concentrations of sand in suspension measured in a real estuary can be reproduced by this means in a computer model (ref. 2). In particular, the time lag between changes in velocity and changes in sand concentration that is an observed characteristic of tidal, as opposed to uni-directional fluvial, sites (ref. 3) is simulated.

8. This form of computer model could now be used to carry out an initial feasibility assessment of sand transport in an estuary, and potential sedimentation in a substantial dredged trench, but it is likely to over-estimate the sedimentation rate for the following reasons. Firstly, the form of the velocity distribution through the water depth may be modified from the classical theoretical form by the presence of appreciable quantities of sediment in suspension, especially near the bed. Recently published work by Soulsby and Wainwright (ref. 4) has shown that the theory then over-estimates the bed shear stress, and hence the sediment transport. This may be corrected using the expressions provided in ref. 4, but at a significant cost in computing efficiency if applied at every computational point and time-step. Secondly, the model has to assume that sand is uniformly available throughout the model area. In practice this is not so, there being spatial variations in size distribution and areas where no sediment supply exists within the tidal excursion of water that passes over the dredged trench during a tidal cycle.

9. Previous engineering project studies carried out by Hydraulics Research have circumvented these problems by establishing the 'in-situ' relationship between tidal flow and sand transportation by direct and detailed measurements

Fig. 3. Small sand waves in the Conwy estuary

at the project site, and using the theoretical frameworks given in ref. 1 to reduce and manipulate the field data to predict the effect of the intrusion of a dredged trench into the estuary regime. This requires purpose-designed field equipment to gain precise measurements of velocity and sediment concentration within 1 m of the bed, as described in ref. 5. Such measurements are still an essential basis for confident prediction of sedimentation, but they could now alternatively be used as the means to validate or calibrate a two-dimensional computer model incorporating sand-transport process functions.

Effect of waves
10. The established sand transport formulae take no account of the additional effect of wave action in increasing sediment transport, and the detailed field measurements made from an anchored vessel must necessarily be carried out under relatively calm conditions. Nevertheless, at a site exposed to waves, or when the water crossing the trench will have passed through an area exposed to waves, the increased sediment mobility and its contribution to sedimentation in the trench needs to be quantified, taking account of the annual frequency of occurrence of wave activity. Note that although the effect of wave action might be very substantial, for a short time and occasional occurrence, in comparison with the regular transportation by tidal currents alone the gross effect of waves in a full year may be relatively small.

GEOTECHNICAL ASPECTS

11. Field measurements in the Thames estuary have provided a simple means of applying a correction to increase sand transport to allow for wave activity, based on the velocity created at the bed by the waves (ref. 6). In addition, recent laboratory experiments (ref. 7) on the combined effects of waves and currents on flat and rippled sand beds have shown that the maximum bed stress calculated from the combination of wave and current velocities can be used against the existing threshold criteria derived from current velocities alone to predict correctly the threshold of motion of sand. Thus there is now a reasonable basis for inclusion of moderate wave effects in sand transport, and hence trench sedimentation, predictions.

Bed movement

12. It is not uncommon to see small sand waves in a dynamic sandy estuary, for example the Conwy estuary shown in Fig. 3. Clearly these represent a potential source of sedimentation if they creep along the bed during the tide. However, to a degree they are implicitly included in the sand transport formulae calculations described above. Many of the transport formulae claim to predict <u>total</u> load (i.e. not just the transport in suspension), and the detailed near bed measurement of sand transport at the site will intercept and include sand particles that are moving in short trajectories as a result of sand wave motion. Moreover, the tidal sediment flux represented by creep of small sand waves under normal conditions can usually be shown to be very small compared with the transport of sand in suspension. Thus unless very large sand waves are found to be moving at the site, the contribution of sand wave creep to sedimentation can be expected to be relatively insignificant.

Slope stability

13. Long-term flume experiments on reversing flows passing across a trench in a sandy bed have indicated that the side-slope of the trench stabilises at a gradient of about 1:20. It is possible to create an initial stable slope at a much steeper gradient, but it is progressively re-graded by the passage of water and deposition/erosion of sediment on its shoulders. In ref. 1, Lean addresses the effect of waves on sand slope stability, and concludes that it is unlikely that the sides of a channel will fail as a result of wave action except in places where it is dredged through a high shoal exposed to breaking waves.

MUDDY SEDIMENTS

14. The characteristics of cohesive (muddy) sediments are quite different from sands, and the procedures and models developed for sandy situations cannot be extrapolated to muddy situations. Detailed information on muds can be found in ref. 8, summarised in ref. 5. It cannot be too strongly emphasised that reliable predictions of sedimentation in

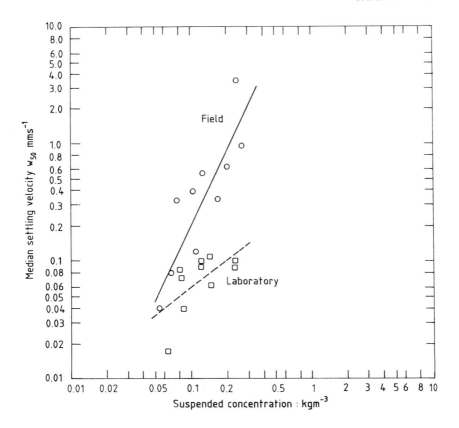

Fig. 4. Comparison of field and laboratory measurement of mud settling velocity

dredged trenches by muddy sediments depend upon the specific characteristics of the local site sediment being determined in an adequate manner. For example, the settling velocity must be measured 'in-situ' at the site under naturally occurring conditions: as is evident from Fig. 4, similar measurement in the laboratory gives significantly different (and erroneous) results.

15. The development of an endless mud carousel at Hydraulics Research (Fig. 5) has provided the means to reproduce and study the deposition of muddy sediments from suspension in flowing water under simulated tidal flow conditions (ref. 9). Experiments show that when the bed shear stress (τ_b) exceeds a minimum value (τ_m) none of the sediment being carried in suspension will settle. Thus during the main run of the tide, a reduction in velocity and hence bed shear stress as the sediment-laden flow crosses the trench will probably not induce deposition. When the bed

Fig. 5. The Hydraulics Research mud carousel

shear stress falls below a lower threshold value (τ_{cd}), all of the sediment will eventually fall out of suspension, at a rate dependent upon the median settling velocity (W_{50}) which in turn depends upon the sediment concentration (c). When τ_b lies between τ_m and τ_{cd}, a proportion of the sediment in suspension will eventually settle out of suspension, that proportion being independent of sediment concentration. Typical values for estuary muds are (ref. 10).

τ_m : 0.5 to 1.0 Nm^{-2}, τ_{cd} 0.06 to 0.10 Nm^{-2}

16. An indication of the rate of settlement (dm/dt) for the intermediate case is given in Fig. 6. It is thus evident that in the time taken for flow to cross a dredged trench during most of the tidal cycle, little or no sedimentation will result. Settlement of sediment in suspension into the trench normally only occurs around slack water periods, or in places where the tidal flow is reduced by other effects such as caissons, spur walls or heavily piled jetties.

Effect of waves

17. Wave action, particularly in shallow water over intertidal mud flats, is now known to be a very prolific source of sediment where such areas exist within the tidal

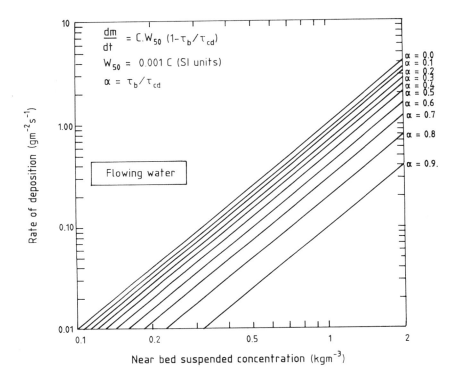

Fig. 6. Rate of deposition of typical mud from flowing water

excursion of the water passing the project site. It may not be seen as a large increase in sediment concentrations in suspension, but can create a very dense 'soup' of sediment, termed "fluid mud", that flows down the slope and along the bed driven by the tide, and which will be wholly intercepted by a trench dredged across its path. The effect of wave action on fluidisation of mud deposits can now be reproduced and quantified in a suitable laboratory facility, but the theory is not yet sufficiently well developed to be applied with confidence. For engineering purposes, if wave action over mud flats appears to exist at or near the project site, its relevance and importance can be quantified in the laboratory.

Bed movement
18. In muddy estuaries, bed movement does not normally occur, but a similar effect is created when "fluid mud" forms. This may be the result of wave action, as noted above, or may be formed by settlement of concentrations of sediments greater than 2500 ppm at tidal slack water times

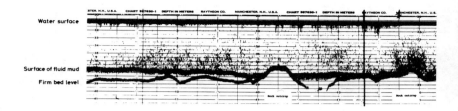

Fig. 7. Fluid mud in a tidal channel

(ref. 11). Once formed, the fluid mud is a dense suspension of sediment particles at a concentration in the range 60,000-120,000 ppm. It tends to flow down gravity slopes into the deepest part of the estuary channels, and will flow along the channel as a separate dense layer under the hydrostatic head imposed by the water slope as the tide reverses. In exceptional cases, fluid mud can be seen on an echo-sounding as in Fig. 7, but more usually needs sophisticated field instrumentation to detect and measure it (ref. 12). It is vitally important to recognise a situation in which fluid mud may occur, and then to be able to quantify it by field measurement, for it represents a potential source of sedimentation in a dredged trench at least one order of magnitude greater than that threatened by sediment in suspension.

Slope stability

19. The critical angle of an underwater slope in muddy sediments can be estimated from Fig. 8 if the undrained shear strength (Cu) and the buoyant weight (γ') of the mud are known. It is most likely that the mud is under-consolidated, so the pore pressure in the bed is greater than the hydrostatic water pressure. The value of $Cu/\gamma' z$ is anticipated to be in the range 0.03-0.15 depending on the degree of excess pore pressure (ref. 10). It is evident from Fig. 8 that the solution of Booker and Davis (ref. 13) for a finite slope, which is appropriate to a dredged trench, predicts a steeper gradient than the solution of Morgenstern (ref. 14) for an infinite slope. This is consistent with observation, that cut slopes in a mud bank will stand more steeply than the mud bank itself.

PRACTICAL CONSIDERATIONS

20. Confident prediction of sedimentation and slope stability of a dredged trench across an estuary depends upon an adequate <u>quantitative</u> knowledge of the estuary sediment and its properties. The effect of the trench on the tidal flow can be calculated either by using the results of previous laboratory and theoretical studies (ref. 1) or by

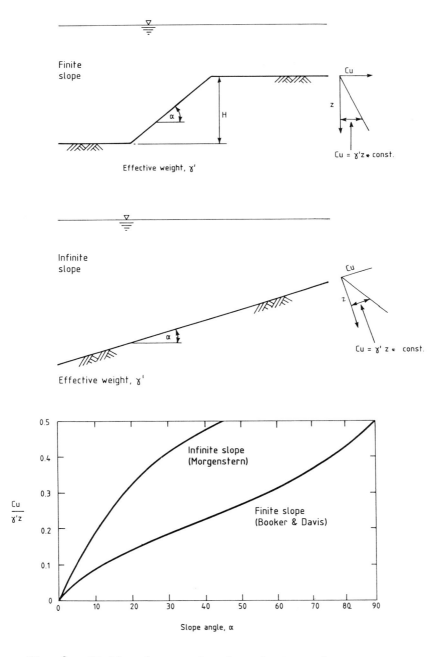

Fig. 8. Stable slope angles in cohesive sediments

creating the effect in a two-dimensional hydraulic or computer model. If the trench will be a major change in estuary bathymetry, and is therefore likely to cause a re-distribution of tidal discharge, a two-dimensional model of the interaction between the trench and the estuary is recommended. If the trench is a relatively small feature in a stable estuary regime, then its effect can reasonably be calculated.

21. Existing sediment transport theories for sand and typical available data for mud will provide a first estimate of sedimentation, but should not be relied upon absolutely. Theories and models need to be calibrated and/or validated by site-specific measurements of the sediments and the sediment supply if the results are to be reliable, and the whole estuary regime should be expertly appraised so as to detect any potentially unpleasant surprises such as the formation of fluid mud or the existence of large, mobile sand bed features.

22. The techniques described in this paper have been applied by Hydraulics Research to a number of engineering project studies in the UK and overseas. They are based on the current state of theory and knowledge, but would benefit from verification at a construction site. The construction of an immersed tube crossing at Conwy, for which extensive detailed studies have been carried out, offers the opportunity for this, but in the present political and commercial climate it seems unlikely that resources will be made available for the purpose.

ACKNOWLEDGEMENTS

23. This paper summarises research and engineering project experience contributed by many colleagues at Hydraulics Research, Wallingford. Much of the research was funded in whole or in part by the UK Department of the Environment through its support for the UK construction industry.

REFERENCES
1. LEAN G.L. Estimation of maintenance dredging for navigation channels. Hydraulics Research Station, Wallingford, 1980.
2. WILD B.R. A numerical sand transport model with time dependent bed exchange. Hydraulics Research, Wallingford, 1988, Report No. SR 148.
3. THORN M.F.C. Deep tidal flow over a fine sand bed. Proc. I.A.H.R. XVIth Congress, Paper A27, Sao Paulo, 1975.
4. SOULSBY R.L. and WAINWRIGHT B.L.S.A. A criterion for the effect of suspended sediment on near-bottom velocity profiles. J. of Hyd. Res., Vol. 25, No. 3, 1987.
5. THORN M.F.C. Physical processes of siltation in tidal channels. Hydraulic modelling applied to maritime engineering problems, Thomas Telford, 1981.

6. OWEN M.W. and THORN M.F.C. Effect of waves on sand transport by currents. Proc. 16th Coastal Engineering Conf., Hamburg, 1978.
7. KAPDASLI M.S. Threshold condition of sand particles under co-directional combined wave-and-current flow. Hydraulics Research, Wallingford, 1985, Report No. SR 69.
8. DELO E.A. Estuarine muds manual. Hydraulics Research, Wallingford, 1988, Report No. SR 164.
9. DELO E.A. Deposition of cohesive sediment from flowing water. Hydraulics Research, Wallingford, 1988. Report No SR 152.
10. DELO E.A. Siltation and stability of cohesive dredged slopes manual. Hydraulics Research, Wallingford, 1988, Report No. SR 180.
11. ODD N.V.M. and RODGER J.G. An analysis of the behaviour of fluid mud in estuaries. Hydraulics Research, Wallingford, 1986, Report No SR 84.
12. WATERS C.B. Measurement of fluid mud layers. Hydraulics Research, Wallingford, 1987, Report No SR 129.
13. BOOKER J.R. and DAVIS E.H. A note on a plasticity solution to the stability of slopes in inhomogeneous clays. Geotechnique, Vol. 22, No. 3, 1972.
14. MORGENSTERN N.R. Submarine slumping and the initiation of turbidity currents. Marine Geotechnique, pp 189-220, Univ. of Illinois Press, 1967.

6. Trench siltation: origin, consequences and how to cope with it

J. H. VOLBEDA, MSc, and G. L. M. van der SCHRIECK, MSc, HAM Dredging, The Netherlands

SYNOPSIS. The authors have recently been involved in the solution of a serious siltation problem during the construction of the Emstunnel in Germany. This resulted in a review of the main factors and phenomena that must be allowed for when using the immersed tunnel technique.

I-INTRODUCTION

When a waterway has to be crossed by a road or railway, immersed tunnel construction is, particularly in lower delta regions of a river, a preferable solution. This is because of the presence of intensive shipping activities. However in these regions the siltation is dominated by the flocculation effects following from salt wedge intrusion, resulting in a strongly increased siltation forming layers of fluid mud. It is this type of siltation that can cause severe problems in the construction phase. In this article the siltation itself, the consequences during construction and possible solutions will be discussed.

II-ORIGIN, QUANTITY AND CONSOLIDATION OF TRENCH SILTATION

1. Origin and quantity of siltation. The amount of silt transport in the upstream region of a river can be measured and predicted reasonably well. Moreover the quantity of silt transport in that region is mostly relatively low. This is why little or no hindrance due to siltation in tunnel trenches has been encountered in upstream regions. Examples of these are some of the tunnels in the upstream region of Rotterdam like the Maastunnel, Heinenoord tunnel, Metro tunnel, etc.

In the lower river region the salt water wedge intrudes along the bottom having a greater density than the fresh river flow .(fig. 1.). The mixing proces of fresh and salt water causes the fresh water silt load to flocculate and have considerably increased settling velocity. This type of increased siltation will occur within the tidal range of the river especially where the saltwedge phenomenon is dominant.

Immersed tunnel techniques. Thomas Telford, London, 1990.

In the bottom layer of the salt wedge the current is in an upstream direction, pushing the silt to the tip of the salt wedge, where it is accumulated and will form a bottom layer of fluid mud. The location of the tip of the salt wedge is not stationary, it will move back and forth with the tide and with the changes in river discharge. Thus
silt that settled at the tip of the salt wedge will be reworked by the currents and partly accumulate and resettle at the new location of the tip. This wandering quantity of accumulated mud is the main origin of siltation troubles in tunneltrenches. Tunneltrenches upstream of the saltwedge normally have very little siltation provided the riverbottom consists of sand. Downstream of the tip of the saltwedge siltation in tunneltrenches can be considerable, due to the combined effects of marine silt brought on by the salt wedge and river silt flocculating and being deposited.

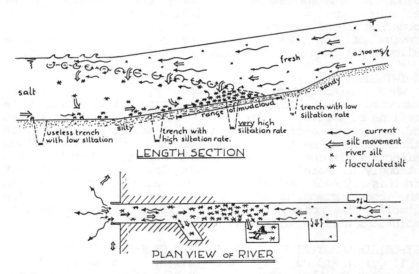

Fig. 1. Flocculation caused by salt wedge intrusion.

The process described above is continuous. It is thus clear that siltation within the saltwedge region can be many times greater than upstream or downstream of the salt wedge. This aspect should be given much attention during the design and construction phase of a tunnel trench.

2. <u>Consolidation of siltation.</u> The fluid mud layer mentionned above will flow into the lower area of a trench and accumulate to a thick layer if no maintenance dredging is performed. Once in the trench gravity induced "consolidation" will form a bottom layer with increasing density and shear stress. The speed of this proces depends mainly on the layer thickness, the percentage of heavy sand grains and the

characteristics of the trench bottom with respect to permeability and groundwater pressure differences in relation to the actual river level. It is especially the latter circumstance that can cause unexpected phenomena. Negative groundwaterpressures in the sands of the trenchbottom (e.g. caused by artificial low water levels in polders alongside the river) will result in rapid consolidation of the mud near the interface with the sandy trenchbottom (Emstunnel). It will be difficult to remove this consolidated mud without disturbing the underlying trenchbottom. On the other hand a positive groundwaterpressure will prevent consolidation of the mud as groundwater is continuously "injected" into the mud (Hamburg).

III-CONSEQUENCES DURING CONSTRUCTION

3. <u>Floatation of tunnel elements.</u> In many cases the tunnel elements are constructed near the site in a temporary drydock. When they are finished the dock is flooded and the elements floated for transport to the trench. If however the siltation in the drydock is high the bottom of a tunnelelement will be sealed to the dock's floor by a mud layer which may prevent it from floating. Regular maintenance dredging in combination with air or waterjets underneath the element at the moment of float out may be necessary in such cases. The presence of an internal piping layout for the final underfilling of the tunnelelements in the trench is a definite advantage here, as this system can be used to break the adhesion between element and dockfloor by pumping water through it.

4. <u>Cleaning of the trench before placing of the elements.</u> If the siltation in the trench has reached a level above the final bottom level of the elements, placing of the elements on their temporary supports can only be achieved by applying sufficient ballast in the elements in order to squeeze away the mud. This is a hazardous operation. Removal of excessive mud just before placing will be necessary. This trenchcleaning operation is an organisational headache as surveying, mudcleaning and underfilling of the previously placed element is all taking place at the same time and in close proximity. Careful planning of anchoring systems and equipment activities is a must to achieve satisfactory quality control. (Ref. 1.) In a busy constricted navigation channel this is even more complicated.

5. <u>Overloading of temporary supports.</u> Most immersed tunnelelements are positioned on temporary supports which enable exact vertical adjustment of the element to the required levels before underfilling will take place. Normally the load on the supports is low and calculated according to Archimedes's law and where necessary kept within the calculated range by pumping limited amounts of water in or

out of special build-in ballasttanks. High siltation rates may effect the supportloads in a very surprising way, well beyond the correction capabilities of the ballasting system. A layer of fluid mud may cause a higher floating force than expected, leading to negative support reactions. Siltation on the roof of the element has the reverse effect. Further consolidation of the mud underneath and besides the tunnelelement will cause variations in the support reactions wich follow the vertical tide. As the mud is no longer a pure (heavy) fluid but a more or less impervious clay the waterpressure at the bottom of the element no longer follows the variation in waterlevel at the top. In the case of the Emstunnel these variations were alarming, however by chance the ballasting system had a wide overcapacity which fortunately could just cope with the situation. (Ref.2.). Dredging silt from the roof of the elements proved necessary in addition to the gredging of the silt from the side of the elements which was a prerequisite for the underfilling.

7. Removal of silt and underfilling. Although the groundpressure of tunnelelements in their final position is low, differential settlements must be avoided. The bed material used for underfilling should therefore have : 1) a constant layer thickness, 2) a low and consistent consolidationcoëfficient throughout. Dredging of the trench bottom must be done within close tolerances. The clearance between trenchbottom and element should be as low as practical conditions allow. Sand is normally used and will, in general, result in reasonable final settlements in the order of 5 to 15 cm. Mud or soft clay with high and unpredictable consolidation characteristics is unacceptable as underfilling material and should be removed before underfilling.

In principal two methods are available for the removal of the silt from underneath the elements. The first one is to erode the silt by waterjets. The second one is to use mechanical means to break the cohesion of the mud.

Eroding the silt by jets is the easiest method in practice. It is often combined with sand underfilling in one operation using the external method as first deployed by Christiani and Nielsen before Wordwar I. To apply this method the mud should be more or less fluid with a very low cohesion, in order to be eroded by watercurrents with a low degree of turbulence.(Fig. 2.). At higher levels of turbulence by high velocity jets not only will the mud be eroded but also the trenchbed itself, resulting in the end in differential settlement.

Siltation with some erosion resistance therefore cannot be removed by jets but has to be treated mechanically. This may be excecuted with a raketype instrument to break the cohesion and more or less transport the mud or by applying a bulldozerblade to shove the mud to the side of the tunnel

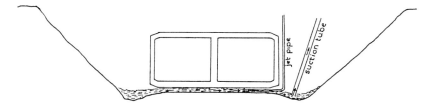

Fig. 2. Silt removal in case of "fluid" mud.

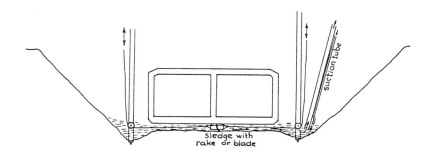

Fig. 3. Silt removal in case of "consolidated" mud.

where it then can be removed by a dredger. Immediately after cleaning and inspection of a small section underfilling of that section with sand must be excecuted to prevent influx of mud.(Fig. 3.).

Underfilling with sand has been carried out in an alternative way via an internal piping system with exits for the sand water mixture in the bottom of the element. When underfilling is completed, the exits are sealed with concrete and later on buried under the ballast layer of conrete in the tunnel. (Fig. 4.). This internal system has definite advantages in operational simplicity, but should only be used if siltation rates in the trench are limited.

IV-SOME OTHER SOLUTIONS

7. <u>Solutions by design adaptations.</u> So far only solutions in dealing with the siltation problem by aiming at the total removal of the silt are considered. There may be other ways to cope with it by making the right adaptations in trench and element design. Although the authors are aware that such

GEOTECHNICAL ASPECTS

Fig. 4. Internal and external sand underfilling system.

design idea's need thorough study and testing some of them are listed below.

a) <u>Foundation layer of very coarse stones.</u> Immediately after dredging the trench a foundation layer of very coarse stones is constructed by means of dumping and very accurate levelling. Because of the great hydraulic stability of coarse stones the use of waterjets to finally clean the layer just before placing the element will be possible. Remaining thin layers of fresh silt will be pushed in the pores between the stones.

b) <u>Foundation on piles.</u> When a pile foundation under the vertical walls of the element is used the special underfilling with sand is no longer necessary.(Ref.3.)

c) <u>Element bottom with vertical ribs.</u> Another way to cope with the silt might be to use vertical ribs under the tunnel elements. After accurate dredging of the trench bottom the bulk of the siltation can be removed just before placement by dredging without touching the original trenchbottom. The thin remaining siltlayer will be penetrated by the ribs squeezing the mud to either side. For reasons of equal spread of load the ribs should be knife edge shaped. Sand underfilling then is no longer necessary. Construction of the ribs in the drydock will be an extra problem however.

d) <u>Use of gravel bags.</u> The use of gravel bags to prevent siltation under the elements immediately after placing is a technique which has been used succesfully at the Eastern Scheld Storm Surge Barrier. Although complicated and expensive it could be applied because it concerned solitary elements. In tunnelconstruction however there are extra complications at the joints of the elements.

V-CONCLUSIONS

The main conclusion of this review is the necessity to recognize the siltation effects as a major item from the very beginning of the design phase. The siltation rate based on only the upstream situation can be far from reality. Special attention should be given to the increased siltation rate from flocculation and the effect of accumulation in the tidal region. When high consolidation rates are to be expected special dredging techniques and/or adaptations in the tunnel design must be considered. The idea's presented here do not pretend to be final solutions but are intended to be a contribution in the search of even more succesfull immersed tunnel construction methods.

REFERENCES

1. HOLSTERS H. Dredging Performance at Great Depth in a Tidal River, Proceedings Wodcon 1968, World Dredging Conference, California, 1986.

2. LINNEKAMP J. and DEN BREEJEN W. The influence of sludge on support reactions during construction of the Emstunnel. Proceedings Modelling Soil-water-structure Interactions SOWAS 88, A.A.Balkema, Rotterdam, 1988.

3. DIJK van H. The Excecution of Dredging for the Construction of the Metrotunnel under the river Nieuwe Maas in Rotterdam, Proceedings Wodcon 1968, World Dredging Conference, California, 1986.

4. STIKSMA K. Tunnels in the Netherlands underground transport connections, National Public Works Department, ISBN 90 6618 536 8

Discussion

MR H.D. OSBORN, HAKA UK Ltd
I was interested in the injection piles used in the Zeeburger Tunnel, described in Paper 1.

In the 1920s and 1930s, screw piles were used to provide bearing capacity in soft alluvium materials. These were used extensively for bridge piers in India and so on. Screw piles were last used in the UK in the 1940s, at the wartime military ports in the West of Scotland. These screw piles were large cast-iron propeller type screws of the order of 3-4 m diameter, with one to three screws per pile shaft.

Is this not a system that could be used in the future to provide bearing capacity in alluvium materials?

MR C.R. WEEKS, Mott Macdonald
On the Second Downtown Elizabeth River Tunnel, referred to in Paper 2, a dyke was used to provide a dry environment for making the final joint between the immersed tube and the in-situ section of tunnel. The dyke is a substantial structure and dewatering was quoted as a problem. Why was the final joint carried out in the dry, rather than being completed under water?

MR R.L. TAYLOR, Acer Freeman Fox
In the case of the Second Downtown Elizabeth River Tunnel, was there not a risk of loss of bedding material from beneath the immersed tunnel unit when the contractor used heavy pumping to dry out the landward tunnel excavation, since he chose not to grout under the unit?

With regard to the Second Hampton Roads Tunnel, also referred to in Paper 2, the Authors relied on sand draining and surcharge to accelerate settlement of the soft clay beneath the new island. How were they able to judge when sufficient settlement had taken place in order accurately to place the immersed tunnel? How accurate were they?

GEOTECHNICAL ASPECTS

PROFESSOR B.A. O'CONNOR, Liverpool University
Would Mr Munfakh indicate why a cable-stay bridge was adopted for the Chao Phya crossing instead of an immersed tube tunnel?

Could Mr Grantz explain why a concrete tube tunnel was rejected in preference to a steel tube tunnel? Was this really on environmental grounds or because of greater experience with steel tunnels?

Would Mr Thorn explain the problems of fluid mud arising from sediment which has settled out on the side slopes of trenches, and also the serious problem of wave reflection from individual dredged trenches?

MR M.N. KELLEY, Kiewit Engineering Company, Omaha, Nebraska
My question concerns the highest current velocities attempted in immersed tube construction. The New York City 63rd Street Tube and Tunnel Project, built between 1969- 1973, had the maximum current velocities experienced by the Kiewit Company. Maximum tidal velocities were in the 3 m/s range and all mooring calculations, for what was known as the standby condition, were based on that velocity. Actual tube placing mooring calculations were based on 1-1/2 m/s with the current dropping. Slack tide at that site lasted about 20 min, with the water surface in reality never coming to a complete slack condition, but with water flowing at minor velocities at the top in one direction and water flowing at greater depths in the other direction. The tubes were placed at a depth of about 30 m. Also, on the San Francisco Trans Bay Tube Project, current velocities were about 1.5 m/s, with a tube placing depth of 40 m. The type of material used for the screeded bed will be influenced by the current. It has been found by experience that graded, rounded gravel of the 37-25 mm minus size works the most satisfactorily. Normal sand, say 5 mm minus size, has been found to be too fine and will easily erode under small bottom currents. On the New York project, even the 37 mm minus material would erode under the maximum tide velocities. To prevent erosion of the screeded bed, it was necessary to use 150 mm minus material and to increase the allowable screed bed tolerances.

MR H.A. KAMP NIELSON, Comar Engineers A/S, Virum, Denmark
Could Mr Thorn expand on the effect of trench slopes and on possible current separation?

Could Mr van der Schrieck explain the type of salt water wedge referred to in Paper 6, and what is meant by a well-mixed estuary?

MR KRUIZINGA, Paper 1
In reply to Mr Osborn, the choice of a pile system depends on

DISCUSSION

the type of soil, the loading and the allowable settlements of the structure (serviceability requirements).

Basically, one can distinguish displacement piles as driven prefab piles, and non-displacement piles as bored piles. The latter show a weaker behaviour in terms of load- settlement characteristics.

Owing to the need for vibrationless piling together with a less weak settlement behaviour during the last decades, screw and/or auger pile systems have been developed with diameters from 0.4 to 0.9 m. The continuous flight auger piles, allowing for decompaction of the soil, are more of the non-displacement type than the other systems with one to three screws just above the pile tip, under the condition that the rate of penetration is as close as according to the pitch. Therefore, a high torque capacity machine is required which is also able to provide a vertical pushing force during screwing in. These machines are now available on the market.

The system with very large cast-iron propeller type screws of 3-4 m diameter seems to be more of the non-displacement type with a more unfavourable price/bearing capacity ratio than the piles mentioned previously.

For the Zeeburger Tunnel, tension piles with a high pull-out capacity were needed; therefore, a displacement type of pile together with a grouted shaft (in the bearing layer) was chosen.

DR MUNFAKH, Paper 2
In reply to Mr Weeks, to meet the right construction schedule, the immersed tube elements had to be placed before construction of the cut and cover section of the project. For this reason, the last joint had to be constructed in the dry.

In reply to Mr Taylor, the bedding material at the Second Downtown Tunnel was designed to meet the filtering requirement with the in-situ soil. Therefore, the loss of material was not a concern. Keeping the excavation dry, on the other hand, was a concern which was handled by the contractor, using heavy pumping.

At the Hampton Roads Tunnel, the long-term consolidation behaviour of the compressible soils was monitored through the use of piezometers and settlement indicators. The horizontal displacements in the clays were monitored by inclinometer readings which verified the analytical predictions. Once the vertical and horizontal movements had been stabilized, the surcharge was removed, a trench was excavated and the end immersed tube was placed and backfilled. The net loading on the native soil at that stage was negative.

In reply to Professor O'Connor, the decision to adopt the bridge option was based on cost and the difficulty in finding a suitable site for building a casting yard near the project location.

GEOTECHNICAL ASPECTS

MR GRANTZ, Paper 3
With regard to the question by Professor O'Connor, my reply is as follows.

Early in the Central Artery/Tunnel (CA/T) Project, we did an extensive evaluation of the two alternatives. In the process, we identified a potential site for concrete tunnel fabrication that would not have involved either environmental impact or an increase in schedule time. There is a large shipyard in Quincy, Massachusetts, which could easily have handled the fabrication of the concrete elements, but it was decommissioned a few years ago. It was decided, however, that the viability of the shipyard for this work was in question since it has been purchased by another agency for other purposes. Further, it was felt that to make this type of tube construction contingent on the use of a single local shipyard would overly restrict bidding competition. In view of this, the viability of two other local sites in which to excavate a graving basin were investigated. The first of these, Lynn Harbor, was found to involve a great deal of dredging for access and was discarded from further consideration. The second site, at Governor's Island Flats, had easy land access and was close to the alignment but involved a considerable amount of rock excavation. Based on this, schedule and cost estimates indicated that if a graving basin had to be constructed, scheduled time would be increased substantially even if there were no delays in the environmental permitting process, and that the concrete element alternative would be more costly. We feel, therefore, that the concrete alternative was given a fair appraisal.

In another portion of the CA/T Project, we are planning a very complicated concrete box tunnel. This was the result of a similar detailed evaluation of alternatives which led to the conclusion that the concrete box element was the only logical choice for the site conditions. In this case, the elements will be fabricated in the approach excavation. This excavation will be landlocked by existing bridges and could not possibly be used for the Third Harbor Tunnel.

MR VAN DER SCHRIECK, Paper 6
Mr Kamp Nielson asks for an explanation of the type of salt water wedge referred to in our Paper.

When a river discharges its fresh water into a salt water sea, the fresh water is lighter than the receiving salt water. Therefore, the fresh water will tend to flow above the salt water, thus forming a fresh water layer on top of the salt water. In the transition zone between both layers, a mixing process will occur, leading to the flocculation of the silt in the fresh water layer. Because of tidal effects, the region in which this phenomenon occurs will travel inland with high tide and seawards with low tide. The salt water body under the fresh water layer is wedge shaped which explains the name 'salt water wedge'.

The meaning of a 'well-mixed estuary' is as follows. The mixing process mentioned above can be more or less active over a larger region, depending on the hydraulic situation. In the case of a narrow river, the volume of the tidal prism is relatively small, resulting in a stratified situation with a salt wedge. However, as the river gradually becomes wider and forms an estuary with a large tidal prism volume, the mixing process becomes more effective, resulting in an estuary in which the salt and fresh water is well mixed with a horizontal salinity gradient.

7. Bosporus railroad tunnel crossing design alternatives

V. TANAL, PE, Parsons Brinckerhoff International, Tapei, Taiwan,
W. C. GRANTZ, PE, and L. W. ABRAMSON, PE, Parsons
Brinckerhoff Quade and Douglas, Inc., Boston and San Francisco

SYNOPSIS. Marine and land borings, geophysical, bathymmetric and hydrographic surveys were performed to study the feasibility and complete the preliminary design for a railroad tunnel crossing under the Bosporus in Istanbul, Turkey. A tube tunnel, entirely immersed below the channel bottom presented an environmentally acceptable, plus technically and economically feasible solution.

INTRODUCTION

1. A comprehensive urban transportation study initiated in 1985 by the Turkish Ministry of Public Works, later taken over by the Ministry of Transport and Communications, concluded that the best means of meeting the long term transportation needs of Istanbul would be to combine an underground metro network with a Bosporus railroad crossing to connect the European and Asian parts of the city (Fig. 1). The study was conducted by Istanbul Rail/Tunnel Consultants (IRTC), a consortium led by Parsons Brinckerhoff International Inc., that included Kaiser Engineers International Inc., PB-TSB Consulting and Engineering Co. Ltd., Temel Muhendislik A. S. and Tumas Muhendislik, ve Muteahhitlik A. S.. The consortium evaluated the technical, environmental and economic feasibility and performed the preliminary designs of both the metro and the Bosporus crossing. This paper presents the results of the engineering evaluations and the preliminary design features of the Bosporus tunnel crossing.

GEOLOGY

2. The Bosporus is a natural channel between the European and Asian continents, that connects the Black Sea and the Marmara Sea and divides the city of Istanbul into the European side and the Anatolian side. The geology of the Bosporus was significantly affected by the geological events during the Upper Miocene and Pliocene periods. Aerial photographs and field measurements reveal the presence of joint sets, faults and shears in the northeast-southwest and northwest-southeast directions.

Immersed tunnel techniques. Thomas Telford, London, 1989.

PLANNING

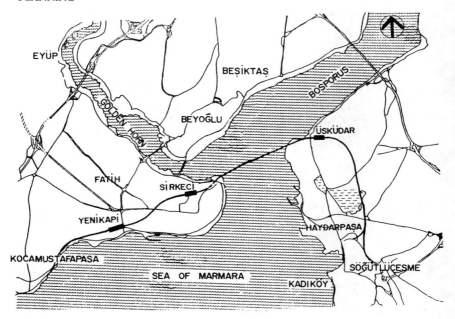

Fig. 1 Bosporus Railroad Tunnel Alignment and Stations

There are three major theories on the origin of the Bosporus: the Bosporus is a graben, it is a drowned valley, or a river which has changed course under the effects of underwater erosion.

3. The geologic evaluation of the area suggests that the Bosporus and the Golden Horn, an inland branch, are ancient valleys that formed a sea connection by drowning during the Pleistocene period. Subsequent changes were caused by surface and underwater erosion or deposition.

SEISMICITY

4. The seismically active Marmara Region encompassing the project site is located on the western edge of the North Anatolian Fault Zone. The area can also be affected by seismic events originating in the adjacent Aegean Graben seismic zone. Earthquake data collected for the project were derived from historical data for the period prior to 1900 and from instrumental data for the post-1900 period. Seismicity was evaluated for the events in the entire region and for events in the Marmara region only. On the basis of these evaluations, a service level acceleration of 0.12 g, velocity of 13 centimeters per second, displacement of 7.6 centimeters, and a survival level acceleration of 0.17 g, velocity of 17 centimeters per second, displacement of 10.2 centimeters was adopted for preliminary design.

The service level was defined as load levels with a 70 percent probability of not being exceeded in a 100 year period. The survival level loads, which may cause damage but not a catastrophic collapse, are defined as loads with a probability of 95 percent of not being exceeded in a 100 year period.

SITE INVESTIGATIONS
5. Field investigations were undertaken by IRTC in 1985 and 1986 to explore the subsurface, bathymmetric and hydrographic conditions of the project site.

Underwater surveys
6. Several types of underwater surveys were carried out over a wide band between Topkapi Point, just south of the Golden Horn and Uskudar, on the Asian side of the Bosporus. These included several bathymetry surveys using three types of survey echosounders. Side scan sonar surveys were used to identify obstructions on the channel bottom. Subbottom reflection and refraction surveys were made and correlated with marine borings.

7. These surveys revealed that the maximum channel depth at the tunnel alignment was approximately 50 meters with the bedrock interface some 40 meters below this. No sunken wrecks were identified by the sidescan sonar although it clearly showed existing telephone cables crossing the alignment. Because of the soft bottom conditions which were found to exist, the subbottom reflection surveys did not penetrate to the bedrock interface except in some limited areas. The refraction survey using a remote hydrophone/transmitter and a compressed air gun gave profiles of velocity in rock and in sediments, which with some speculation, could be correlated with the borings to extend the bedrock interface.

Geotechnical investigations
8. Geotechnical investigations included borings, laboratory testing, and geologic mapping. A total of 62 borings were drilled along the proposed railroad tunnel alignment. Seventeen of these borings, or 746m, were drilled on the European side of the alignment, twenty five or 772 m were drilled in the Bosporus, and twenty or 598 m were drilled on the Asian side of the alignment. The twenty five borings in the Bosporus were drilled off a floating barge held in place by 4 to 5 heavy anchors. Because of the relatively great sampling depths below water and the equipment and expertise at hand, standard penetration tests were difficult to carry out and often produced very poor sample recovery. Grab samples were obtained from the channel bottom by divers; several

undisturbed samples and numerous split spoon samples and rock cores were recovered from the borings. Laboratory tests were conducted on selected soil and rock samples to determine the physical properties of the subsurface materials. Geologic mapping was conducted along the alignment in the nearby rock exposures and construction excavations to investigate the orientation and characteristics of discontinuities.

Subsurface conditions

9. The immersed tube tunnel alignment can be divided into four major sub-sections according to location and subsurface conditions: the western end section in rock; the section founded on marine clays on the west side; the section founded on marine sands on the east side; and the eastern end section in rock (Fig. 2). Subsurface conditions at the west shore of the Bosporus where the tube tunnel will be founded on rock consist of marine sediments ranging in thickness between 10 and 20 meters. They consist of very loose to very dense, medium to course sand and gravel with silt, clay, brick fragments, coal, shells, and cobbles. Subsurface conditions at the west side of the Bosporus where the tube tunnel will be founded on marine clays consist of three major soil stratifications. The upper zone is approximately 10 to 17 meters thick and consists predominantly of very loose to very dense, coarse to fine, silty, clayey sand with shell, coal, and brick fragments. The middle zone consists of very soft, sandy, silty clay with shell fragments and ranges in thickness between approximately 5 and 20 meters. The lower zone above bedrock consists predominantly of clayey gravel with a mixture of sand, silt, and shells and is between 3 and 16 meters thick.

10. At the east side of the Bosporus, the tube tunnel will be founded on marine sands. There are two major soil units at this side. The upper zone is similar to the upper zone on the west side of the strait and consists of loose to very dense, silty and clayey, coarse to fine sand with gravel, shells, and brick fragments. This zone is between 10 and 24 meters thick and density generally increases with depth. The lower zone, between 5 and 9 meters thick, is similar to the lower zone on the west side of the strait, consisting of very dense, coarse to fine, sandy gravel and cobbles with clay, shells, and occasional sandstone boulders. Subsurface conditions near the east shore consist of marine sediments ranging in thickness between 7 and 18 meters. They consist of very loose to very dense, coarse to fine sand and gravel with silt, clay, shells, coal, and rock fragments.

11. Bedrock consists of gray medium to fine grained sandstone, siltstone and claystone. The upper zones generally consist of highly altered gray sandstone and siltstone with core recoveries of 5 to 25 percent and RQDs of zero. Along the shorelines bedrock is intensely fractured but sufficiently strong to provide good foundation support for the end tubes. At the west shore it is encountered at about elevation -31 meters and dips toward the Bosporus at an angle of approximately 15 degrees. Bedrock near the east shore is encountered at about elevation -6 meters and dips toward the Bosporus at and angle of approximately 5 degrees. At a distance of approximately 400 meters off-shore, bedrock drops about 20 meters to elevation -45 meters along a fault zone. Bedrock was not encountered in borings toward the middle of the Bosporus which were drilled up to 46 meters below sea bottom to the elevation of -90 meters.

Hydrographic surveys

12, Because of the significant depth of the Bosporus the vertical profile of the tunnel was an important consideration. The deeper the profile, the longer or steeper the approach tunnels would have to be. In earlier studies it was planned that at least in the deepest part, the tunnel elements would be supported and protected in an embankment which would extend above the channel bottom.

13. The Bosporus is characterized by a two-layer flow. Fresh water runoff into the Black Sea finds its way to the only outlet at the Bosporus. Salty Mediterranean water migrates toward the Black Sea underneath this outgoing fresh flow. The result is an environmentally sensitive balance for the Black Sea. Changes in salinity of small magnitude could potentially have long-term effects on the ecology of the Black Sea and its fishing industries.

14. The factors affecting the currents in the Bosporus are the level difference between the Black Sea and the Marmara Sea, the atmospheric pressure difference, the wind direction and velocity, and the difference in salinity (therefore densities) between the two seas. The Sea of Marmara has a higher concentration of salt and therefore a higher density. When interfaced with the lower density of Black Sea water, a two-layer, two-directional flow results.

15. Measurements and studies were carried out for the simultaneous evaluation of flow characteristics, level differences, meteorological conditions and water density. Vertical profiles of depth versus temperature, salinity and Ph were made at specific stations along the Bosporus. Static strings of current meters were placed and left to record temperature, salinity and current for a month at a time.

PLANNING

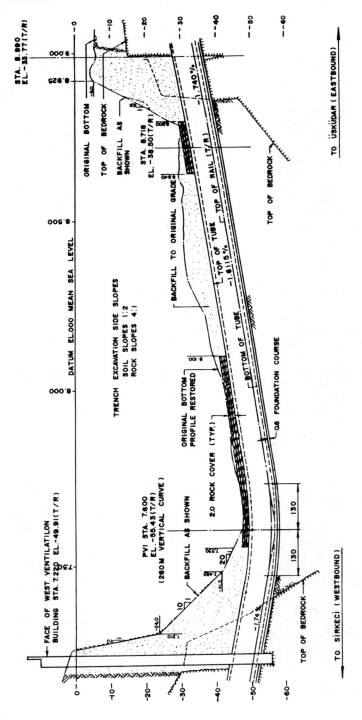

Fig. 2 Tube Tunnel Profile

16. The long term current measurements revealed that for more than 95 percent of the time there is a two-layer flow system. However, there are periods in which there is a flow in only one direction from the Black Sea to the Marmara. During a storm the interface between salt water and fresh water was discovered to be almost driven entirely out of the Bosporus to the south; a phenomenon which had been suspected but never actually measured before.

17. The measurements showed that the top layer has a salinity of above 18 parts per thousand and the bottom layer about 57 parts per thousand. The middle, or transition layer, is vertically stratified with varying properties. The layer is up to 10 meters thick at the southern end and about 2 meters thick at the northern end of the Bosporus. Current measurements made at the proposed tunnel location indicate that the interface is 13 to 20 meters below the surface; therefore, if the tunnel were to project above the channel bottom, it would be subject only to the lower current flow. The recorded velocity of the lower layer flow at this site did not exceed 1.0 meter per second.

DESIGN ALTERNATIVES
Vertical and horizontal alignment considerations
18. Since the rail tunnel is to handle freight trains as well as commuter trains (the largest traffic component) maximum gradients are set to be 1.74 percent by the Turkish Railway criteria. Based on this, the depth of the tunnel at the crossing determined the minimum depth of the stations on both sides of the Bosporus. Horizontal alignment was investigated in earlier studies and set further to the north than was ultimately arrived at due to better geotechnical information.

19. Borings on the European side indicated that competent rock was not available along the first alignment and that a mined tunnel would not be feasible. This left the alternatives of either constructing by cut-and-cover methods through a very densely built-up area, or shifting the alignment to one where rock was available to the beginning of the immersed tunnel. On the Asian side it was also decided to shift the alignment to provide for stations in locations more favorable to patronage, in the busy center of Uskudar where the major ferry traffic currently operates.

Bored tunnel
20. A bored tunnel alternative to the immersed tube crossing was studied. A bored tunnel would have to be constructed at great depth in the permeable marine

sediments. To mine a tunnel by compressed air techniques in the soft ground below the 50 meter water depth would require air pressures that are more than twice the value considered feasible. Use of pressurized face tunnel boring machines at such depths is unprecedented and would be considered experimental.

21. Since sound rock along the tunnel alignment occurs at a depth of more than 100 meters, a rock tunnel would have to be mined at a depth of about 130 meters. It would place the Sirkeci and Uskudar Stations at depths of over 100 meters below sea level. This would require steep grades, a greater length of tunnel and difficult patron access. The resulting higher construction and operation costs removed the bored tunnel from consideration.

Immersed tube tunnel

22. Early during the planning stages, the most economical immersed tube tunnel profile was established such that at the deepest portion of the Bosporus, the tunnel would have been located 4 to 6 meters above the original bottom. This would have required the construction of an underwater embankment or the underwater installation of pile foundations to support the tubes.

23. Pile supported tubes would have required precision pile driving below 50 meters of water. More importantly, the differential settlements between the pile supported and "at grade" sections of the tubes were of concern. Additionally, there would be a significant risk of damage to the tubes by dragging anchors or tackle. These issues eliminated the pile supported alternative from further consideration.

24. The embankment alternative is easier to construct. However, settlement and potential liquefaction of the upper soft and loose zones in the foundation materials presented design problems. Densification of the foundation materials at such great water depth, although unprecedented, was not considered impossible but would have been very costly. Further disadvantages of the embankment alternative were potential ecological impact on the Bosporus regime and safety considerations from sinking vessels.

25. To overcome primarily safety but other concerns as well, it was decided to immerse the tube below the bottom of the channel. This required a deeper dredged trench, at 60 meters below sea level, and a deeper ventilation building on the European side. However, potential long term ecological effects and potential hazards were thus eliminated.

Steel vs concrete elements

26. Both steel shell elements and concrete box elements were considered and preliminary designs were prepared for each. However, the steel shell tunnel was recommended because it utilized established local shipyards, involved a shorter schedule, involved a better structural efficiency and watertightness, and the costs were competitive. Additionally, the use of steel shell elements for the immersed tunnel was considered to have the least impact on the environment because the elements would be fabricated in existing shipyards in an industrial area. Concrete elements would require extensive dredging and filling to create a graving basin for fabrication.

DESIGN ANALYSES
Design parameters

27. Because of the difficulty in sampling the Bosporus soils, the design parameters can only be generalized. The two types of foundation soils in the Bosporus were categorized as either marine sands or marine clays. Most of the immersed tube will be founded on marine sands except at the west side of the Bosporus where it will be founded on marine clays.

28. The marine clay is intermixed with occasional layers of clayey sand. The saturated unit weight of the material is 1.76 grams per cubic centimeter. The cohesion values obtained from field and laboratory tests range between 0.05 and 0.20 kilograms per square centimeter. The material is relatively compressible. Consolidation test results indicate an initial void ratio (e_o) of 1.12, a compression index (C_c) of 0.311, a recompression index (C_R) of 0.043, and a coefficient of consolidation (C_V) between 0.0002 and 0.0004 square centimeters per second, depending on the effective stress range.

29. The marine sand is intermixed with gravelly layers. The saturated unit weight of the material is 1.85 grams per cubic centimeter. The angle of internal friction is 28 degrees and cohesion ranges between 0.3 and 0.4 kilograms per square centimeter.

Trench slope stability

30. Slope stability analyses were conducted using the Modified Bishop method of slices. On the west side of the Bosporus the dredged trench slopes will generally be within the very soft silty clay, with cohesion values between 0.05 and 0.2 kilograms per square centimeter. On the east side, the trench will be within the sandy soils having an angle of internal friction between 12 and 28 degrees. Even though the soil conditions are different on each side, an average trench slope of 2H:1V for the entire Bosporus

crossing is feasible for design and payment. Localized sloughing in the upper weak soils will occur during dredging. However, steeper slopes may be achieved in the stronger materials.

Liquefaction potential

31. A preliminary assessment of earthquake-induced liquefaction potential was made on the basis of the seismic hazard analysis and the subsurface exploration results. The results indicated that, for the service level event, the soils would not liquefy if they have a standard penetration resistance greater than 16 and 9 for fine contents of 5 and 35% respectively. For the survival level event, standard penetration resistance must be greater than 24 and 16, respectively. A comparative evaluation of the analysis with the boring data indicated that there is a potential for liquefaction of small isolated pockets during a service level event and of somewhat larger zones, in a survival level event. However, during difficult drilling in the Bosporus, the sampling techniques employed often resulted in significant loosening of the in-situ soils and loss of fines during sample recovery in deep water. These sampling imperfections probably resulted in underestimated blow counts, and underestimated fine contents, both of which lower the evaluated liquefaction resistance. It is concluded that liquefaction would not present a design problem, but this should be confirmed by further analyses during final design.

Settlement

32. At the west side, the immersed tube sections will transition from bedrock and onto marine clays. The net foundation pressure will generally be less than the unload pressure of the soil removed by dredging. Assuming that compression of the foundation course will be on the order of 5 centimeters, the maximum estimated settlement is 16 centimeters, and occurs in the section of tunnel where the middle clay soil zone is the thickest. The anticipated angular distortion (settlement per unit length of tube) is estimated to be 0.0002. Settlement is expected to be negligible at the west end founded on rock and up to 5 centimeters in the middle of the strait where the tubes will be founded on cohesionless, granular materials.

33. Since the clay is very sandy and relatively free draining, it is difficult to assess the time rate of settlement. Conservatively assuming that there are three equally spaced sand lenses within the clay zone, primary consolidation settlement will take up to 20 years. Most of the primary settlement will occur within the first year. Secondary consolidation rates were not estimated but should be relatively minor. The estimated settlements of the tube tunnel are within tolerable limits.

ENVIRONMENTAL IMPACT

34. Since the tube tunnel will be completely immersed below the natural channel bottom, it will not cause any change in the cross section of the Bosporus. It will therefore not have any adverse effects on the ecological or hydrographic characteristics of the Bosporus. However, it is inevitable that during dredging, foundation course placement, and backfilling operations, adverse environmental impacts may be experienced.

35. It is estimated that 1.13 million cubic meters of material will be dredged for the trench tunnel. Turbidity, sedimentation and water quality impacts will affect the marine life during construction. Turbidity and decreased dissolved oxygen levels will affect the distribution and migration of fish, especially bluefin tuna. Sedimentation, estimated at 2 centimeters thick within 500 meters of the construction, will particularly affect the mussels. Additionally, man-made underwater noise during construction may affect migration patterns of fish schools.

RAIL OPERATION AND SAFETY MEASURES

36. The tunnel includes facilities for ventilation and environmental control as well as for safety and fire fighting. The ventilation buildings located on either side of the Bosporus will be supplemented by three mid-tunnel ventilation shafts, two to the west and one to the east of the respective ventilation buildings. Normally air velocity will be controlled by draft relief shafts with turnarounds. The maximum horizontal air velocity on the platform is 5 meters per second. During emergencies tunnel air velocity increases to a maximum of 11 meters per second. The station and tunnels will be ventilated by purging the tunnels with sufficient air so that carbon monoxide end nitrogen oxide concentrations will not exceed 150 parts per million and 5 parts per million, respectively.

37. Electrification for vehicle traction and auxiliary equipment will be provided by an internal mid-voltage distribution system from the local power company. The traction power system utilizes a 25KV AC catenary service matching that of the existing railroad.

38. Train signaling ad communications are designed to permit reliable and safe realization of high operating speeds and low headways. Minimum operating headway for the year 2005 peak hour will be 2 minutes.

39. Operations plans include allowance for regional, intercity, and freight trains, in addition to more frequent suburban commuter service. The double-track tube tunnel

can accommodate projected traffic for 2005 but will require coordination of train operations and maintenance activities. Achievement of future operating goals requires improvements to facilities beyond the project limits, and improvement to existing vehicles.

CONSTRUCTION METHODS
Dredging and spoil disposal

40. Dredging for the tunnel will involve both rock dredging and soft ground dredging. Rock dredging will be necessary for the areas where the immersed elements would be near shore. On the west end a trench will be needed for the elements to reach the deep west ventilation building. On the east end, a trench will be needed to reach a point where a portal into a mined tunnel can be reliably established. The rest of the dredging will be across the overburden materials of sands, gravels and silty clays.

41. It is estimated that about 50 percent of the 1.13 million cubic meters of the dredged materials will consist of sand and gravel, suitable for use as backfill. Dredging can be accomplished by a dedicated clamshell dredge. Comparative cost estimates indicate that the utilization of large cutterhead suction dredges, capable of dredging to the project depths of greater than 50 meters, would not be economical for the relatively small quantity of dredging.

42. Rock dredging would be done by conventional drill and blast methods using barge-mounted line drilling rigs. Some excavation of the more weathered rock may be done without blasting to strip the surface of loose rock as much as possible before starting the drilling operation.

43. The usable portion of the dredged spoil may be moved to a temporary contained stockpile area in shallow water for later use as backfill over the tunnel. The clays and silts can be either disposed of in the deep areas of the Marmara Sea or used as fill for land reclamation.

Placement of elements

44. While it is deemed feasible to carry out placement operations using conventional methods, estimates of the forces on the tunnel elements and placing equipment indicate that this will be an operation requiring great care and planning. It is estimated that lateral anchor forces of 1,000 tons can be expected and, as the element passes through the transition zone between the upper and lower water flows, hydrodynamic forces may vary between +100 tons to -300 tons. It was recommended that a detailed model study be carried out during final design.

Mined tunnel transitions:
45. On both ends of the immersed tunnel, a connection must be made to a mined tunnel. On the west end, at Topkapi Point, this is done via the ventilation building. On the east end it is accomplished by means of a tremie concrete connection of the easternmost element to a trench cut in sound rock. Tunnels are mined out to meet this connection and the mining is carried to the bulkhead of the tunnel element.

46. The ventilation building is constructed in the dry in a very deep excavation carried below the tunnel invert. It is equipped with an adaptor to which the first tunnel element can "made up". The building is equipped with wings tied into the cofferdam so that the exterior face can be flooded without affecting the interior face. This allows the rock excavation to be carried up to the excavation already done in the front area of the cofferdam. Once this excavation is completed and the screeded bedding is placed, the first tunnel element can be joined to the building.

CONCLUSION
47. The feasibility studies and preliminary design for the Bosporus railroad tunnel crossing were completed in 1986. The studies indicate that the immersed tunnel is the proper method of construction and that steel shell tunnel elements are likely the best choice. When constructed, this will represent the first tunnel link between two continents. It will improve the regional transportation, and when integrated with the future Istanbul metro, will provide much needed commuter rail service between the two sides of the city. As designed, there will not be any long term environmental impact from the tunnel construction.

ACKNOWLEDGMENTS
48. For the Turkish Ministry of Public Works, and the Ministry of Transport and Communications, Mr. Zafer Ozerkan was the Project Manager. For IRTC, T. R. Kuesel was technical advisor, W. W. Hitchcock was principal in charge and R. Shellmer followed W. C. Grantz as Project director.

8. Immersed steel railway tunnel across the eastern channel of the Great Belt — Denmark

A. GURSOY, PE, Parsons Brinckerhoff International, Inc., Storebaelt, Denmark

SYNOPSIS. This paper summarizes the results of development of Tender Design documents for an immersed steel rail tunnel across the Eastern Channel of the Great Belt in Denmark. Although this design was not selected for implementation, tenders received for construction demonstrated that it was economically competitive and may have other applications in Europe. The project layout, basis of design and principal design features are described.

INTRODUCTION
1. The rail tunnel between Zealand and Sprogø will be part of the fixed link across the Great Belt, connecting the main Danish islands of Funen and Zealand.
2. The construction of the fixed link is based on the political agreement made on 12 June, 1986 between the Social Democratic party and the government coalition. The agreement has been followed up by the Public Works Act for the fixed link across the Great Belt, passed by the Danish Parliament on 26 May 1987. In the agreement and in the Act, the main principles of the fixed link have been set out:

(a) The fixed link across the Great Belt shall include a low level bridge across the Western Channel between Funen and the small island of Sprogø in the middle of the Great Belt. The bridge shall carry road as well as rail traffic.
(b) The railway link shall be carried in a tunnel across the Eastern Channel between Sprogø and Zealand to be constructed either as a bored tunnel or as an immersed steel or concrete tunnel.
(c) The road link across the Eastern Channel shall be carried out either via a bridge or via a tunnel.
(d) The rail link will be opened in 1993 and the road link in 1996.

Immersed tunnel techniques. Thomas Telford, London, 1989.

3. Tender documents for the project were completed and made available to prequalified bidders in February, 1988. Documents covered three alternate tender designs; an immersed steel tunnel, an immersed concrete tunnel and a bored tunnel.

4. In October 1988, the bored tunnel alternative was selected by the Danish government for the eastern rail crossing, primarily on environmental considerations. The economic evaluation, which included allowances for Owner's Risk concerning possible construction time and cost overruns, did not indicate a decisive difference between the immersed and bored tunnel alternatives.

5. The construction cost of the total fixed link across the Great Belt will be approx. DKK 13bn (1986 prices) exclusive of VAT and interest. The construction cost of the immersed steel tube tunnel alternate for a railway connection across the Eastern Channel has been established as approx. DKK 3.4bn (1988 prices).

6. Bid prices for the steel alternate were within six percent of the engineer's estimate. Bid prices for the concrete alternate were about 10% higher than for the steel alternate. A total of four construction consortia were selected as prequalified bidders. Each bidder was required to bid for each alternate (steel and concrete immersed tubes, and bored tunnel).

PHYSICAL CONDITIONS

7. The fixed link will be approx. 18 km long. On the Zealand (east) side, the link will start north of the existing car ferry terminal at Halsskov Odde. Approx. 6 km into the Great Belt, the link passes 200 m north of Sprogø. The water depth in the Eastern Channel between Halsakov and Sprogø is up to 60 m.

8. From an artificial island north of Sprogø, the link will continue onto the Western Bridge. The link will end on the Funen side just north of the Knudshoved ferry terminal. The Great Belt is approx. 6 km long between Sprogø and Funen, and the water depth is up to 27 m.

ALIGNMENT

9. The landfill of the railway link on the Zealand side will be a man-made peninsula north of the Halsskov ferry terminal and it will be connected to the present railway just west of Svenstrup. At Sprogø, the railway link will be connected to the combined road-and-rail bridge across the Western Channel.

10. The alignment under the Eastern Channel is determined by the railway requirements of the longitudinal

profile, the soil conditions, the sea bed and the construction method of the tunnel. The length of the immersed tunnel is about 5,600 m from portal structure to portal structure. At the deepest point the rails would be at a level of -49 m. The maximum gradient will be 1.7% and the minimum horizontal radius about 4,500 m. The maximum design travelling speed is 160-200 km/h.

DESIGN BASIS
11. The tender design is based on the Danish standards for concrete and steel construction of DS 409, DS 410, DS 4111 and DS 412.

12. The tender design satisfies:

(a) Serviceability limit state
(b) Ultimate limit state and
(c) Applicable other codes and standards

13. The project's physical risk factors were studied, including accidental loads due to shipping traffic on the Eastern Channel (e.g. dredging, dropped anchors, and sunken ships, or submarines settling on the tunnel). The design life of the project is assumed to be 100 years.

IMMERSED TUNNEL TYPES FOR THE PROJECT
14. The tender documents specified two alternative types of immersed tunnels. One is an immersed concrete tunnel (European type), the other is an immersed steel tunnel (American type). For both types the tunnel element would be placed in a trench dredged in the sea bed. The geological strata through which dredging would be made consist mainly of moraine deposits and marl. A rockfill on top protects the immersed tunnel against the impact of ships and the like.

15. In the tender documents the immersed concrete tunnel has been specified to comprise 38 tunnel elements of a length of 148 m each. The two train passageways have an inside height of 6.1 m above the rails and a width of 6.5 m.

16. In the tender documents the immersed steel tunnel has been specified to comprise 39 tunnel elements of a length of 144 m each. Each tunnel element contains two train pasageways having an inside height of 6.1 m above rail and an inside width of 6.9 m maximum.

PORTAL RAMPS AND ISLANDS
17. The 1,300 m long ramp on the Zealand side starts

Table 1. Principal quantities

		Immersed Tunnel Concrete	Immersed Tunnel Steel
Immersed Tube Tunnels			
Trench Excavation	(m^3)	3,700,000	3,600,000
Backfill	(m^3)	1,600,000	1,400,000
Rockfill	(m^3)	230,000	310,000
Steelwork	(t)	13,000	43,000
Formwork	(m^2)	450,000	285,000
Reinforcement	(t)	40,000	9,200
Concrete	(m^3)	410,000	300,000
Ramps and Portal Structures			
Excavation	(m^3)	400,000	400,000
Backfill	(m^3)	2,900,000	2,900,000
Rockfill	(m^3)	650,000	650,000
Sheetpiling	(t)	7,800	7,800
Formwork	(m^2)	17,000	17,000
Reinforcement	(t)	900	900
Concrete	(m^3)	7,800	7,800
Road construction	(m^2)	42,000	42,000
Rail construction	(m)	5,700	5,700

at a level of +1.0 on the coast north of Halsskov ferry terminal and leads down to the portal structure, where the top of rail is -12 m. Halsskov peninsula would be extended westwards, by up to 1.2 km long and about 0.5 km wide.

18. Protective rock fills and peripheral dikes in combination with sheet piles would be used for the construction of the ramps and islands, where dikes serve as access and service roads for the finished portal structures. On the Sprogø side a 2 km long ramp would be terminated with the rails at a level of +5.5 m where the railway would be connected to the bridge over the Western Channel. Construction of artificial islands and ramps is similar to that on the Halsskov side. The ramps at Halskov as well as at Sprogø are designed to be constructed as open excavations with flat slopes and with a permanently lowered water level. The lowering of the water table by an extensive system of subsoil drains and a filter system with overflow wells is part of the ramp design.

19. Surface water and ground water from the permanently drained ramps are collected and pumped into the Great Belt.

PORTAL STRUCTURES

20. The portal structure at each end consists of a portal building, a pump station, an oil and petrol separator and a fire reservoir. The buildings are designed to contain rescue and emergency supplies, fire pumps, and control rooms for the surveillance of the railway services and the tunnel installations, as well as exhibition and recreational rooms.

FOUNDATION

21. The tunnel elements are designed to be supported by a screeded gravel foundation, placed in a dredged trench. Over a length of 1 km at the deepest section of the Eastern Channel, the top of the element would protrude above the existing sea bed. The dredging depths are 6 to 53 m. The upper stratum consists of unsorted moraine and the bottom stratum of stiffer marl. At certain places the moraine may be covered by post-glacial deposits. The bottom width of the trench is about 20 m and the slope of the trench at the moraine stratum is at least 1.5H:1V, while it may be as steep as 0.5H:1V for the marl. The tunnel foundation may consist of an approx. 60 cm thick screeded gravel base on which the tunnel element would be placed directly. After the tunnel element has been placed in correct position, a locking fill would first be

placed to half the height of the tube and followed by suitable backfill material to a level of 2 m above the element. Various types of protection layers would be used along the profile.

IMMERSED STEEL TUNNEL
22. The total height of the steel shell is 9.9. m inclusive of an outer ballast box of 1 m. The total width is 16.5 m. The tunnel elements consist of a composite structure with an outside steel shell plate and an inside lining of reinforced concrete. It was assumed that the total steel of each element including the outer steel shell plate, transverse stiffeners, longitudinal centre truss and bulkheads will be prefabricated at a shipyard and launched as a watertight floating structure. The bottom slab of the concrete lining (keel concrete) would be cast before launching. The inside concrete lining can be cast while the element is moored at an outfitting pier. After the casting of the concrete lining was finished, concrete hatches would be closed by steel plates welded to the steel shell. The tunnel element then would be towed to its position in the tunnel alignment, where it would be lowered by the immersing equipment after placement of the gravel ballast. Other interior concrete, such as track bed, service walks and structural concrete within the joints would be cast after the tunnel elements are placed, joints are made and backfilling is completed.

CROSS SECTION
23. The load-carrying structure consists of a steel shell with an inner reinforced concrete lining. The steel shell is fabricated of a 10 mm plate with transverse stiffeners at intervals of 1.5 m consisting of ½ IPE 360 profile. A longitudinal centre bracing system is provided to stiffen the flat top and bottom shell plates against launching and towing stresses. It consists of a truss with flanges of ½ IPE 360 profiles, verticals of HEB 200 profiles and diagonals of angles 150 x 100 x 12 mm. A ballast box of a 6 mm plate braced with a 6 mm plate at intervals of 1.5 m is welded to the top of the tunnel tube. The steel shell can be fabricated in 12 m modules using 4 m or 3 m plate widths, common to most shipyard operations, which facilitates fabrication of sub-assemblies.

24. The reinforced concrete lining consists of a 1,000 mm thick bottom slab, 800 mm thick outer walls, and 800 mm thick top slab. The concrete central partition wall is 1,100 mm thick due to the space requirements of the electrical installations. The space will

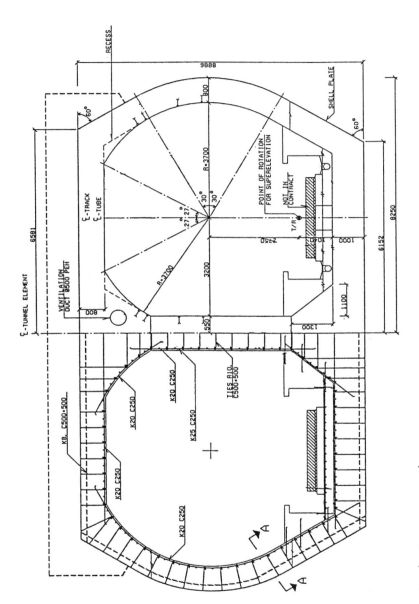

Fig. 1. Typical cross-section

PLANNING

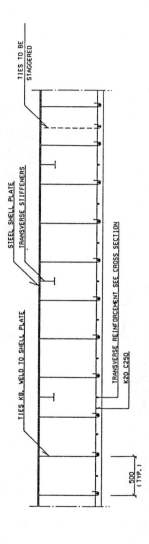

Fig. 2. Section A-A

also be used for sump and pump rooms at the low point and quarter points.

JOINTS

25. To facilitate the joint construction, each end of standard tunnel elements is designed to have an annular ring, containing rubber gaskets for initial seal and steel wedges for alignment correction. The annular ring and the end bulkheads are installed before tunnel elements are launched, one end of the element has a flat bearing surface, and the other end has a rubber gasket ring. When two element ends are jointed after the lowering, the rubber gaskets are compressed thus forming a temporary waterproof joint so that the gap between the end bulkheads of the two elements can be drained, and the final joint construction can be carried out in the dry from the inside. The final watertightness of the joint is secured by welding steel closure plates between the two elements, and the gap behind the closure plates is filled with concrete grout.

26. The tunnel placement schedule was optimized by placing elements from both the Halsskov and Sprogø and meeting at a closure joint at about a quarter of the tunnel length from Sprogø.

INSTALLATIONS AND EQUIPMENT

27. Installations and equipment in the railway tunnel comprise railway systems, electrical systems and mechanical systems. In the tunnel tubes the tracks, overhead traction power system, block signals, fans, lighting, cable trays and catch basins are installed. Control and electrical panels as well as mechanical systems are placed in niches in the central partition wall. All systems are designed to be controlled from control rooms in the portal structures. Furthermore, the design incorporates the control and surveillance installations to be adapted for remote control from the centre in Copenhagen as well as from future regional remote control centres.

28. The tunnel elements are protected against corrosion by an impressed current cathodic protection system.

VENTILATION AND SAFETY ASPECTS

29. The Great Belt railway tunnel presents several interesting challenges. There are no restrictions on the goods to be carried nor on the possibility of a goods train to be on the same track and in the tunnel at the same time as the passenger trains. No emergency closures are permitted at the ends of the tunnel to assist emergency ventilation, and no central service/e-

mergency gallery is provided. The passenger trains will be diesel powered. The freight trains will be powered from a 25kV catenary.

30. Calculations show no requirement for fan ventilation during normal operating schedules. Approximately 100 reversible jet fans are provided in each trackway for emergency duties. The jet fans will be similar in design to the 630mm fans used in road tunnels throughout the world. However, the high train speeds at high blockage ratios will require several fan modifications to increase the endurance of these fans which spend almost their entire life cycle in a windmilling mode. The fans can deliver emergency ventilation air at a rate of five metres per second with three stopped trains in the tunnel on the same track.

31. Fire mains are provided for each trackway with hydrants every 75m at the cross passage doors each of which are fire rated. Sumps are located at the single low point and the two quarter points of the tunnel. The pumping capacity at the sumps is ample to remove the water introduced by fire fighting operations. Because of the unlimited types of goods to be carried through the tunnel, the sumps must cope with potentially corrosive and/or flamable materials. The sumps for each track way are physically isolated to preclude a situation in one tunnel spreading to the other, provided with a fire suppressent system, and include a means for introducing dilution water. The drainage water is pumped immediately to a holding tank in the ramp area for hauling away, or treatment, or discharge into the waters of the Great Belt.

9. Planning and design of the Guangzhou immersed tube road and railway tunnel

AU SHIU KIN, MSc, David Au and Associates, Hong Kong,
J. D. C. OSORIO, FICE, Posford Duvivier, London, JIN FENG and
CHEN SHAO ZHANG, Guangzhou Underground Railway
Preparatory Office, Guangzhou, China

SYNOPSIS. Recently the "open" coastal cities of China have developed rapidly. Guangzhou, at the head of the Pearl delta is the economic centre of southern China, the capital of a fertile province and the nearest chinese city to Hong Kong, Macau and the adjacent special economic zones, has been particularly affected by this trend. Further progress is now hampered by lack of an adequate transport network and the proposed tunnel is urgently needed; it will allow development to take place on the south bank of the river which is at present served only by a foot and bicycle ferry for commuters and two rail lines and two carriageways for inter urban traffic. The tunnel will be designed and built almost entirely by local resources. The paper describes the important features of the project.

BACKGROUND
1. Guangzhou (formerly known as Canton) is at the upstream limit of navigation for pre-19th Century ocean going ships with draughts of up to 5m, and is the natural site for an old port city. Ships drawing 9m can reach Huangpu a short distance down stream but large modern ocean going ships with draughts of up to 17m can only get just upstream of Hong Kong. Between Hong Kong and Guangzhou there is a great delta of soft, low lying fertile land and shallow waterways which has traditionally been served by inland waterways barges and across which it is difficult and costly to construct roads and railways.
2. The position of the city relative to Hong Kong, Shenzen, Huangpu, Foshan, Zhuhai and Macau which are the other centres of development around the delta is shown on figure 1. The proposed tunnel is designed as part of the transport network inside the city. Other crossings will be constructed to facilitate regional transport.

URBAN TRANSPORT IN GUANGZHOU
3. The central urban area is shown on figure 2. the population figures are conflicting but the most recent (1987) are as follows:-

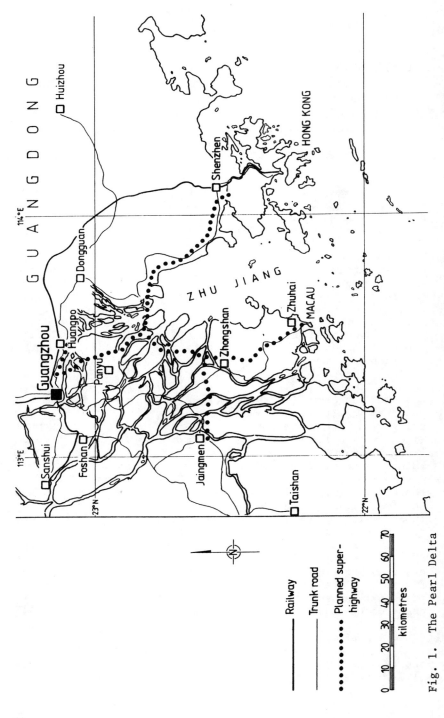

Fig. 1. The Pearl Delta

	Area Km^2	Population Millions
Administrative area	6400.0	5.65
Total Urban districts	1443.6	3.42
Central area	54.4	2.19
Non resident		1.1

It is not clear whether the non residents are in addition to the "population". The population density rises to 158,000/km^2 in the old city.

4. Available figures for the number of vehicles is as follows:-

	1952	1986	1987
Motor vehicles	2,000	123,000	158,000
Bicycles	---	1,550,000	1,920,000
Other non motor vehicles	---	80,000	76,000

The number of motor cars, taxis and motorcycles on the roads relative to the number of bicycles appears to be increasing much more rapidly than these figures suggest.

5. The modal split for urban passenger transport in 1986 was:-

	% of journeys
Foot	38
Bicycle	30
Bus and trolley bus	26
Ferry	2
Taxi	2
Minibus	1
Other	1

Almost all journeys by residents are by bus, bicycle or on foot.

6. The public transport system consists (1985 figures) of:-

	Routes	No. of vehicles
Bus	40	}
Trolley bus	7	} 1836
Minibus	7	}
Express bus	40	}
Taxi		7000
Ferries	23	

This system is unable to cater for the demand at the morning and evening peaks.

7. In 1987 there were 536km of highway with an average width of 8.4m. The roads have been substantially improved by widening, grade separation and traffic control but the average speed of urban vehicles has decreased to 10km/hr.

PLANNING

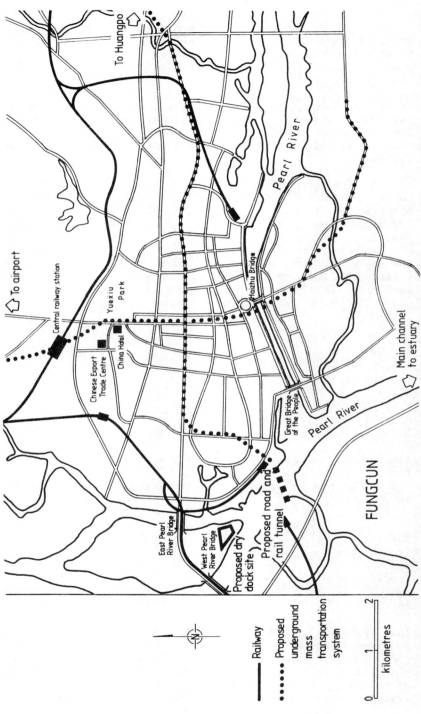

Fig. 2. Guangzhou central districts with initial mass transit railway system

PLANS FOR METRO

8. It is accepted by the Municipal government that the only way to solve the traffic problems is by the construction of a rapid transit railway. The preparations made for this are:-
1984 - establish the "Guangzhou preparatory office of underground railway".
1984 - approval by the state council of a project to construct 21km of railway with 21 stations by 2000.
1985 - Present. Social and economic studies.
1985 - Compilation of available borehole data.
A feasibility study is in progress.

9. The proposed routes are shown on figure 2. The initial east-west route is to be 14km long with 14 stations to be extended later to Huangpu in the east and Fungcun in the west. The initial north south line is to be 10km long with 9 stations for subsequent extension to Xinshi in the north and Chigan in the south. It is proposed that high rise buildings for housing and offices will be constructed over the stations and depots.

PLANNING THE TUNNEL

10. The tunnel will connect the central and most congested part of the Guangzhou with the relatively undeveloped newly established city district of Fangcun. Fungcun has a population of 120,000 in an area of 40.8km^2. Existing industries include ship building, pharmaceutical, building materials and machinery. It has a long frontage on navigable waterway with warehouses and a railway station. All the traffic from Guangzhou to the centres on the west side of the delta (Foshan, Zhuhai, Zhongshan, Jaingmen and Macau) passes through Fungcun. The tunnel will therefore facilitate the development of Fungcun and, until the proposed outer ring road is constructed, also serve regional transport.

11. The tunnel alignment has been chosen to connect the inner ring road of Guangzhou and the main thoroughfare of Fungcun. It avoids significant existing development. Longitudinal and transverse sections are shown on Fig.3. The dimensions of the crossing are:-

Total length of structure	1130m
Length of tunnel	685m
Immersed tube section	403m
Gradient	3.0%
Horizontal curve radius	300m
Vertical sag curve radius	1000m
Vertical hog curve radius	3000m
Outside dimensions of units	33.6m wide
	8.05m high
	100m long
Internal dimensions of units:-	
Traffic lane	9.55 x 4.9m
Service tube	1.5 x 5.7m

PLANNING

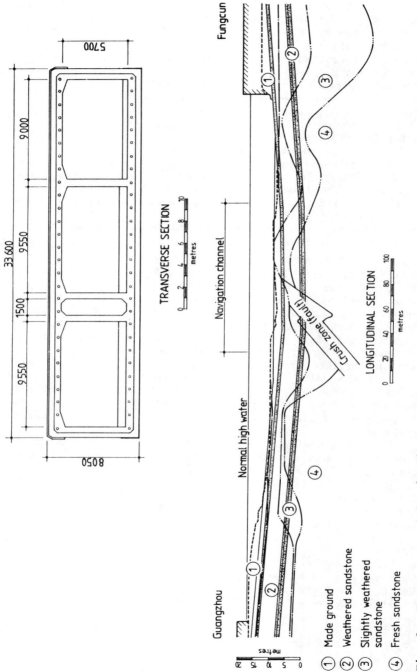

Fig. 3. Transverse and longitudinal sections

Lane for bicycles and subsequently
rail   9.55 x 4.9m
Traffic lane   9.0 x 4.9m
Expected speeds and capacities are:-
Design vehicle speed   30km/hr
Maximum vehicle speed   60km/hr
Capacity in each of 4 lanes   705 vehicles/hr
Capacity in non motor vehicle tube   10,000 bicycles/hr

12. The whole of the immersed tube section of the tunnel will be below the navigable waterway. The main channel is 120m wide and 6m deep at low water. Normal tides rise approximately 2m and the extreme (1:100 year) flood level is 4m above navigation level. Therefore there will not be sufficient water to float the units without dredging.

13. The bed of the river consists of red, cretaceous sandstone. The depth of decomposition varies from 1 to 4m. There is a thin layer of mud near the banks but the rock is exposed in mid river. The "average" current velocity is said to be 0.8m/sec and Guanghzou is near to the tidal limit: annual average fresh water dishcarge from the Pearl River system ranges from approximately 100 million m^3/day in December and January to over 1,000 million m^3/day in June and July. Therefore it will probably be necessary to place the units on high tides between November and February when the currents will be small. Moving the units from the dry dock in October or March would involve more dredging as the tides are significantly lower.

14. No studies of other methods of constructing the crossing have been reported. However given:-
(a) the need to provide an underground railway crossing,
(b) that the land on both banks is flat and densely developed,
(c) the river is used by coastal shipping,
(d) the whole of the crossing slopes downward from the shores to mid river,
it is evident that a high level bridge, deep bored tunnel or opening bridge would not compete with an immersed tube.

15. Finance for domestic loans in China requires the repayment of capital and interest within 10 years. The rules are:-
- repayment in 7 years at estimated traffic
- repayment in 10 years at 3/3 estimated traffic
- repayment in 13 years at 1/2 estimated traffic.
The interest rate is not stated. Tolls will be charged to cover the domestic loan. Foreign currency will be required for the procurement of gaskets, sheathing materials, specialised mechanical/electrical equipment and advice. Tolls will yield very little foreign currency so that the foreign currency requirements must be minimised. It is hoped that these will be recovered by constructing a multi functional building adjacent to the north ramp of the tunnel. This site is adjacent to the White Swan Hotel and

PLANNING

the Esso offices on the east end of the island which was occupied by foreign merchants during the 19th Century.

DESIGN

16. The units will be constructed in a drydock formed in an island 2km upstream of the site. The existing ground level on this island is at about low water level. It is cultivated with small farmers houses clustered along the flood banks. The dock will therefore be about 6m deep below existing ground level. Boreholes are not available but the river channel bed is irregular so the base of the dock will probably be either rock or weathered rock.

17. 66 boreholes were sunk along the proposed alignment of the tunnel. Most boreholes reached either fresh or slightly decomposed sandstone with an average core recovery of 60%. The tunnel will be entirely in rock and will be founded on fresh or slightly decomposed rock except in mid river where there is a fault zone 20m wide dipping to the north at about 20° to the horizontal. For the loads applied by an immersed tube tunnel these rocks are effectively rigid.

18. Studies carried out locally recommend the following criteria for the design earthquake:-
- intensity magnitude 7
- maximum ground surface acceleration 0.1g
- shear wave amplitude 0.004m

It is considered that risk of significant displacement occurring at the fault is negligible. These figures have been questioned.

19. Sand flow, screeded crushed rock and grout were considered for bedding the units. Chinese marine construction groups have shown themselves to be extremely proficient at screeding gravel beds manually by standard suit divers; this method is commonly used for the construction of quay walls which are precast in large elements and placed by floating cranes. However it is considered that even with an abundant supply of skilled divers this method will not be sufficiently accurate for bedding 33m x 100m units. The use of mechanical screeding equipment could not be justified for such a short wide tunnel.

20. A thin grouted bed should be the cheapest and quickest for a tunnel founded on rock due to the substantial reduction in the volume of rock to be excavated. A grout bed cannot liquify in an earthquake. However it is not considered that local construction teams have sufficient experience of grouting to attempt to pump such a large volume of material, and it therefore proposed that the units be bedded by the "sand flow method" with secondary grout injection to tighten the bed and fill any voids.

21. Tests on samples of the available materials and the charts prepared be Seed and Idriss indicate that the bed should have a relative density of 25% to obviate the risk of

liquifaction in the design earthquake. The initial density of this sand when placed in a large scale trial is much less than 25% and the Chinese authorities consider that the density should be higher than 25%. The very low densities observed in the large scale trial, cannot be reproduced in the laboratory or accurately and reliably measured at the site where the trials are being carried out. However, it is known that tunnel units placed on a sand flow bed 1m thick settle by about 70mm soon after laying. These uncertainties and conflicts have not yet been resolved.

22. The preparatory office have not provided any information on the design loads and methods to be used for the reinforced concrete units. Initially they stated that they required all the intermediate joints to be locked before the tunnel was put into service. All subsequent movements would b e accepted at the two joints between the approach ramps and the immersed section; with a temperature range of 20°C and a rock backfill in a narrow rock trench this resulted in unmanageable axial stress in the units. It is now agreed that all the joints will be flexible with gina gaskets and omega seals. The movements expected from earthquakes are small compared with temperature strains and shrinkage and there is no reason to use an unconventional arrangement. The amount of shrinkage and the casting and laying tolerances still have to be confirmed. It is expected that shear keys will be provided in the walls to prevent vertical movements due to uneven live load and partial loss of support from the bedding.

THE CURRENT SITUATION

23. Detail design work started in Dcember 1988. The arrangements that have been made for the project are as follows:-

o   The preparatory office will be the leading design organisation and project managers. They will be advised by local design institutes and Urban Transit Consultants from Hong Kong.

o   The preparatory office are holding 20% of the shares and the remaining 80% are held by a leading finance firm established in Guangzhou and Hong Kong.

o   The project will be initially financed by the Provincial Government, they are planning to recover their costs by selling sites on both sides of the tunnel and fabrication yard, tolls and tax.

o   The detailed design of the foundation and joints will be carried by Urban Transit Consultants in Hong Kong.

o   A number of local contractors have already been pre-qualified for the work.

PLANNING

o The Guangzhou economy is continuing to grow rapidly. The gross domestic product increased 3.82 times between 1977 and 1987. Personal earnings increased by 3.57 times in the same period.

o The city circular road has just been completed. The Shenzhen, Guangzhou and Zhuhai super highway is being built. Roads to south of Guangzhou i.e., Foshan and Panyu are also being constructed.

FUTURE CROSSINGS
24. There are probably many sites in the Pearl Delta where an immersed tube will be preferred to a high level bridge when the expertise necessary is available locally and this project should therefore be the first of many to be constructed in Southern China.

10. The Sydney Harbour tunnel — influence of local conditions on the route and on the marine and land tunnels

N. SAITO and A. M. NEILSON, The Sydney Harbour Tunnel Joint Venture, Australia

SYNOPSIS. The Sydney Harbour Tunnel is located in a well-developed and environmentally sensitive area bordering the Central Business District of Sydney. Its design has been greatly influenced by the many constraints peculiar to this site, with the adopted proposal providing an economical solution to Sydney's cross-harbour traffic problem with surprisingly few adverse environmental effects. The long lengths of land tunnels on each shore demanded that the design of the immersed tube and the land-sea interfaces recognise additional constraints in the interest of overall economy of the Project.

CURRENT STATUS
1. Construction commenced on the Project in early 1988 after an offer to design, construct, finance and operate the tunnel, by a Joint Venture comprising construction companies Transfield Pty Ltd of Australia and Kumagai Gumi Co., Ltd of Japan, was accepted by the New South Wales (NSW) State Government. This proposal was the latest of the many tunnel or bridge concepts proposed to alleviate Sydney's cross-harbour traffic congestion. It was conceived in 1984, and after exhaustive study of alternative routes and design options, a firm offer was submitted to the Government in December 1986. Detailed design commenced on the project in July 1987 and was completed in December 1988. It is anticipated that the Tunnel will be open to traffic in September 1992.

EXISTING HARBOUR CROSSINGS AND ROAD NETWORK
2. Sydney Harbour is an attractive, natural deep water port around which the city of Sydney has developed. The entire area around the harbour, extending 30 kilometres from the Heads to Parramatta and beyond, is highly developed with residential, commercial or industrial facilities. This development also extends some 20 kilometres to the north and south of the Harbour. Consequently there is a high demand for highway traffic to cross the Harbour generally, and in particular, to access the City centre.

PLANNING

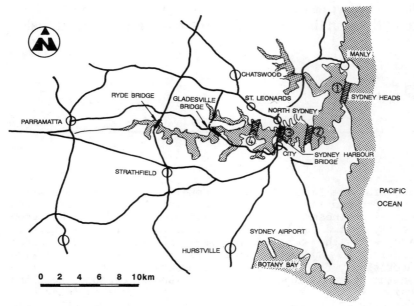

Fig. 1 Road bridge crossings

3. Several road bridge crossings exist. (Fig. 1)

(a) The Sydney Harbour Bridge, providing eight road lanes and two rail tracks, is located some eight kilometres from the Heads and was opened in 1932.
(b) The Gladesville Bridge, providing six road lanes is located some 14 kilometres from the Heads and was opened in 1964.
(c) The Ryde Bridge, originally providing three road lanes and recently supplemented with a new four lane bridge is located some 20 kilometres from the Heads.
(d) There are also several crossings of the Parramatta River further to the west.

4. The existing major crossings have generally been located where the width of the Harbour is narrowest and where construction costs would be least. They were established many years ago and have been significant influences on the growth of the road network and of development in general.

5. The road and freeway network in the area east of the Ryde Bridge generally radiates from the City centre with the predominant Harbour crossing being the Sydney Harbour Bridge. The bridges to the west of Gladesville, and to some extent Gladesville Bridge, form part of a "ring-road" system serving north-south traffic which is not destined for the City.

6. The Sydney Harbour Bridge, being the only crossing for 14 kilometres from the Heads, also carries significant traffic which wishes to by-pass the City itself.

7. The well-developed Warringah Expressway distributes traffic to the northern suburbs on the northern side of the Harbour Bridge. On the southern shore there are two expressways, one to the west of the City, the Western Distributor, and the other to the east, the Cahill Expressway. These carry traffic which wishes to access the southern end of the City or to by-pass the City completely.

8. There is congestion on all main cross-harbour arteries with long delays experienced on the approaches to the Harbour Bridge in peak hours because of the limited capacity of the Bridge itself.

POSSIBLE HARBOUR CROSSING CORRIDORS
9. Many ideas have been floated to increase the cross-harbour road capacity. (Fig 1). These have included

(a) a crossing over the Heads (1)
(b) a crossing between Mosman and Darling Point (2)
(c) additional capacity on, or in the vicinity of the Harbour Bridge (3)
(d) a crossing between Greenwich and Balmain. (4)

10. The crossings east of the Harbour Bridge have not been considered seriously. They would require long cross-harbour bridge or tunnel structures, the construction of approach roads through developed areas, and generally would satisfy only a limited traffic demand.

11. The last two options have been the subject of exhaustive study in recent times. A crossing between Greenwich and Balmain would be well-located to satisfy traffic demand and the Harbour crossing itself would be economical. However the construction of approach roads would require severe disruption to well-established residential areas and the resumption of large numbers of homes. The price, both in terms of construction costs and environmental disturbance, was considered too high.

12. It is for this reason that many studies have concentrated on the corridor in the vicinity of the Harbour Bridge. Such schemes have included the addition of extra traffic lanes to the existing Bridge. The study which led to the acceptance of the Transfield-Kumagai proposal considered supplementing the Harbour Bridge using an additional crossing located in the general vicinity of the Bridge.

PLANNING

FUNDAMENTAL CONCEPTS
13. This proposal was developed around four fundamental criteria which were considered essential for the Project to be accepted by the people of Sydney and by the Government of NSW. These were

(a) It must offer an effective solution to the cross-harbour traffic problem.
(b) It must be constructed without major disruption to communities along the route - no resumption of private land should occur.
(c) It must be constructed and operated without significant adverse environmental effects.
(d) It must be constructed and operated at a cost which allows it to be funded privately without drawing on the State's finances.

TRAFFIC STUDIES
14. Examination of historic data on traffic volumes has shown a steady growth of traffic across the Bridge. This growth has continued in spite of severe congestion and is expected to continue. In 1986, the average annual weekday traffic was 196,000 vehicles per day and is expected to reach 240,000 vehicles per day by the year 2000. The Bridge is at maximum capacity at both morning and evening peaks and the growth would be accommodated by an increase in the length of the peak periods.

15. On the northern shore, the Warringah Expressway effectively collects and distributes all cross-harbour traffic. However on the southern shore, the traffic breaks into three streams: one to by-passing the City to the west via the Western Distributor; one by-passing the City to the east via the Cahill Expressway; and that destined for the City itself. (Fig. 2)

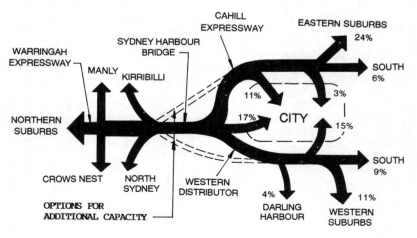

Fig. 2. Traffic flow distribution

16. Some 55% of traffic crossing the harbour wishes to by-pass the City completely and of this traffic some 60% wishes to by-pass the City to the east.

17. Also, the Cahill Expressway is currently at capacity during peak hours and some 30,000 vehicles per day divert through City streets to avoid it.

DESIGN INFLUENCES
18. The corridor of investigation was located in the most conspicuous part of Sydney and hence the environmental issues became of prime importance. Issues of particular relevance were

(a) The Harbour itself, one of the most beautiful in the world.
(b) The Sydney Harbour Bridge and the Sydney Opera House world-renowned structures which could not be disturbed or visually interfered with.
(c) The northern shore parkland.

19. In fact the need to maintain the established beauty of this region demanded that a new bridge could not be tolerated within this corridor.

20. As well, Sydney is a busy commercial port and shipping could not be disrupted. The busy Circular Quay ferry terminal is also located within the corridor. Obviously the already congested cross-harbour road traffic must be maintained during the construction period.

21. There were many utilities and services located within the corridor which had to be avoided if possible. The most notable were

(a) high voltage electrical cables across the Harbour just to the east of the Harbour Bridge
(b) major telephone cables across the Harbour just to the west of the Harbour Bridge
(c) several sewer and stormwater tunnels on both the northern and southern shores
(d) railway lines, the most significant influence on the final selected route being the City Circle Railway tunnels on the southern shore.

ALTERNATIVE ROUTE STUDY
22. Many routes were investigated within the corridor of investigation established. All satisfied the same fundamental criteria previously outlined and linked the Warringah Expressway on the northern shore with the Cahill Expressway or the Western Distributor on the southern shore. By use of existing approach freeways, major disruption to the established areas on each shore could be avoided.

23. Routes which satisfied the criteria were identified one by one and then examined to determine the structure of the land and marine tunnels, the method of connecting to the existing expressways, and services requirements (most significantly the ventilation), so that preliminary construction costs and risks could be established. In all, eight routes were identified, each with a variety of connections to the existing expressways, particularly on the southern shore. (Fig. 3)

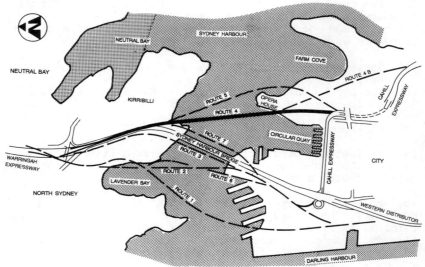

Fig. 3. Route investigation

24. The traffic studies indicated that it was preferable to provide additional capacity to the eastern side of the City and to either expand or by-pass the Cahill Expressway across Circular Quay to eliminate that restriction on the approaches to the existing Bridge.

25. Consequently the final selection became a choice between Route 4, which provided an almost direct link to the eastern side of the City, and Route 7, a more indirect route paralleling the existing Bridge across the Harbour and requiring upgrading of the Cahill Expressway across Circular Quay. Route 7 was estimated to have a capital cost of $80 million less than Route 4. However it was some 800 metres longer than Route 4 and also required an additional ventilation station on the southern shore.

26. Route 4 best suited the traffic demands of the future and the additional capital cost was justified by savings in time and fuel costs associated with the shorter route. Consequently Route 4 was selected as the preferred route for the Sydney Harbour Tunnel and defines the Project currently under construction.

THE SELECTED ROUTE

27. The selected route linked the Warringah Expressway at North Sydney with the Cahill Expressway on the eastern side of the City in almost a straight line. It required a total of 2300 metres of tunnel, approximately 900 metres on the north shore, approximately 1000 metres of marine tunnel and approximately 400 metres on the southern shore.

28. As well as the many natural and man-made features and requirements previously noted, the site geology significantly influenced the design. Hard sandstone bedrock occurred at a shallow depth on both sides of the Harbour. This rock extended into the Harbour at the depth of the Tunnel for 200 and 100 metres on the northern and southern shores respectively. In the centre of the harbour, the rock was up to 45 metres below sea level and overlain by alluvial silts and clays on the south side and sands on the north side. The maximum depth of water over the alluvial deposits was approximately 15 metres.

29. As the route alignment was determined by the need to satisfy the many constraints in the corridor, it is quite surprising that such a straight alignment was possible. However, the grading was greatly influenced by these features, particularly in the southern shore where a grade of 8% over a 220 metre length of tunnel was necessary. Here the tunnel must rise from the shoreline, where adequate depths of water in the Harbour must be maintained, to above the City Circle Railway Tunnels at the southern end of the route.

30. The grading on the northern shore was influenced by the geometry of the Ventilation Station located on the shoreline which had to be completely buried for environmental reasons. Across the Harbour the grading was dictated by the shipping channel clearances. (Fig. 4)

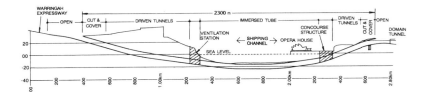

Fig. 4. Longitudinal section

PLANNING

THE TUNNEL DETAILS
31. The facility provides two high clearance lanes in each direction. Each pair of lanes is located in one roadway void, which provides a width of 7.0 metres between New Jersey barriers. A walkway 0.75 metres wide is provided on one side of the roadway.

32. The maximum vehicle height is 4.5m and a minimum of 4.8m clearance is provided from the road surface to the underside of the ceiling to allow for light fittings and clearance above the nominal vehicle height. There is limited capacity to provide for any cross-harbour services in addition to those required for the operation of the Tunnel.

THE TUNNEL VENTILATION SYSTEM
33. Environmental constraints demanded that no ventilation station could be located on the southern shore. Consequently the only Ventilation Station is located on the northern shoreline. The supply air station is located over the tunnels and the exhaust air station is located in the adjacent North Pylon of the Bridge. The exhaust air extraction ports over the Tunnel are linked to the Pylon by a 100 metre long Exhaust Air Tunnel.

34. The ventilation system is termed a modified semi-transverse system, in which the supply air is ducted through the tunnel and distributed to the roadway. The vitiated air is then drawn along the tunnel roadway to the exhaust air extraction points where it is ducted to the top of the North Pylon and discharged at a high level where it is safely dispersed into the atmosphere.

35. The tunnel ventilation system has a great influence on the tunnel structure in that air ducts must be provided to suit that system. These ducts must be continuous from the ventilation station to the vicinity of the Tunnel portals. The size of the ducts is a function of the prescribed air quality and the distance from the ventilation station to the portals. In the case of the immersed tube and southern shore tunnels, a duct cross-section area of 13.5 square metres for each roadway was required.

THE LAND TUNNELS
36. The land tunnels are driven through sandstone bedrock with each two-lane roadway contained in one tunnel. The two tunnels are separated to provide a rock pillar between them, preferably at least seven metres wide, to ensure stability of the surrounding rock mass. Since the depth of rock near the tunnel portals is low, "cut and cover" tunnels, constructed in an open excavation, are used in these locations. The most economical shape for the land tunnels locates the air duct above the roadway and forms it by use of a suspended ceiling. (Fig. 5)

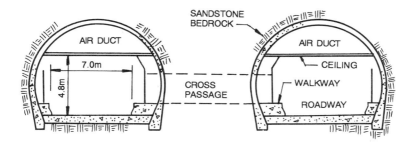

Fig. 5. Driven tunnel section

THE MARINE TUNNEL
37. The deep alluvial deposits in the Harbour, together with grading requirements, demanded that the marine tunnel be constructed using the immersed tube technique. Studies concluded that the most economical immersed tube section for Sydney was formed using a rectangular reinforced concrete unit configured with the roadway voids located beside each other and with the air ducts located beside them at the extreme edges of the unit. (Fig. 6)

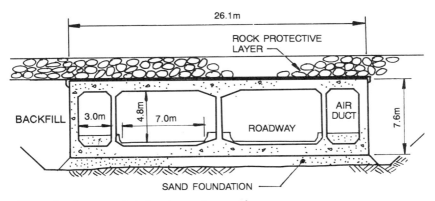

Fig. 6. Immersed tube tunnel section

INTEGRATION OF THE IMMERSED TUBE TUNNEL (IMT) WITH THE LAND TUNNELS
38. There were two major inconsistencies between the requirements for the IMT and land tunnels. By its very nature the IMT requires the roadway voids be located parallel to each other and separated only by the central wall of the IMT. On the other hand, the driven tunnels required that the roadway voids be separated to maintain the

PLANNING

integrity of the rock mass. Also, the air duct in the IMT tunnel was located beside the roadway void, whereas in the land tunnels it was located over the roadway.

39. There was also a requirement for a breakdown bay to be located at the base of the uphill sections of roadways. Neither the driven land tunnels nor the IMT could readily and economically accommodate the widening required for this breakdown bay.

40. Fortunately the need for special structures at the interfaces allowed these conflicts to be readily resolved. On the northern shore the Ventilation Station was interposed between the driven land and IMT tunnels. On the southern shore, the construction using "cut and cover" techniques under the Opera House Forecourt provided a suitable transition between the land and IMT tunnels.

41. On both shores, the roadways diverge immediately on the landward side of the IMT tunnel interface to provide as much separation as possible where the driven tunnel portals occur. This deviation and the breakdown bay, can be readily accommodated within the insitu concrete structures of the Ventilation Station and the Concourse Structure. At the adjacent driven tunnel portals, the rock pillar between the driven tunnels was still insufficient to support the overlying rock, and this pillar required strengthening using "filler-wall" techniques.

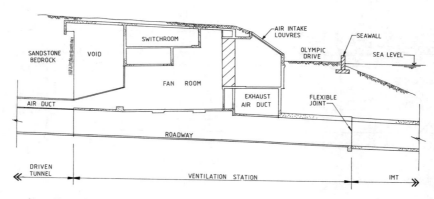

Fig. 7. North shore interface

THE NORTH SHORE INTERFACE (Fig.7)
42.   The Ventilation Station has been planned to be completely underground with the only evidence of it on the surface being the intake louvres fifty metres long by seven metres wide located on a sloping bank.

43. Since the supply air system is divided into zones to the south and north of the Ventilation Station, there was no need to provide continuity between the supply air ducts in the IMT and driven land tunnels at this location. In fact, the different location of the air ducts, to the side of the roadway voids in the IMT, and above the roadway in the driven land tunnel, facilitated the compact design of the Ventilation Station.

44. An interesting feature of the Ventilation Station is that its southern face forms a permanent cofferdam between the sea and the cavity in which the structure is constructed. This seawall is socketed into the sandstone bedrock. Water flows through the sandstone are so low that the Ventilation Station need not be designed for any uplift forces and, since the rock extends virtually to the surface, there are no lateral hydrostatic or earth pressures on the structure.

45. The cavity between the excavated rock face and the Ventilation Station structure is well drained to prevent buildup of water. Depressurisation wells, extending up to eight metres into the rock, are used to relieve any ground water pressures which could cause heave of the base rock under the structure.

46. The design requires that the first IMT Unit will be placed after the first stage of the Ventilation Station is constructed.

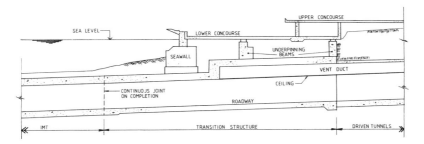

Fig. 8. South shore interface

THE SOUTH SHORE INTERFACE (Fig. 8)

47. The construction on the southern shore generally, and under the Opera House Concourse area in particular, is considered the most difficult work on the Project. The NSW Government has just completed a major construction on the shoreline in this area, comprising a new seawall, a two-

PLANNING

level restaurant and promenade. Work on the Tunnel must be completed without interruption to the use of this new facility.

48. The seawall crosses the Tunnel at an angle of about 50 degrees to the centreline. The area of the rock above the Tunnel roof is very low near the seawall but increases rapidly to the south. To the south of the new concourse structure, driven tunnels are feasible, but the rock cover is still considered low in tunnelling terms.

49. The recently completed Concourse Structure and the seawall itself require underpinning before excavation can proceed to roadway level.

50. Consequently, a "cut and cover" construction technique was used under the Concourse Structure, extending northward under the seawall to the IMT interface. To the south of the Concourse Structure, driven tunnels were used.

51. On the southern shore, continuity between the supply air ducts in the IMT and driven tunnels was required. This was achieved by deflecting the air through three 90 degree bends in a streamlined air duct incorporated in the "cut and cover" Transition Structure.

52. The design requires that the IMT unit be placed before the major construction work under the Concourse is commenced. This required that the "final joint" be placed at the northern end of this IMT unit.

53. The selected system involved the use of a skew-ended IMT unit which finally joined to the insitu structure to provide complete continuity with that structure.

CONCLUSION

54. The Sydney Harbour Tunnel has been designed to satisfy the peculiar traffic, environmental and other requirements of the City of Sydney.

55. Like all other tunnels around the world the technical concepts and details have been developed to accommodate the local site conditions. The Designers have taken the experience of similar previous projects, and applied sound engineering principles to develop a "state of the art" design for this Project.

ACKNOWLEDGEMENTS

56. The authors recognise the contributions of the following consultants to the design of the components of the project discussed in this paper

Wargon Chapman Partners:

— Design Manager for the Project, and Designers of the route and of the Ventilation Stations.

Macdonald Wagner-Freeman Fox:

— Designers of the IMT.

John Connell-Mott Hay Anderson:

— Designers of the land tunnels.

Gutteridge Haskins Davey, Maunsell Partners, Parsons Brinckerhoff International:

— Designers of the tunnel services.

Devine Erby Mazlin:

— Architects for the Project.

Coffey & Partners:

— Geotechnical Consultants.

11. The Sydney Harbour tunnel — evaluation of form of construction for an immersed tunnel

M. W. MORRIS, MSc, and D. G. MORTON, BSc, FICE, Acer Freeman Fox (Far East) Ltd, and L. GOMES, BSc, BE, PhD, Macdonald Wagner Pty Limited

SYNOPSIS. The Sydney Harbour Tunnel will be the first immersed tube traffic tunnel in Australia. In the absence of established practice, an investigation was carried out as part of the Sydney Harbour Tunnel Feasibility Study to determine the most appropriate cross section and form of construction. The Paper briefly describes the methodology of that Investigation.

INTRODUCTION

1. The Sydney Harbour Tunnel will, when completed in 1992, run from new road connections at the Warringah Expressway on Sydney's North Shore to a point immediately south of the Opera House and Botanic Gardens on the opposite shore (Figure 1). The Tunnel will be about 2.2 km between portals and the central section under the Harbour, about 1 km long, will be constructed by the immersed tube method.

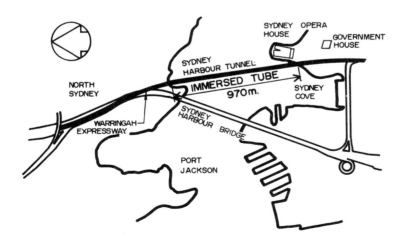

Fig. 1. Location Plan

PLANNING

2. The preliminary design of the immersed tube section was developed during a Feasibility Study undertaken in 1986 by the proponents of the Tunnel, the Transfield-Kumagai Joint Venture. Macdonald Wagner in association with Freeman Fox, were consultants to the Joint Venture responsible for the immersed tube section and its interface connections to the approach tunnels. The detailed design of this tunnel has been reported elsewhere (1).

3. In countries where a number of immersed tubes already exist (Holland, North America, Japan, Hong Kong), established working practice and availability of materials and labour skills may predetermine the cross section configuration and use of steel or concrete as the principal structural material. The Sydney Harbour Tunnel was the first immersed tube traffic tunnel to be proposed within Australasia although a small immersed tunnel was constructed in 1981 as part of a cooling water intake canal system for Eraring Power Station, Newcastle, New South Wales (NSW). There was therefore no established practice for design or construction ; nor did availability of construction materials impose any constraint.

4. In these circumstances, a detailed investigation of a number of cross section configurations in both steel and concrete was undertaken as part of the Feasibility Study in order to determine the most cost-effective arrangement in conditions specific to the Sydney area. This Paper presents an overview of the methodology of this Investigation.

STRUCTURAL TYPE

Options
5. There were two primary structural options :

(a) The circular steel shell section with a composite concrete lining, two such tubes often being joined by diaphragms to form a binocular section ;
(b) The rectangular concrete section, being either reinforced or prestressed longitudinally and transversely depending on circumstances.

6. The circular steel shell has always been more popular in North America, the rectangular concrete section being more popular in Europe. The reasons probably relate to relative cost of materials and availability of labour skills.

Relative Merits
7. Factors affecting the decision betwen these two structural types are summarised in Table 1.

CONSTRUCTION MATERIALS

8. Major sources of construction materials in NSW are shown in Figure 2.

Table 1. Concrete vs Steel – Advantages and Disadvantages

REINFORCED CONCRETE SECTION	STEEL SHELL SECTION
ADVANTAGES o conventional construction methods using readily available materials; o construction carried out in ideal conditions in the dry enabling a high degree of supervision and quality control; o conventional methods and working conditions enable realistic competitive bidding and hence lower prices; o rectangular shape enables efficient tailoring of structure to suit traffic and ventilation requirements, thereby minimising use of materials, minimising tunnel depths below navigation requirements and hence dredging and backfill costs; o the unit must have a minimum deadweight for stability against flotation. It is sensible to use concrete for the deadweight since it is efficient in cost, density and structural terms. **DISADVANTAGES** o reinforced concrete, by its nature, and if it is to be used efficiently, cracks when strained, whether due to flexural, drying or early thermal shrinkage. These cracks must be controlled (typically to 0.2mm flexure, 0.1mm shrinkage) to minimise width and make the structure durable. Typical reinforcement content 120-150 kg/m^3. (Alternatively, prestress is used to achieve a Class 1 structure, no tensile stresses). A waterproof membrane is often specified as a second line of defence; a high standard of workmanship is required throughout; o structural elements must be designed for transverse bending rather than membrane compressive action as in a steel shell unit, thereby increasing reinforcement weights; o a major casting facility is necessary to enable handling of the completed units by flotation; the units are too heavy for launching by slipway.	**ADVANTAGES** o elaborate casting facilities are not required, thereby saving time and cost; o the inner steel shell provides a permanent impermeable membrane when protected by structural concrete on both sides; o the outer steel shell permits placing of final ballast concrete from outside, thereby minimising concrete to be placed inside after sinking; o the steel shells have a secondary role in providing longitudinal bending strength; o the circular shape enables hydrostatic load to be carried efficiently in ring compression. **DISADVANTAGES** o steel fabrication is a specialist activity, thus reducing the range of potential tenderers; o slipway and fitting out berth facilities are still required. Significant time savings may only occur where existing facilities can be used/taken over with minimum start-up time; o the requirement for minimum negative buoyancy in the unit means that the steel is largely an additional cost; o all concrete, except keel concrete, must be placed whilst the unit is floating to a carefully controlled sequence; o all concrete within the steel shell must be placed under difficult working conditions; transport costs increase and workmanship and supervision suffer; o the circular shape fits the essentially rectangular traffic gauge less well, resulting in extra duct volume; this can be minimised by the use of a "horseshoe" cross section.

PLANNING

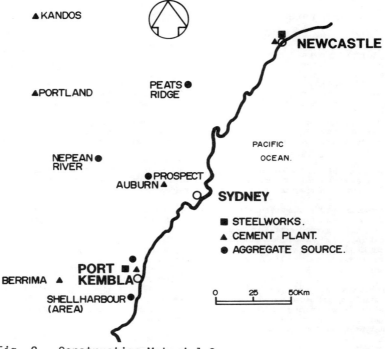

Fig. 2. Construction Material Sources

Steel
9. Australia's largest steel making company, BHP, has major steel works at Newcastle and Port Kembla, both within 120 km of Sydney. Port Kembla has an annual output of nearly 1.5 million tonnes of plate and coil strip products, whilst Newcastle produces 0.5 million tonnes of reinforcing bars per annum.

Cement
10. The two principal cement companies, Blue Circle Southern Cement and Australian Cement have works throughout NSW with a total annual capacity of over 2 million tonnes.

11. These plants, together with a limited amount of cement imported from Tasmania, offer a full range of cements; ordinary portland, high early strength, low heat, sulphate resisting. Blended cements consisting of ordinary portland cements and ground granulated blast-furnace slag (ggbfs) or pulverised fuel ash (pfa) are also available as standard products, although the level of cement replacement is low. Both ggbfs and pfa are freely available from the steel works and power stations in the area.

Aggregates
12. The Sydney area has a good supply of high quality aggregates sourced from quarries within a 100 km radius. River

gravels, generally crushed for use in concrete, are sourced from the Nepean River system 60 km west of Sydney. Metallurgical slags are also available as concrete aggregates.

CONSTRUCTION FACILITIES

General
13. An important factor in choosing between steel or concrete units for the Sydney Harbour Tunnel was the availability of a suitable construction site. The sites considered are shown in Figure 3.

Existing Shipyards
14. Carrington Slipways Pty Ltd of Newcastle prepared a budget estimate based on preliminary drawings of a steel shell tunnel. At the time of the Study, they were in a position to make available a slipway large enough for tunnel unit construction within the time frame envisaged in the project programme. However, construction of the Tunnel was dependent on approval of the Feasibility Study by the NSW Government and it was impossible for the Joint Venture to make any commitment at that stage. There was therefore, some uncertainty about the availability of the shipyard's facilities.

Existing Dry Docks
15. Only two dry docks in Sydney were large enough to be considered. The Captain Cook Dock at Garden Island was under the control of the Royal Australian Navy. The No 1 Dock at Cockatoo Island was barely large enough. Both were ruled out on grounds of cost and programme restraints generated by their current order books.

16. A small former graving dock known as Woolwich Dock was considered. However, the dock would have required enlargement and further investigation showed that it was listed by the National Trust and could not be altered in any way.

Other Sites in Sydney Harbour
17. A number of other sites around around Sydney Harbour and the Parramatta River were considered. These were mostly disused industrial sites ; berths, power stations, coal gas plants with waterside access for deep draught units. They were all rejected for one or more of the following reasons:

(a) high rockhead making casting basin excavation costly;
(b) suspected toxicity of existing landfill;
(c) impact of construction traffic or noise on residential areas.

18. It was concluded that there was no suitable site for construction of units by whatever means within Sydney Harbour.

PLANNING

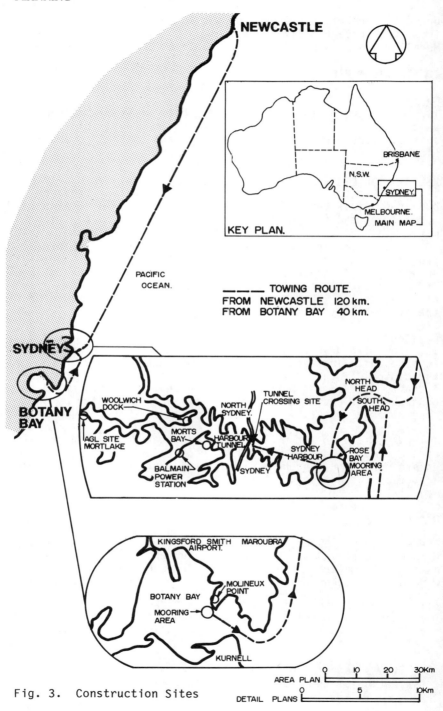

Fig. 3. Construction Sites

South of Sydney
19. The Maritime Services Board the port authority for NSW, offered to make available a 12 ha site in Botany Bay about 15 km south of Sydney. The site had water frontage for a side-launch slipway, but was also big enough to contain a 3 or 4 unit casting basin.

20. Perceived difficulties were seen as:

(a) keeping dewatered an excavation 8m below sea level in permeable sands with open sea frontages;
(b) a 40 km open sea tow to Sydney;
(c) settlement effects on adjacent industrial structures.

21. It was believed that these problems could be overcome but it was clear that if it became necessary to tow a concrete unit, further investigation of its wave induced behaviour would be necessary.

Conclusion
22. In the absence of any suitable site for construction of steel or concrete units within Sydney Harbour, either the use of the Carrington Slipways facilities at Newcastle (for steel units) or Botany Bay (for steel or concrete units) were the only choices. Of these, only Botany Bay could be expected to be available. The review of alternative tunnel types was therefore made on the basis of construction at Botany Bay.

CROSS SECTION REQUIREMENTS

Vehicle Gauge
23. The vehicle gauge adopted for the Sydney Harbour Tunnel is shown in Figure 4. A range of overall carriageway widths (between barrier profiles) was considered:

(a) for 2-lane carriageway 7.0 to 8.3m;
(b) for 3-lane carriageway 9.9m.

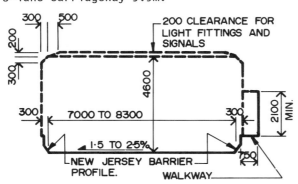

Fig. 4. Sydney Harbour Tunnel Vehicle Gauge

PLANNING

24. A raised walkway at least 750mm wide was required faced with a New Jersey Barrier profile. A similar profile was required on the other side. The standard vertical clearance was 4.6m but because the existing Harbour Bridge provided an alternative route for heavy vehicles and buses, consideration was given to a low vertical clearance tunnel (3.5m) suitable only for cars, light commercial vehicles and small buses.

25. A further 200mm vertical clearance was allowed for signs and lighting and 150mm for formation of vertical curves in the road surface, and for any deviation of the tunnel units from their intended position after laying.

Ventilation Requirements
26. At an early stage, the electrical and mechanical consultant was able to establish a basic requirement for semi-transverse ventilation. Preliminary calculations showed that for this purpose an air duct cross sectional area of 15m^2 would be required for each road duct.

Ballast
27. For each cross section developed, provision was made for the placing of internal or external concrete ballast to provide permanent negative buoyancy after sinking.

CROSS SECTION DEVELOPMENT FOR THE SYDNEY HARBOUR TUNNEL

Scope of the Study
28. Eight basic cross sections in concrete and steel were originally proposed for review. This Paper will only consider the last two stages of the Investigation by which time some of the sections had been eliminated as unworkable or uneconomic. These stages will be referred to here as Stages 1 and 2. For clarity, only 2-lane, standard vertical clearance versions of the cross sections will be reviewed.

Stage 1 Cross Sections
29. The alternatives considered at this stage are shown in Figure 5. The concrete alternatives differ only in the positioning of the ventilation ducts ; centre/overhead (C1), totally overhead (C2), and outboard (C3). The steel alternatives comprise the traditional binocular form (S1) and an attempt to reduce the depth and extra duct-space of that section by truncating the base (S2).

Points of Comparison
30. Whether the cross section is steel or concrete, either type must have the same mass for any given cross sectional area in order to meet buoyancy requirements. Since the mass of the steel shell is small in relation to the mass of concrete (about 1:20), the steel shell cost is largely an "extra" to be offset against the cost of extra reinforcement in the concrete section and, of course, the casting basin. This applies equally to comparison of concrete cross sections.

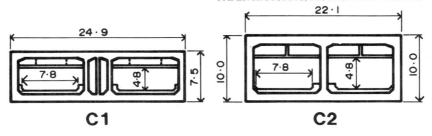

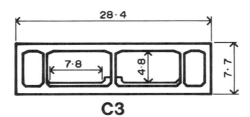

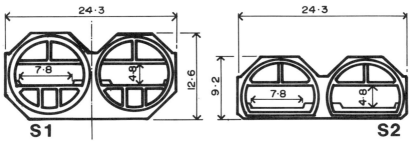

Fig. 5. Stage 1 Cross Sections

Since alternatives C1, C2 and C3 had much the same overall cross section areas, they contained similar quantities of steel and concrete ; direct costs were therefore similar. Cross section alternatives were therefore also assessed subjectively on how well they satisfied a series of requirements which would affect overall project cost, but might not be reflected in direct costs as obtained from unit rates.

31. The requirements were :

(a) External physical dimensions — affects dredging quantities and casting basin requirements.
(b) Ease of construction — multiple ducts require complex formwork and slow down construction.
(c) Ventilation ducts — efficiency (in terms of shape and relationship to traffic ducts) and accessibility.

PLANNING

(d) Utility provisions – space for tunnel and utility services.
(e) Extraneous space – related to (d) ; extraneous space may be used for utility provisions.
(f) Drainage
(g) Cross connection between traffic ducts – important for operational and emergency circumstances.
(h) Interface with bored approach tunnels – bored tunnels must have minimum separation and will have overhead ventilation ducts.

32. It should be emphasised that, whilst some of these requirements have general application, others are project specific. Perhaps unusually for an immersed tube, the approach tunnels and part of the immersed tube are in rock, Hawkesbury Sandstone. This material is hard but also abrasive because of its high quartz content. This makes it difficult to dredge by cutter suction methods, and adverse public perceptions of blasting made minimisation of rock dredging critical.

33. The approach tunnels are of horseshoe cross section with overhead ventilation ducts. Environmental considerations have dictated a single ventilation building on the north shore of the Harbour at the interface between the north approach tunnels and the immersed tube. At the south interface, which is immediately adjacent to the Opera House, the ventilation ducts must be transferred from their locations within the immersed tube to an overhead location within a transition structure to be built under the Opera House Concourse. The configuration of the ducts within the immersed tube had to facilitate this.

34. The results of this screening exercise are summarised below. As in most subjective assessments, the results were difficult to interpret.

35. <u>Alternative C1</u> was a compromise between the greater depth of C2 and the greater width of C3. It had the advantage of a reduced number of ducts (the dividing wall in the central duct may be constructed of blockwork and the overhead duct separated by a precast ceiling at a late stage of construction), and a common drainage trunk line, although this required reverse crossfall on the road surface. Cross adits obstructed the central ventilation duct but if these were faired to smooth airflow, small plant rooms could be sited there. The overhead ventilation duct was ideal for emergency smoke extraction, but was too small for convenient personnel access. The necessity to combine the central and overhead ducts made connections to the bored tunnel duct configuration difficult.

36. <u>Alternative C2</u> was probably best in this latter regard, but to supply ventilation air at traffic level (which was

preferred) required air ducts down the tunnel walls. Emergency smoke extraction via ceiling grilles was straight-forward. There was a lack of cabling/plant space.

37. Alternative C3 had an additional duct in comparison to C1, but had shallowest overall depth (which minimised dredging and backfilling costs). Ventilation air could be supplied at traffic level but emergency smoke extraction panels at high level were required. The outboard ducts enabled normal crossfall and facilitated greater central sump storage capacity since this could be split between the two ducts.

38. Alternatives S1 and S2 shared the disadvantages of a deep cross section and consequent high dredging and backfilling costs. Ventilation duct transition to the approach tunnels was straight-forward. In Alternative S1, the lower ventilation duct was extraneous space unless used for cabling/plant but access in operation was difficult.

Cost Comparison
39. A comparison of direct (tunnel structure) and indirect costs (dredging and backfilling) is presented in Table 2. Since these were approximate estimates, they are shown in the form of the ratio of the cost of each alternative to the cost of the cheapest alternative (C1/C2/C3 for direct costs, C3 for indirect costs). In addition, the effect on C1/C2/C3 of adding in the cost of the casting basin is shown.

Table 2. Stage 1 Cost Comparison

	ACTIVITY	COST RATIO Section Width (m) x Height (m)				
		C1 24.89 x 7.48	C2 22.13 x 10.01	C3 28.37 x 7.66	S1 24.27 x 12.63	S2 24.27 x 9.23
A	Tunnel Structure (Direct Cost) (1)	1.00	1.00	1.00	1.70	1.45
B	Tunnel Structure (Direct Cost) (1) plus Cost of Casting Basin	1.00	1.00	1.00	1.40	1.19
C	Marine Works (Indirect Cost) (2)	-	1.15	1.00	1.65	1.20
	RATIO OF TOTALS (A+C) RATIO OF TOTALS (B+C)	- -	1.08 1.07	1.00 1.00	1.68 1.51	1.32 1.19

NOTES:
1. Direct costs includes all costs of basic units as floated except temporary fitments, bulkheads, finishes, joints.
2. Indirect costs includes dredging, backfilling and rock armour but not towing, sinking, handling.
3. Costs based on 970mm immersed tube length at June 1986 prices.

PLANNING

40. Unit rates were based on typical Sydney construction costs for June 1986 ; budget costs were obtained from international contractors for major items like dredging. The fabricated cost/tonne of the steel shell used for S1/S2 included slipway/launch costs. It was based on :

(a) the budget estimate provided by Carrington Slipways Pty Ltd, including the use of all existing shop and launching facilities;

(b) a fabricated cost/tonne for steel work of this kind provided by Transfield Pty Ltd to which construction costs for a temporary slipway were added.

41. Alternative C1 was rejected by the Joint Venture on the basis of the subjective assessment whilst preparation of indirect costs was in progress. These are not therefore available for C1. Alternative S1 had the highest direct and indirect costs and was rejected. It appeared that S2 had the capacity to be competitive and was carried forward to Stage 2.

Stage 2
42. The four alternatives for study in Stage 2 are shown in greater detail in Figure 6. Alternatives C2, C3 and S2 were brought forward from Stage 1. Alternative S3 was a hybrid section having a reinforced concrete base as a means of reducing the structural steel content whilst retaining a launch capability. It should be emphasised that structural aspects of this alternative were not fully investigated.

43. These four cross sections were developed to preliminary design stage and detailed bills of quantity prepared. A cost comparison of both direct and indirect costs in shown in Table 3. A lump sum cost for the casting basin was included for "C" alternatives. This sum is in fact very close to the actual cost of the casting basin which is complete at the time of writing. The costs do not include towing, handling, sinking, jointing or finishes and are quoted only in "units of cost" to preserve commercial confidentiality.

44. Table 3 shows that the concrete alternatives have a considerable cost advantage over Alternative S3, even after allowing for the casting basin. This advantage results not only from direct cost, but from substantial savings in dredging and backfill. Alternative S2 was competitive but there were doubts as to whether the reinforced concrete base would make it too heavy to launch. This alternative would have broken new ground in a number of ways and there was insufficient time to develop it further. It was concluded that, even though the question of the sea tow of a concrete unit had to be further investigated, a concrete "C" alternative should be chosen.

45. Alternative C3, whilst having virtually identical direct

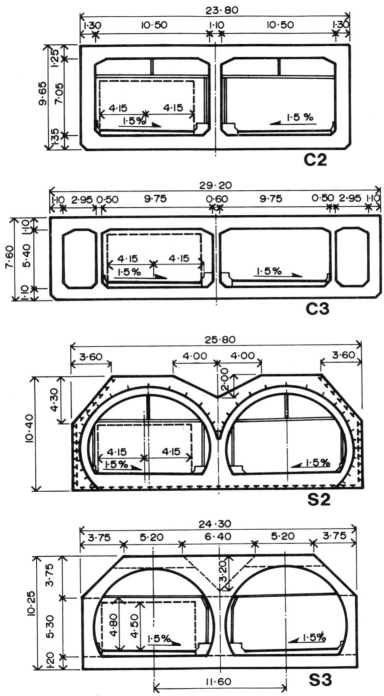

Fig. 6. Stage 2 Cross Sections

cost to Alternative C2, showed significant savings in dredging and backfilling costs, becoming the cheapest overall. Since it was expected to show cost savings from the provision of a shallower casting basin, and also to exhibit more stable handling during the tow from the casting basin to Sydney Harbour, Alternative C3 was selected as the preferred cross section.

Table 3. Stage 2 Cost Comparison

	ACTIVITY	UNITS OF COST Section Width (m) x Height (m)			
		C2 23.80 x 9.65	C3 29.20 x 7.60	S2 25.80 x 10.40	S3 24.30 x 10.25
A	Tunnel Structure (1)	10.0	10.05	16.29	12.31
B	Dredging (in rock)	7.95	6.70	9.04	9.04
C	Dredging (in other than rock)	3.77	3.40	4.10	4.10
D	Sand Foundation	0.21	0.26	0.23	0.23
E	Ordinary Backfill	4.47	3.79	5.40	5.40
F	Special Backfill	0.05	0.05	0.05	0.05
G	Rock Armour	0.36	0.36	0.36	0.36
H	Casting Basin (3)	2.60	2.60	- (4)	- (4)
	Sub Total	29.41	27.21	35.47	31.49
	Contingency 10%	2.94	2.72	3.54	3.15
	TOTAL ($ million)	32.35	29.93	39.01	34.64
	COST RATIO	1.08	1.00	1.30	1.16

NOTES:

1. Tunnel structure includes all costs of basic tunnel units as floated except temporary fitments, bulkheads, joints, finishes.
2. Activities (B-G) are those making up the Indirect Costs shown in Table 2 with the addition of Activity D : Sand Foundation.
3. The slight difference in Casting Basin costs between C2 and C3 has been neglected for purposes of this comparison.
4. Cost of tunnel structure for S2 and S3 include slipway/launch costs.
5. Costs based on 970m immersed tube length at June 1986 prices.

CONCLUSIONS

46. This Paper has described the determination of the cross section and form of construction of the Sydney Harbour Tunnel. The cross section chosen has now been designed in detail and construction of the units had just begun at the time of writing. However, the casting basin site has been moved to Port Kembla about 100 km south of Sydney because of concern, which arose subsequent to the Feasibility Study, about effects of dewatering at Botany Bay.

REFERENCES

1. GOMES L and MORRIS, M W, The Sydney Harbour Tunnel – Design Aspects of the Immersed Tube Structure, Proc. 2nd Australasian Port, Harbour and Offshore Engineering Conference, Brisbane 1988.

12. The Conwy tunnel — scheme development and advance works

G. W. DAVIES, BSc, and G. CRAMP, BSc, FICE, Travers Morgan, and H. KAMP NIELSON, MSc, formerly Christiani and Nielson A/S, now COMAR Engineers A/S, Copenhagen

SYNOPSIS. The A55 trunk road is the major traffic artery along the North Wales Coast and for some years has suffered from major traffic congestion particularly where the road passes through the town centres. The worst bottleneck of all occurs at Conwy where the road passes through the ancient town walls and along the narrow streets of the town centre. This paper describes the studies leading to the selection of a new bypass for Conwy which incorporates a 1km long immersed tunnel beneath the Conwy Estuary. The paper should be read in conjunction with Paper No.22 'Conwy Tunnel - Detailed Design' by Messrs. Stone, Lunniss and Shah.

INTRODUCTION
1. The A55 trunk road passes along the North Wales Coast from Chester to Bangor, a distance of 91km (figure 1). It is the main traffic artery in the area and forms a vital communication link between the North Wales Coast and the rest of Britain.

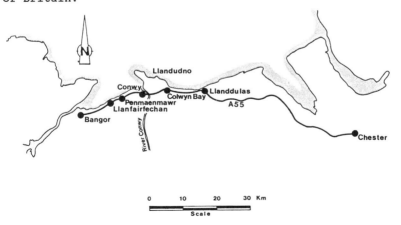

Fig. 1.   The existing A55

Immersed tunnel techniques. Thomas Telford, London, 1989.

PLANNING

2. During the 1960's it became increasingly clear that serious traffic congestion was occurring along major lengths of the A55 and that extensive and costly improvements would soon be required. The particular problems were the tortuous alignment, narrow road widths and numerous at-grade junctions where the road passed through the town centres. In the summer months there is a major influx of visitors to the North Wales Coast and between June and September traffic conditions were becoming intolerable.

3. In 1969 the Welsh Office commissioned Travers Morgan and Partners, Consulting Engineers, to undertake a feasibility study for a new or improved route between St. Asaph and Aber and over forty five alternative routes and permutations of routes were examined. Extensive consultations were carried out with those affected and in July 1972, the Secretary of State announced his preferred alternative. This was to construct a new dual two lane carriageway route to follow the coastline for most of its length but to pass inland over the central section where the road crossed the Conwy Estuary. A low level bridge was favoured for the estuary crossing and the new road was to pass immediately alongside and just to the south of Conwy Castle and continue behind the town in a new bored rock tunnel.

4. Some minor modifications were made as a result of further development work and in May 1974 draft highway orders were published for the route between Llanddulas and Llanfairfechan a distance of 27km, this being divided into three separate sections known as Stages 1, 2 and 3 (figure 2).

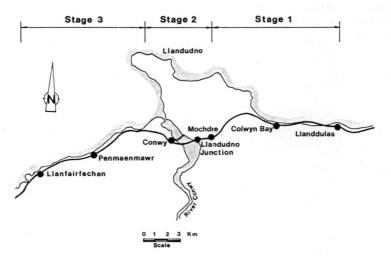

Fig. 2. The published route

5. Over 600 objections to the orders were received and all objectors were given the opportunity of presenting their views to an independent Inspector at a Public Inquiry in Llandudno between May 1975 and February 1976. At the time this was the longest Inquiry into a road scheme ever held in the UK.

6. Most of the objections to the Conwy Crossing (Stage 2) proposals were concerned at the change which would be caused by building a major trunk road of such scale and dimensions alongside the ancient town of Conwy with its medieval Castle and town walls.

7. After studying the Inspector's report, the Secretary of State announced his intention to proceed with Stages 1 and 3 (subject only to certain minor amendments) but to defer a decision on Stage 2 until further detailed studies of alternative estuarial tunnel routes had been undertaken and compared with the low level bridge.

8. Travers Morgan and Partners were appointed to undertake these studies with Christiani and Nielsen A/S acting as specialist sub-consultants.

CONWY ESTUARY

9. Conwy Castle dates from 1283 in the reign of Edward 1 and is an outstanding example of medieval military architecture. The town walls are the finest left in Britain. The town is a major tourist attraction and is filled with visitors in the summer months.

10. The first road crossing of the Estuary was completed by Thomas Telford in 1826. He constructed a wrought iron suspension bridge with a long approach embankment or causeway (figure 3). The suspension bridge has now been bypassed by a replacement bridge built in 1959. The Chester and Holyhead Railway shares the causeway and crosses the estuary in a wrought iron tubular bridge built by Robert Stephenson in 1848.

11. The estuary follows a south-east to north-west alignment and is flanked on both sides by high ground. On the east side there is extensive development and residents of Deganwy enjoy fine views of the Conwy Estuary and the Snowdonia National Park. To the west are the treelined slopes of Bodlondeb and further north Conwy Morfa is a flat sandy area with a fringe of dunes. It has a well established golf course, and areas of housing, caravans and light industry.

12. At low tide large areas of mud and sand are visible and in several locations these have developed into salt marshes, attracting a wide variety of wildlife. A large number of yachts are moored in the lower estuary and a small fishing fleet and a number of pleasure boats are based at Conwy Quay.

13. The ground conditions in the area are extremely complex and the underlying bedrock consists of sandstones, siltstones and mudstones of the Silurian and Ordovician periods.

PLANNING

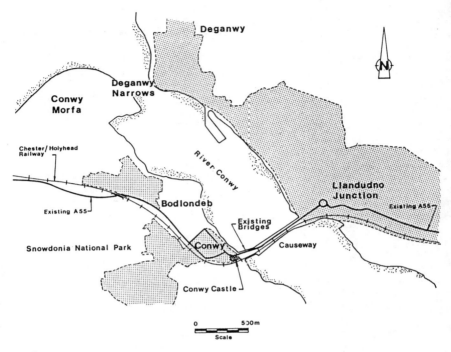

Fig. 3. The Conwy Estuary

14. Glaciation has led to the deposition of considerable depths of boulder clay which is over 20 metres thick in parts. The bed of the estuary is formed of soft alluvial clays overlain by varying depths of sand and sediment.

15. Deganwy Narrows, a channel some 170m wide, forms the entrance to the estuary and in this area current velocities can reach 2m/s. A further restriction occurs at the bridges near Conwy Castle and this also produces high current velocities which have scoured the bed producing the deepest water (approximately 19 metres at mean tides). The tidal range is 6.5 metres at Spring tides and substantial quantities of sediment are in transit with the Upper Estuary (ie to the south of the bridges) continuously accreting sediment brought in by the tide. The lower estuary is substantially in balance with most of the sediment brought in by the flood being removed on the ebb.

THE ROUTE ALTERNATIVES

16. These are illustrated in figures 4 and 5 and all have common termination points with Stage 1 on the eastern side of Glan Conwy Interchange and with Stage 3 at a point on Conwy Morfa to the west of Conwy Town. There are three alternative tunnel routes known as the Brown, Green and Blue routes and the previous 'published' route is also shown crossing the estuary on a low level bridge.

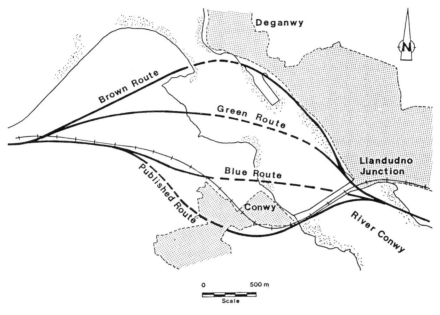

Fig. 4. Alternative routes

17. Various types of tunnel construction were studied for each route and an extensive site investigation involving over 150 boreholes was undertaken to provide information on the ground conditions. It was concluded at an early stage however that only immersed tunnel schemes were feasible with cut and cover lengths constructed in reclaimed areas on either bank. All forms of bored tunnels were rejected with the exception of the western length of the blue route passing in a bored hard rock tunnel beneath the high ground known as Bodlondeb on the west bank.

18. The Brown route is the most northerly option and involves a 580 metre long tunnel crossing immediately to the south of Deganwy Narrows. The Green Tunnel is similar in principle although it involves a longer 1030m tunnel, crossing the estuary more or less diagonally and positioned mid-way between the Brown and Blue alternatives.

PLANNING

19. The Blue route involves a 1200m long tunnel passing beneath the estuary and continuing under Bodlondeb.

20. The 'published' route crosses the estuary on a low level bridge on the south side of Conwy Castle and various forms of bridge, including cable-stayed and variable depth concrete options, were investigated. (Figures 6 and 7) The published route also involves a 1000 metre long bored tunnel known as 'Cadnant Tunnel' behind Conwy Town. This is a rock tunnel comprising two one-way two lane tunnels passing beneath the eastern limits of Snowdonia National Park. With the published route there is an option of providing an additional interchange in the Gyffin Valley area (immediately next to the Castle) and this provides some additional traffic benefit not available with any of the other schemes.

Fig. 5. Alternative routes - aerial view

COMPARATIVE EVALUATIONS

21. The feasibility studies were completed in March 1980 and a summary of the principal findings is given in Table 1. Full details are available in 'The Conwy Estuary Tunnel - Feasibility Reports' published in March 1980. The studies were undertaken by Travers Morgan in association with Christiani and Nielsen A/S with the following organisations providing specialist advice :-

- i) The Hydraulics Research Station (hydraulic modelling and surveys)
- ii) Howard Humphreys & Partners (rock tunnelling at Bodlondeb and Cadnant Park)
- iii) Nutall Geotechnical Services (geotechnical investigations)

Fig. 6. Published route - cable stayed scheme

Fig. 7. Published route - variable depth scheme

TABLE 1 - COMPARISON OF PRINCIPAL FEATURES

Group	Aspect	Published Route	Brown Route	Green Route	Blue Route
Road Users	Length Length in tunnel Number of bridges Traffic Design Year Peak Hour (low forecast and high forecast assumptions)	6.00 kms 1.01 kms 11 Glan Conwy to Llan.Junc (2240-3680) Llan.Junc. to Gyffin Valley (2130-3380) Gyffin Valley to Conwy Morfa (1820-2850)	6.36 kms 0.58 kms 10	5.92 kms 1.03 kms 11 Glan Conwy to Llan.Junc (2240-3690) Llan.Junc. to Conwy Morfa (1790-2820)	5.67 kms 1.39 kms 11
	Percentage relief to existing A55 Conwy Bridge Conwy Town Bangor Road	Low Forecast / High Forecast 72 / 72 49-55 / 50-59 86 / 84		Low Forecast / High Forecast 57 / 57 7-52 / 47-56 85 / 83	
Residents	Houses Demolished Curtilage Affected Houses affected by traffic noise greater 5 dB(A) Houses subject to significant visual change	18 4 81 65	3 7 65 124	4 7 24 64	3 7 91 23
Visitors	Change in character of area	Significant effect on Gyffin Valley and view of Conwy Castle from south	1. Significant effect due to road passing along east side of lower estuary 2. Significant effect due to reclamation north of Deganwy Pier. 3. Significant effect on Conwy Morfa	1. Significant effect due to reclamation on both sides of the river 2. Less significant effect on Conwy Morfa	-

TABLE 1 (Cont.) - COMPARISON OF PRINCIPAL FEATURES

Group	Aspect	Published Route	Brown Route	Green Route	Blue Route
Commerce and Industry	Loss of facility (or effect so bad that business could not continue)	8 (3 in Gyffin Valley on Conwy Morfa)	7 (5 on Deganwy Pier)	2	5 (4 on Conwy Morfa)
Those involved in recreational activities	Golfers	None	Major Impact on layout of course. Area lost 4.8 ha	Lesser impact on course Area lost 3.3 ha	None
	Yachtsmen and Sailors	Loss of facilities in Gyffin Valley	Loss of facilities at Deganwy Pier	Loss of 12.7 ha. of lower estuary	None (in permanent case)
	Anglers	No significant change	1. Disturbance to fish life during construction 2. Reclaimed area may afford better access to water at high tide	1. Disturbance to fish life during construction 2. Reclaimed area may afford better access to water at high tide	Disturbance to fish life during construction
	Others	Loss of cricket ground bowling green, recreation ground and Playing Field in Gyffin Valley			Effect on landscape gardens at Bodlondeb

TABLE 1 (Cont.) COMPARISON OF PRINCIPAL FEATURES

Group	Aspect	Published Route	Brown Route	Green Route	Blue Route
Ecological Effects	Salt Marshes (Area permanently taken by scheme)	6 ha in upper estuary	45 ha in upper estuary	45 ha in upper estuary 5 ha in lower estuary	46 in upper estuary
	Fish life (salmon and other migratory fish)	–	←――――――――― Change in water quality due to sediment disturbance during dredging ―――――――――→		
	Invertebrates	–	Loss of 3.5 ha of primary mussel beds	–	–
	Birds	–	Loss of feeding areas east side of estuary	Loss of feeding areas on both sides of estuary	Loss of feeding area on east side of upper estuary
Local Government (Planning Opportunities	Opportunities to encourage tourism	1. Access to Conwy improved 2. Dramatic new view of Castle from new	1. Access to Conwy improved 2. Some new views of Castle and	1. Access to Conwy improved 2. Some new views of Castle and estuary	1. Access to Conwy improved 2. Some new views of Castle and upper estuary from new road
Central Government	Estimated construction cost (Nov.1979 prices)	£61.6 m (excluding Gyffin Interchange) £62.6 m (including Gyffin Interchange)	£66.9 m	£76.7m	£83.3 m

TABLE 1 (Cont.) COMPARISON OF PRINCIPAL FEATURES

Group	Aspect	Published Route	Brown Route	Green Route	Blue Route
Ecological Effects	Salt Marshes (Area permanently taken by scheme)	6 ha in upper estuary	45 ha in upper estuary	45 ha in upper estuary 5 ha in lower estuary	46 in upper estuary
	Fish life (salmon and other migratory fish)	–	–	Change in water quality due to sediment disturbance during dredging	–
	Invertebrates	–	Loss of 3.5 ha of primary mussel beds	–	–
	Birds	–	Loss of feeding areas east side of estuary	Loss of feeding areas on both sides of estuary	Loss of feeding area on east side of upper estuary
Local Government (Planning Opportunities	Opportunities to encourage tourism	1. Access to Conwy improved 2. Dramatic new view of Castle from new road	1. Access to Conwy improved 2. Some new views of Castle and estuary	1. Access to Conwy improved 2. Some new views of Castle and estuary	1. Access to Conwy improved 2. Some new views of Castle and upper estuary from new road
Central Government	Estimated construction cost (Nov.1979 prices)	£61.6 m (excluding Gyffin Interchange) £62.6 m (including Gyffin Interchange)	£66.9 m	£76.7m	£83.3 m

PLANNING

iv) British Rail (track diversions and Underline Bridgeworks)
v) Institute of Terrestrial Ecology (salt marsh ecology and plant life)
vi) University College of North Wales (birds, insects and marine invertebrates)
vii) Unit of Coastal and Estuarine Studies (river hydrology)
viii) Freshwater Biological Association (freshwater fisheries)

22. Clearly the comparative evaluation picture is complex with value judgments having to be made at every stage as to the relative importance of each of the effects and principal characteristics of the different schemes.

23. In traffic terms all the routes provide substantial relief to the existing A55 and form an effective bypass of Conwy. However the Published Route is considerably cheaper than any of the tunnel alternatives (see Table 1). The Published Route, however, has a major environmental impact particularly on close views to and from Conwy Castle on the south side and this was undoubtedly a major consideration at the Public Inquiry. Distant views from the south (Glan Conwy) are marginally affected, although the main form of Castle is free from obstruction. The tunnel routes also have an affect (even if only during construction) on the Lower Estuary which is an important element in the setting of the Castle within a larger framework of Benarth Hill, Conwy Mountain and the old walled town.

24. In reaching a decision, the Secretary of State clearly had to decide whether the net impact of the Published Route on the Castle and the introduction of a major road into the small scale of the Gyffin Valley (with consequent effect on the environment of property overlooking the area) was worth the £5 to £15 million (1979 prices) additional cost needed for a tunnel scheme. All the tunnel schemes would make a significant change in the present form of the Lower Estuary (as seen by visitors and those living on the Deganwy shore) and lead to some loss of part of the present golf course on Conwy Morfa. In the event he decided that a tunnel alternative was preferable.

25. Of the tunnel schemes available, the Brown Route was the least costly, and had a lesser effect on the form of the Lower Estuary than the Green Route. The Green Route however had less effect on those living in Deganwy overlooking the estuary and less effect on the golf course. The Blue Route had the least effect of all but it is the most costly and the interchange behind Conwy Town is a major adverse feature due to residential properties and recreational parkland.

26. The Secretary of State selected the Green Tunnel Route and his decision was announced in July 1980.

THE SELECTED SCHEME

27. The Green Route is 6km long and commences at the western limit of Stage 1 where a viaduct is to be constructed over Glan Conwy Interchange (figure 8). The road passes westward over the Blaenau Ffestiniog branchline onto the mudflats south of Llandudno Junction. Here a grade separated junction is provided linked to the existing road network by a viaduct over the railway. The new A55 turns northwards and descends to pass beneath the Causeway in an underpass. North of the Causeway the road remains at low level and enters the tunnel which crosses the estuary in a slow curve to emerge on the Morfa between the golf course and the housing on Morfa Drive.

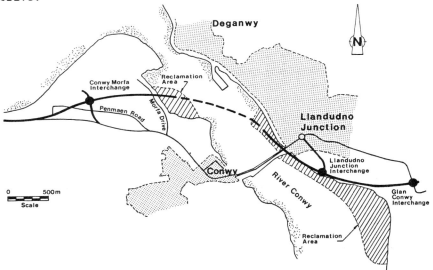

Fig. 8. The selected scheme - plan layout

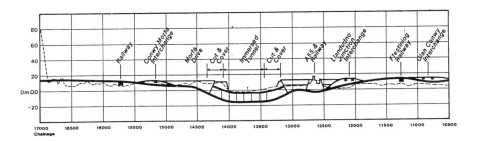

Fig. 9. The selected scheme - longitudinal section

PLANNING

28. The tunnel approaches are constructed as permanent open cuttings through two reclaimed areas within the estuary, one on the mudflats north of the Causeway and the other on the salt marshes on the western bank. Bunds at the sides of the road protect against tidal inundation.

29. The total length of the tunnel is 1030 metres and comprises three separate parts: an eastern cut and cover tunnel (200m), the immersed tube (710m) and a western cut and cover tunnel (120m). The immersed tube consists of six prefabricated units each 118m long. (Figure 9).

30. On the west side the road continues beneath Morfa Drive and crosses Conwy Morfa. There is an interchange on the Morfa and a link road over the railway provides a western access to and from Conwy. Beyond the interchange the road rises from below ground level to pass over the railway and join the Penmaenbach Tunnel section of Stage 3. A Tunnel Monitoring and Maintenance Building is provided on Penmaen Road and a tunnel services building to house the tunnel E&M equipment is provided near each portal. The Monitoring and Maintenance Building also serves the Penmaenbach and Pen-y-Clip Tunnels on the A55 further to the west.

31. Traffic flows in 2005 are predicted to reach a maximum of 34,000 vehicles per day (2 way) and dual 2-lane carriageways each 7.3m wide, are provided throughout. 1.0m wide hardstrips are provided either side of the carriageways and these reduce to 0.5m within the tunnel and Causeway Underpass. The 1.8m wide central reserve contains lighting columns and open box beam safety fencing.

32. From Llanddulas to Conwy Morfa the new A55 is a Special Road with usage restrictions similar to a motorway. Hazardous loads are permitted through the tunnel.

ORDER PROCEDURES AND PARLIAMENTARY BILL

33. Main line and side road orders for the scheme were published on 28th May 1981 using powers in the 1980 Highways Act. The orders comprise main line and side road orders dealing with that part of the new scheme between Glan Conwy and Conwy Morfa Interchanges (which is designated a Special Road) and main line and side road orders for the part to the west of Conwy Morfa Interchange which is designated a trunk road.

34. These draft orders give the Secretary of State powers to construct the new scheme in its entirety including powers to divert the navigable waterway and to construct the tunnel beneath the estuary. Also published at the same time was a detrunking order for the existing A55.

35. Relatively few objections to the orders were received since the tunnel scheme had the support of the overwhelming majority of those in the local community. However detailed 'holding' objections were registered by Gwynedd County Council and by Aberconwy Borough Council. Both sought extensive undertakings from the Secretary of State as to the way in which the powers were to be exercised.

36. Detailed discussions with both authorities then ensued. Those with Gwynedd County Council (who are the local highway authority) resulted in various modifications to the line and side road orders.

37. The discussions with Aberconwy Borough Council were also very detailed and ranged from such diverse issues as the compensation provisions for estuary users disrupted by the tunnel construction to the landscaping proposals and after use treatment of adjacent areas. As a result of these detailed discussions and also representations and discussions with other local organisations it became clear that additional powers to supplement those in the Highways Act would be required and that these could only be obtained from Parliament. A Conwy Tunnel (Supplementary Powers) Bill was therefore prepared and this was submitted to Parliament in November 1982.

38. The Bill gives the Secretary of State powers to undertake remedial work to mitigate the adverse effects of construction. These include the provision of replacement moorings in the Harbour, the re-construction of a fish pass at Conwy Falls (to open up spawning grounds for salmon parr), the possible restocking of the river with salmon and sea trout (and also the restocking of the mussel beds should that be necessary) and powers to regulate navigation and shipping.

39. The Bill also provides for compensation to persons and businesses losing income as a result of tunnel construction activities (mainly trawlermen, musselmen, boatmen and salmon nets men using the Harbour area) and also to the Harbour Authority itself.

40. Powers to acquire land for the tunnel works were also included in the Bill in respect of land within and immediately adjacent to the estuary.

41. The Conwy Tunnel (Supplementary Powers) Bill successfully completed its passage through both Houses in near record time and received overwhelming all-party support. Two new clauses were added during the Committee stages, one to enable the Secretary of State to pay compensation for the cost of having to move boats from moorings displaced by the construction works and the second to secure, as far as practical, maximum employment of persons normally resident in Gwynedd and Clwyd. Royal Assent was granted in the Spring of 1983.

42. As a result of further development work, various minor amendments to the original side road orders were found to be required and Variation Orders were published in November 1983. At the same time Compulsory Purchase Orders for land outside the area of Parliamentary Act powers were published together with an order converting the two link roads at Llandudno Junction and Conwy Morfa Interchanges to trunk roads. This latter order also contained provisions to de-trunk the eastern part of the existing A55 between the Glan Conwy and Llandudno Junction Interchanges.

PLANNING

43. Both Aberconwy Borough Council and Gwynedd County Council registered objections to various parts of these orders and 16 other objections were also received. A Public Inquiry was therefore held in June 1984. Following consideration of the Inspector's Report, one or two minor amendments were undertaken and the orders were made in May 1985.

SCHEME DEVELOPMENT AND PRELIMINARY DESIGN

44. The ground conditions at Conwy are complex and despite the fact that over 150 boreholes and seismic investigations had been undertaken during the feasibility study, a further detailed investigation was necessary to provide information for detailed design. A contract was let to Soil Mechanics Ltd and a major programme of cable tool borings and rotary core drillings were carried out on land and over water together with an extensive programme of field and laboratory testing. Fieldwork was commenced in November 1981 and the final results and factual reports were completed by the end of 1983. A total of over 150 boreholes were sunk in this contract and additional seismic investigation and other specialist testing work was undertaken.

45. As the results became available, a deep glacial channel was revealed crossing the line of the western approaches in the vicinity of Morfa Drive and it became rapidly apparent that the original feasibility study scheme based on permanent open side slopes would have to be abandoned. The channel is filled with permeable sand and gravel materials reaching to 35 metres below ground level in places and in direct contact with open water in the estuary. Any attempt at ground water cut off or ground treatment would have been prohibitively expensive.

46. A new scheme was therefore developed based on retained approaches in a reinforced concrete U-box construction with ballast concrete to prevent uplift; the overall height of the walls being reduced using an impermeable geo-membrane (see Paper No.2 for details). This major change involved the need to steepen the approach ramp gradients (from 4% to 5%) and to divert Morfa Drive eastwards to cross the U-box on a new overbridge.

47. This change was not possible within the powers of the existing highway orders and Variation (No.2) Orders were published in April 1985. Fortunately most of the objections received were concerned with matters of detail and could be resolved in negotiation and discussion without the need for a Public Inquiry. The orders were eventually made in January 1986.

48. On the east side of the estuary the ground conditions were more or less as predicted. However in the vicinity of he eastern tunnel portal the permeable 'lake deposits' comprising interbedded silts, sands and clays were found to

extend further eastwards than anticipated and the disposition of the rockhead in the vicinity of the eastern tunnel portal led to a decision to extend the tunnel length by some 60 metres and to move the portal eastwards. The original position for the revetment around the eastern reclaimed area was preserved however, and the space above the extended eastern cut and cover tunnel earmarked for landscaping.

49. The construction of the cut and cover tunnels on each side of the river relies on dewatering deep temporary excavations in which the reinforced concrete box sections can be built. The solution envisaged in both instances was to form these excavations with open sideslopes making extensive use of well pointing and other dewatering techniques. These excavations are protected from the estuary by temporary bunds.

50. In order to confirm the feasibility of dewatering the excavations and to provide factual data for tenderers, a series of elaborate pumping out tests were carried out in a further site investigation contract let to Osiris-Cesco Ltd in July 1984. Additional 'pressurisation' tests in the boreholes were carried out by Hydraulics Research Ltd.

51. The results confirmed the original assumptions and demonstrated the feasibility of the construction method.

ADVANCE WORKS

52. The Conwy Estuary is one of the most popular centres of boating activity along the North Wales Coast and during the summer months, large numbers of boats can be seen in and around the estuary. Most of these are kept at moorings and there are over 400 licensed moorings in the lower Estuary of which the majority are directly affected in some way by the construction activities.

53. Following consultation with the Harbour Authority, with the local sailing and yachting clubs and with the boat owners themselves, an extensive programme of relocation works was carried out (figure 10). This involved :-

i) The provision of a new deep water pontoon next to Conwy Quay to provide space for up to 100 deep fin boats which require deepwater at all states of the tide.

ii) The provision of 130 moorings on the mud flats next to the Causeway. Although boats using these moorings can rest on the bed at low tide, some dredging work to remove 70,000m^3 of material was required to provide extra water depth.

iii) The provision of other moorings next to Deganwy Foreshore and in the Upper Estuary to provide additional mooring accommodation.

54. The above works were carried out in 4 separate advance contracts between September 1985 and July 1986.

PLANNING

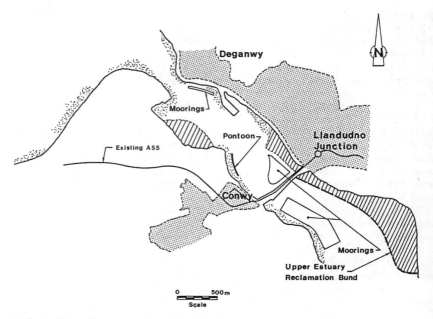

Fig. 10. New mooring facilities

55. The total volume of excavation required for the entire tunnel scheme is about 4.0 million cu.m, with about 2.3 million cu.m being excavated by dredger. Up to 2.7 million cu.m is likely to be re-used as fill the rest having to be disposed of as unsuitable.

56. During the design of the scheme it became apparent that it would be impractical to dispose of all of this unsuitable material by transporting off site. The Upper Estuary Reclamation Area was therefore conceived to take much of this material and also to provide a site for temporary stockpiling of sands and other suitable material. The reclamation area was needed early in the construction period and since the bund between the reclamation site and the estuary required time to consolidate, it was decided to construct this also in an advance contract let to John Howard PLC in October 1985.

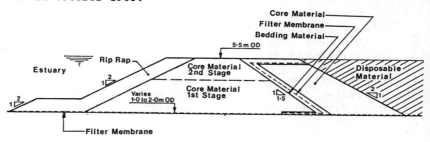

Fig. 11. Upper estuary reclamation bund

57. A total of 130,000 cu.m of stone rip-rap material was imported to construct this bund and construction was completed in August 1986 (Figure 11). The bund has a total length of 1.8km and rests on geo-membrane material to prevent mixing of the fill with the underlying alluvium.

58. Other relatively minor works constructed in advance include some noise insulation of adjacent property, some fencing and 'accommodation works' for adjacent landowners.

59. Negotiations with British Rail were also undertaken with a view to them providing points and a signalling system for a new rail siding on Conwy Morfa for the importation of material by rail and for a new level crossing over the Llandudno Branch line to allow road access to the east portal. Both of these were carried out in advance although in the event the level crossing construction overlapped with the start of the main contract.

CONSTRUCTION ACCESS PROVISIONS AND NOISE CONSTRAINTS

60. Throughout the whole of the planning process for the Conwy Crossing scheme there has been a major concern that the construction activities within the site and the movement of large numbers of construction vehicles to and from the site will severely disrupt residents living nearby and will cause considerable congestion on the adjacent road network.

61. Studies to predict the numbers of construction vehicles likely to use the road network were carried out as were studies of the effects of imposing restrictions of various types and various degrees of severity.

62. The final measures taken and the major constraints imposed on the contractor can be summarised as follows :

1. A complete embargo on all overnight dredging in the estuary between the hours of 2200 and 0700 and severe restrictions on the maximum noise levels permitted at other times.

2. A complete restriction on the use of the existing A55 through Conwy of all construction vehicles over 1.5 tons laden weight during morning and evening peak hours, during summer weekends and during Bank Holiday weekends and limitations on the numbers which can use this road at other times.

3. A complete restriction on the use of the A55 through Conwy at any time, of vehicles engaged in delivering bulk cement.

4. Major restrictions on the use of certain other roads. Many residential roads being completely prohibited with others only being available for vehicles engaged in specified activites and at certain times.

63. In addition to the above, land was made available on Conwy Morfa for the construction of a rail siding for the importation of material by rail thereby avoiding much of the need to use the road network. Negotiations with the Crown Estate Commissioners were also carried out to enable the contractor to excavate various shoal areas in the estuary to recover sand and gravel materials, thereby reducing the amount which might otherwise need to be imported.

64. The above measures will do much to reduce disruption during the construction period and hopefully residents will not be unduly disturbed. Clearly it is impossible to build works of this scale without any disruption but the inconvenience suffered during the construction period should be largely rewarded when the tunnel is opened and Conwy is relieved of its heavy burden of traffic.

ACKOWLEDGEMENTS

The permission of Welsh Office - Director of Highways Mr K J Thomas - to publish this Paper is gratefully acknowledged.

13. Development of immersed tunnels in Canada and Hong Kong

P. HALL, MSc, FICE, Per Hall Consultants Ltd, Hong Kong

SYNOPSIS. The paper is prepared based on the authors experience of certain conditions and requirements which had a decisive influence on planning, engineering and costruction of four immersed tunnels including the Deas Island and Lafontaine tunnels in Canada, the Cross Harbour and Mass Transit Railway Tunnels in Hong Kong which were build over a period of 25 years between 1956 and 1980. Brief references are also made to a sequence of developments in immersed tunnel techniques which were pioneered on these projects and have subsequently gained general acceptance.

INTRODUCTION

1. Design and construction techniques developed for immersed tunnels built in Canada and Hong Kong are well documented in technical papers previously presented to Engineering Societies in United States, Canada, England, Denmark and Hong Kong.

2. In this paper, a brief review is made of conditions which had a decisive influence on planning and the design development of four particular tunnels which followed on from each other over a period of nearly twenty five years.
    (a) Deas Island Tunnel (1956 - 59)
        under Fraser River at Vancouver
    (b) Lafontaine Tunnel (1962 - 67)
        under St. Lawrence River at Montreal
    (c) The Cross Harbour Tunnel (1966 - 72)
    (d) and Mass Transit Railway Tunnel (MTR) (1970 - 80)
        both under Victoria Harbour in Hong Kong

3. All four tunnels were designed to facilitate construction, in units approximately 100m long. Each unit was designed when equipped with temporary end bulk heads to float with the minimum freeboard consistent with towing requirements and to facilitate sinking in place under controlled conditions in a pre-dredged trench within a tolerance of a few centimeters. Based on these primary design parameters the most cost effective tunnel configuration would appear to be a shallow concrete box.

4. However the cross section of each tunnel was developed to accommodate the predicted flow of traffic with appropriate space for ventilation and other mechanical and electrical installations. Moreover the design and method of construction of each tunnel were governed by location-specific requirements such as soil conditions, seismic activities, climate, hydraulics and navigation.

5. Consideration was also given to such matters as land use, environmental impacts, civil defence, social and industrial interests, methods of financing and political requirements. It is explained in the following section that each of the four tunnels referred to in this paper was developed with different configurations to satisfy specific local requirements.

DEAS ISLAND TUNNEL - VANCOUVER BRITISH COLUMBIA
BACKGROUND (Ref.1)

6. The city of Vancouver is surrounded by water and mountains, Georgia Strait lies to the west, the Selkirk mountains to the north and east and the Fraser River to the south (Fig.1).

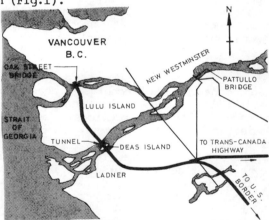

Fig. 1. Deas Island tunnel location plan

7. The Deas Island tunnel is part of a freeway which was build in the fifties for the Provincial Government of British Columbia (B.C.).

8. The tunnel replaces a ferry service across the Fraser River Delta at Ladner and supplements the capacity of the Pattullo Bridge upstream at New Westminister. It has effectively removed the confining influence of the Fraser River on the development of metropolitan Vancouver by opening large tracts of farm land south of the river for industrial, commercial and residential growth.

9. The tunnel also provides for a direct connection to the Trans Canada Highway, to the main highway to U.S.A. and to a short and quick ferry crossing to Victoria and Vancouver Island.

10. In 1955 when the B.C. Highway Department commenced investigation of the feasibility of a fixed crossing of the Fraser River Delta it was recognized from the outset that a bored tunnel would be impractical and that an immersed tunnel would be the only viable alternative to a long span highlevel Bridge.

11. However at that time only a few subaqueous multilane traffic facilities had been build as immersed tunnels and only the 4-lane tunnel under the Maas River in Rotterdam under comparable conditions. The Maas Tunnel had been built nearly 20 years earlier, by Christiani & Nielsen (C&N) a Danish firm of Engineers and Contractors, in joint venture with Dutch and German contractors.

12. To benefit from experience gained on the Maas Tunnel the B.C. Highway Department consulted C&N who entered into a joint venture with Foundation of Canada Engineering Corporation Ltd. (FENCO) to undertake a comprehensive engineering study of a fixed crossing of the Fraser River Delta.

13. From the outset the two parent companies disqualified themselves from bidding competitively for the construction of the project.

14. In the course of the study new concepts in immersed tunnel technology were developed including the "Hydrostatic Joint" between tunnel units which subsequently has gained general acceptance and a semitransverse ventilation system which has proved to be a practical and also a economical arrangement for medium length tunnels.

15. On completion of the study it was concluded that a 4-lane rectangular concrete tunnel (cross section shown on Fig 2) could be safely supported in the soft marine silt formations of the delta. Furthermore it would be both cheaper and faster to construct than a bridge which would have been required to be of adequate span and height to satisfy river navigation requirements downstream from New Westminster. This conclusion was reached with the appreciation that design of the tunnel and methods and timing of construction would be complicated by severe restrictions imposed by site conditions which were extremely difficult and are summarised below:

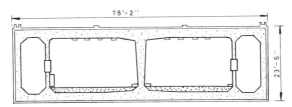

Fig. 2. Cross-section

(a) RIVER FLOW which can vary from a low of 20,000 cu.sec. during the winter season to a max of 550,000 cu.sec. in the spring and early summer when snow in the mountain is melting and causing flood conditions which precludes work on the river and restricts dredging operations and placing of tunnel units to a five month period from early November to early April which is further reduced by frequent occurrences of dense fog in November and December. It was therefore necessary to place the tunnel units in a short period of three months from January to early April to save the high cost of extending construction of the project by another year.

(b) TIDAL FLUCTUATION of the water level in the river (as much as 11 ft. at spring time required continuous adjustments of anchor lines for the sinking rig to ensure accurate placing of tunnel units in the trench. Intrusion during flooding of salt-water from the sea under the fresh river water causeed variation in water density which made it necessary to provide for adjustable water ballast to maintain proper trim of a tunnel unit during the sinking operation.

(c) SEASONAL BED LOAD in the flood season sand dunes on the river bottom were observed reaching a height of 15 ft. and a length of 200-300 ft. travelling downstream at a speed of as much as 10 ft. per hour which was another compelling reason for restricting dredging and sinking operations to the winter season. During the flood season scour holes were eroded in the river bottom to considerable depth, a short distance downstream from the tunnel alignment.

16. The tunnel was designed with sufficient flexibility to allow for seismic activity (zone 3) and for alternating consolidation and rebound of the supporting soil formations caused by variation in load from the shifting sand dunes.

17. Substantial scour protection was required and provided over the whole length of the tunnel by an articulated blanket of reinforced concrete slabs. The river banks for a distance of 200 ft. up-and downstream from the tunnel was protected by a revetment of riprap.

18. The report on the feasibility of the fixed crossing project recommending a 4-lane immersed tunnel under the Fraser River at Deas Island was accepted by the Department of Highways. The new tunnel, the Deas Island Tunnel was to be constructed in six units each 344 ft long with adjoining approaches. A joint venture of C&N and Fenco were appointed to undertake the engineering and manage construction of the Deas Island Tunnel to ensure completion of the project in 3 years ready for Traffic by 1 July 1959.

19. To carry out this contract efficiently and on time a Joint Venture office was established in Vancouver in July 1956 to complete site investigations and, with guidance from specialists in the two head offices, prepare basic plans and specifications in sufficient detail to call for competitive fixed price tenders and award contracts in time to enable the successful bidders to mobilize and proceed with construction by early January 1957.

20. This phase of the design work was completed and separate contracts awarded for construction of the casting basin, the two approaches with ventilation buildings and the 6 tunnel units. Contractors however were not prepared to contract for handling and placing of the tunnel units in the Fraser River which left the J.V. with the responsibility for the development of construction mehtods and sinking and to take over the duties of the Contractors for this phase of the project.

21. At this stage the New Westminister Port Authority which is responsible for navigation on the lower Fraser River raised an objection. They held that an option should be preserved to increase the navigable depth of the river by 10 ft. and therefore requested the corresponding clearance over the tunnel. It was however established that the high cost of dredging to maintain a 10 ft deeper channel in the river could not be justified. A compromise was accepted to lower the tunnel by 5 ft. extra depth under the existing navigation channel. On this basis permission was received to proceed with construction in March without however postponing the opening date for the tunnel.

22. To make up for this critical delay at the start of construction, the area of the casting basin was made sufficiently large to build all six tunnel units in one batch.

23. Construction of the basin and also excavation for the approaches and installation of cofferdams for construction of the approaches were carried out over a period of 4 months. Concurrently work proceeded with preparation of detail design of the tunnel and development of construction methods. The design office was moved to the site to expedite exchange of ideas and information between designers and construction engineers.

24. As a result of these measures the six tunnel units were completed and ready to be floated out of the casting basin only 8 months after construction was started. By mid November a special sinking rig developed by the J.V. was ready for a "dry run" on the river and a number of 'Dutch Anchors' were installed in the river bottom to facilitate safe warping and accurate positioning of the sinking rig. No. 1 tunnel unit was subsequently warped out of the casting basin and moored at the "fitting out wharf" to be equipped for the sinking operation.

25. The first tunnel unit was placed by mid January 1958 and the sixth on the 17th April 1958 just before the spring freshet prevented any furhter work on the river and a mere 13 months after the start of construction. The Deas Island Tunnel was completed in the spring of 1959 and opened for traffic 1 May two month earlier than planned.

LAFONTAINE TUNNEL - MONTREAL P.Q.
BACKGROUND (ref.2)
26. Montreal City is located on an Island in the St. Lawrence River and connected to the South Shore by three bridges and since 1967 also by the Lafontain Tunnel (Fig 3).

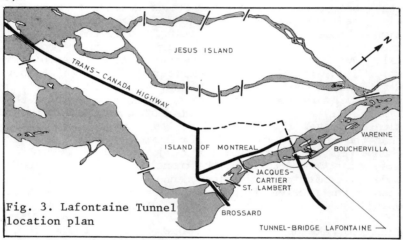

Fig. 3. Lafontaine Tunnel location plan

27. The Quebec Government decided in 1960 that a fourth crossing of the St. Lawrence River was required to supplement the capacity of the three existing bridges and form part of a network of urban express routes to be build in time for the celebration of Canadas Centenary and the staging of "Expo 67" in Montreal. The crossing would also provide the last link in the Trans-Canada Highway the nearly 5,000 miles long 4-lane Highway from the Pacific to the Atlantic coast. As a part of the Trans Canada Highway it qualified for Federal Government subsidy of 90% of the cost of 4-lanes. The balance of 10% and the cost of additional traffic lanes required in urban areas of Montreal was funded by the Provincial Government.

28. The new crossing was to be located 5 miles downstream from the Montreal inner harbour and build through Long Point, a suburb in Montreal East in a previously reserved right of way from where it would cross the river to Charron Island and continue over an arm of the river to the south shore of the St. Lawrence River at Boucherville.

TECHNICAL CONSIDERATIONS

29. Six 12 ft wide traffic lanes designed - in accordance with Trans Canada Highway standards were required to serve the combined volume of long distance and local traffic. A 2,400 ft. wide shipping channel with 150 ft. height above highwater level and a depth at low water of 40 ft was provided to satisfy maritime requirements on the river downstream from Montreal.

30. Conceptual designs were developed in compliance with above requirements for a suspension bridge and the alternative of an immersed tunnel of the same type as the recently completed Deas Island Tunnel. The possibility of two parallel tunnels bored in rock under the river each wide enough for three lane traffic was considered impractical. Engineering studies based on preliminary site investigations lead to the conclusion that an immersed tunnel could be build under the river at less cost than the suspension bridge. It could also be completed in the available time.

31. The cost of operation and maintenance was estimated to be of the same order for bridge and tunnel but the latter would avoid the impact of long approach structures of the flat shore areas and was also preferred from the viewpoint of shipping and civil defence.

32. In the autumn of 1962 the Quebec Roads Department retained the services of three firms of Consulting Engineers i.e. Brett & Qulett, Lalonde et Valois and Per Hall Associates. Their brief was to jointly assume responsibility for planning, engineering and project management to provide for safe and economical construction under the St. Lawrence River of the 6-lane vehicular tunnel, which subsequently became known as the Lafontaine Tunnel.

33. The project was scheduled for completion by March 1967. In addition to the tunnel it included the construction of about 3 miles of dual-3 lane highway with two large interchanges and a 1500 ft. long bridge between Charron Island and the South Shore.

VENTILATION

34. The configuration of a tunnel is developed primarily to serve a specific flow of traffic and to provide for ventilation of the traffic compartments to keep pollution below an acceptable level. A suitable ventilation system must therefore be selected before the tunnel cross section can be designed.

35. Based on experience gained from the Deas Island Tunnel a modified semi transverse ventilation system was considered particularly well-suited for the Lafontaine Tunnel because it was possible to arrange for fresh air intake through ventilation buildings located reasonable close to the quarter points of the 4561 ft. long tunnel

from where fresh air could be distributed over the entire length of the tunnel through a central duct. Vitiated air could be exhausted directly from the traffic lanes by fans in the ventilation buildings.

36. The recommended semi transverse ventilation system was tested in a full scale model of a 100 ft. long tunnel section and found to operate efficiently in a 3-lane tube with the anticipated volume and mixture of vehicular traffic.

TUNNEL CROSS SECTION

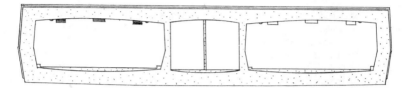

Fig. 4. Cross-section

37. When the design of the ventilation system was advanced sufficiently to size the center duct and the two traffic compartments the tunnel cross section was designed as shown on Fig 4. To provide space for dual-3 lane traffic new concepts were developed for the design and construction of the wide traffic lanes. Transverse prestressing of roof and bottom slabs was required to resist the hydro static pressure at a maximum water depth of 90 ft and still keep the weight of the tunnel within limits to provide for adequate buoyancy to float the tunnel units into place. To counter-balance the force of prestressing and preserve structural integrity of a tunnel unit under all load conditions, bottom and roof slabs were held together along the center line of the traffic compartments by temporary ties. These were gradually relaxed as the hydrostatic pressure increased during sinking operations and subsequently removed after the tunnel units were in place and permanently loaded.

North Ventilation Building

38. To save time and the high cost of a 90 ft. deep cofferdam which would be requied for construction in the dry of the substructure of the north ventilation building in 80 ft. deep water it was prefabricated in the casting basin as a buoyant concrete caisson which was floated across the river and sunk into position on a prepared seat on top of tunnel unit No.7. The joining of the caisson to the tunnel unit follow the same principle as the joining of the tunnel units themselves by means of a rubber gasket which under the weight of the caisson establishes water tightness so that all subsequent work can be carried out in the dry from the inside.

CASTING BASIN (Ref.3)
39. Information obtained from preliminary field investigations indicated that a casting basin could probably be constructed on the Charron Island. Further site investigations confirmed that the most advantageous location would be in the river behind an earthdyke, on the north side of the Island within which the tunnel units could be prefabricated.
40. The casting basin was build large enough to provide space for the construction in one batch of seven tunnel units each 360 ft long 119.5 ft wide and 25 ft high together with the north ventilation building. Additional area for the construction in situ of the south approach was included with the casting basin (Fig 5).

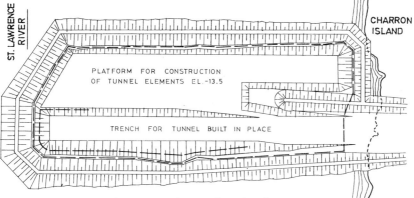

Fig. 5. Lafontaine Tunnel casting basin

41. Work in the casting basin was completed by May 1965. Prior to flooding explosives were placed to blow up a short stretch of the dike to provide an exit to the river.
42. With this arrangement the length of the immersed tunnel was reduced to 2520 ft. and major economies realised in construction time and cost.
43. Trial dredging of a short stretch of trench to full depth indicated that the trench for the tunnel could be dredged and remain stable with 1:4 slopes.
44. The tunnel units were outfitted alongside a wharf in the casting basin near the exit to the river. By pre-sinking and suspending the tunnel units under the sinking rig at the wharf they could be warped out and safely placed in their final location in the trench sheltered against river currents and the wash from frequently passing ocean vessels. The first tunnel unit was sunk in place by early September 1965 and the last of the 7 units during the first week of April the following year. The Lafontaine Tunnel was completed and opened for traffic in March 1967 a month before the opening in Montreal of Expo 67.

PLANNING

CROSS HARBOUR TUNNEL - HONG KONG
BACKGROUND
45. The idea of a fixed crossing of Victoria Harbour between Hong Kong Island and Kowloon was talked about before World War II but it was in the late fifties that the demand for a vehicular crossing was growing at a sufficiently fast rate to justify serious consideration of the project. A few years later Wheelock Marden a large local firm, took the initiative to promote the project and with the support of other local firms presented the Government with an offer to finance, build and operate a tunnel for vehicular traffic under the harbour.

PREPARATORY INVESTIGATIONS
46. Preliminary studies were carried out by consulting engineers Scott Wilson Kirkpatrick & Partners jointly with Freeman Fox & Partners and a conceptual design of a stayed girder bridge was developed for consideration. The Local Marine Department objected however, to the presence of bridge piers in the harbour which would present a hazard to shipping during typhoon conditions. The Civil Aviation Department also entered an objection in their case to the high towers of the bridge, which encroached on airspace required for safe operation of Kai Tak Airport. For these reasons the proposed highlevel bridge was ruled out in favour of an immersed tunnel, between Causeway Bay and Hung Hom (Fig. 6), for which a franchise was granted early in 1965 by the Government to the Cross Harbour Tunnel Co. Ltd. which had in the meantime been formed by the firms that promoted and developed the project.

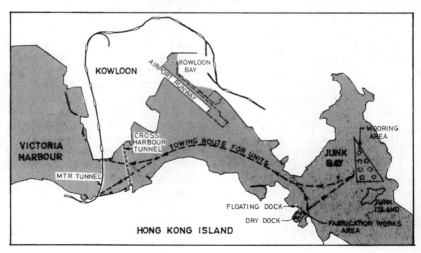

Fig. 6. Cross Harbour and MTR Tunnels location plan

154

DESIGN DEVELOPMENT

47. In order to apply experience gained in Canada on the Deas Island Tunnel and the Lafontaine Tunnel under construction in Sept. 1965 the Joint Engineers retained Per Hall Associates a Canadian firm of Consulting Engineers, to advise on immersed tunnel techniques.

48. Available information indicated that physical conditions in Victoria Harbour were well-suited for the construction of an immersed tunnel in the proposed alignment.

49. In anticipation of the rapid growth of traffic of a variety of motor vehicles which would be generated by a fixed harbour crossing, it was decided to build the tunnel for 4-lane traffic in separate 2-lane compartments with sufficient height for double decker buses and extra width for narrow service side walks and provision for a suitable ventilation system. The optimum choice of site for the casting basin on open land adjoining the Kowloon Landfall of the tunnel would have allowed substantial economies through common excavation and execution of all construction at theo tunnel site. Unfortunately, however the land was not available at the time and a remote site at Kwai Chung had to be accepted for the recommended scheme.

50. To satisfy these requirements and to benefit from the shallowest possible alignment it was concluded that a rectangular reinforced concrete tunnel would be cost-effective because the compact cross section could be developed to achieve minimum displacement of the tunnel units and achieve corresponding economies in quantity of materials.

51. A semi transverse ventilation system arranged for air intake and exhaust through ventilation buildings located in short distances from the tunnel portals and distribution over the entire length of the tunnel through a central duct was found suitable and economical for the length and selected configuration of the Cross Harbour Tunnel. The cross section shown on Fig. 7 was recommended.

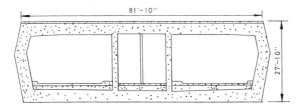

Fig. 7. Concrete tunnel cross-section

CONSTRUCTION

52. When tenders were received from prequalified international Contractors, a single steel shell tunnel alternative (Fig.8) was offered by a British led group. It was accompanied by a financing package for 75% of the contract value at low interest rates and 7 years grace period, a United Kingdom Export Credit.

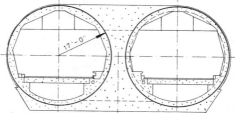

Fig. 8. Steel tunnel cross-section

53. Furthermore the open area at the Kowloon Landfall of the tunnel (para 48) was made available for a yard required for fabrication of steel shells and launchways for the tunnel units and other construction facilities.

54. The combined advantages derived from exceptionally favourable financing and the availability of land to carry out all construction activities at the tunnel site enabled the Group of Contractors who submitted the steel shell alternative to offer the Cross Harbour Tunnel Co. the most economical package for construction of the tunnel. Their tender was accepted.

55. Following 2 years delay during the period of civil unrest construction of the project was started in Aug 1969. Three years later the project was successfully completed and the Cross Harbour Tunnel opened for traffic in Aug. 1972.

MASS TRANSIT RAILWAY TUNNEL - HONG KONG (ref.4 and 5)
BACKGROUND

56. In 1970 barely a year after construction was started on the Cross Harbour Tunnel the Hong Kong Government decided to proceed with the development of a Mass Transit Railway for Hong Kong and retained consulting engineers Freeman Fox & Partners (Far East) to plan and engineer a suitable system.

57. To cope effectively with a project of this magnitude and complexity Freeman Fox retained the services of five firms of specialised consultants including Per Hall Associates who were charged with the planning and engineering of a twin track railway tunnel under the Victoria Harbour.

Alignment

58. Available information indicated that ground conditions throughout the Harbour were favourable for immersed tunnel technology, as already recognized when the design was developed for the Cross Harbour Tunne. This

was subsequently confirmed in the initial phase of construction of that tunnel which was under way at the time the MTR tunnel was planned.

59. The MTR Harbour Tunnel connects the east-west aligned railway along the north coast of Hong Kong Island at the waterfront of Wanchai with the north-south aligned Nathan Road Line at the waterfront in Kowloon a distance of 1400m.

60. To reduce curvature of the railway at the Wanchai Landfall the harbour tunnel is build in a curve with a radius of 2800m and each precast tunnel unit is constructed with the same curvature. The alignment of the tunnel is indicated on Fig. 6.

TUNNEL CONFIGURATION

61. A conceptual design for a binnocular shaped reinforced concrete tunnel with longitudinal prestressing was developed and recommended as the most cost-effective arrangement. A steel tunnel alternative was considered but not found competitive.

62. A steel tunnel alternative was considered but not found competitive. Ventilation is provided by a longitudinal forced air system. Intake an exhaust fans are located in ventilation buildings at the shore ends of the tunnel with ducts streamlined to move air into the train tubes in the direction of train movements.

63. The interior circular shape of the tubes provides extra space for overhead conductors, power cables, a service and emergency side walk and helps to reduce drag and air resistance on fast moving trains (Fig. 9)

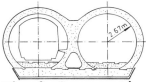

Fig. 9. Cross-section

64. The tunnel is constructed in 14 units each 100m long. Twelve of the units are of standard design with the two end units specially designed to support the ventilation building and to facilitate the connection with bored tunnels on land. A flexible joint is provided below the seawall to compensate for differential settlement caused by the weight of the ventilation buildings and load of backfill behind the seawalls. Floodgates are provided for the train tubes at the ventilation buildings from where they can be operated.

65. The tunnel units are sunk in place on a screeded gravel mattress in a trench dredged across the harbour and subsequently connected with hydrostatic joints and locked into position relative to each other by means of a reinforced concrete shear ring cast in in place in a recess between the units.

PLANNING

66. It was planned to sink tunnel units in sequence from Wanchai. However to make up for delays encountered in the construction of the casting basin for prefabrication of the tunnel units arrangements were made to sink units no. 13 and 14 from the Kowloon side and establish the final closure between units no. 12 and 13 by means of a tremie concrete joint.

VENTILATION BUILDINGS

67. The caisson method of construction developed for the Lafontaine Tunnel was adopted to save time and to avoid the high cost of temporary cofferdams more than 30m deep which would have been required for construction in the dry of the underwater portion of substructures for the ventilation buildings.

68. The substructures for the ventilation buidings were build in a small floating dry dock in Chai Wan to the height required for floating them into position over the tunnel end units at Wanchai and Kowloon. The caisson walls were subsequently build up to the extent that the caissons could be sunk on to a prepared seat on top of the tunnel and connected to the unit by means of a hydrostatic joint arrangement similar to the joint used between tunnel units.

JUNCTION WITH BORED TUNNELS

69. The tunnel end units were constructed with enlargement of a short section at the shore end to provide space for the shield for the bored tunnels. The bored tunnels were driven under compressed air pressure on both shores. To provide for easy tunneling into the immersed tunnel the end units were surrounded by a 2m thick layer of a cement bentonite sand grout to form an airtight cover over which the rest of the backfill was placed to provide weight to resist the compressed air pressure.

70. The enlarged end sections of the tubes were equipped with a second temporary bulkhead to form an airlock compartment inside the end bulkhead of the tunnel unit in which air pressure could be equalised with the compressed air pressure in the tunnel drive and thereby create conditions under which the shield could safely be moved through the end bulkhead and facilitate completion of the junction with the immersed tunnel under compressed air pressure.

71. The MTR Harbour Tunnel was successfully constructed in 33 months and completed on schedule in December 978.

CONCLUSIONS AND FUTURE TRENDS

72. It is characteristic for immersed tunnel technology that the design of the work is always influenced by construction methods and the availability of land and plant and sometimes by the method of financing.

73. Since the basic techniques of prefabricating tunnel units and floating them into place in a prepared trench

was first used in the early part of the century, immersed tunnel techniques have been improved and developed from project to project by engineers and contractors who have jointly pioneered new ideas and concepts in design and construction of immersed tunnels.

74. Notwithstanding that for years only few immersed tunnels were built many design and construction concepts which have gained general acceptance were developed on the early projects, amongst which the following have probably been the most significant i.e.

(a) The rectangular tunnel configuration for multilane traffic and sandjetting
Maas Tunnel 1941
(b) The hydrostic joint between tunnel units
Deas Island Tunnel 1959

(c) Configuration for dual 3-lane Traffic
Lafontaine Tunnel 1967

75. The early projects did however create great interest in immersed tunnel techniques and more than 50 tunnels have been built during the last 25 years, which have caused a rapid increase in improvements of existing techniques and development of new concepts - a trend which is likely to be continued on future immersed tunnel projects particularly on projects where planning and design are developed in close cooperation between engineers and contractors.

76. Immersed tunnels have been built to serve as subaqueous ducts for utilities, disposal of waste water and cooling water intakes for thermal and nuclear power plants. Other civil engineering works such as jetties, wharfs, ferry terminals bridge piers and foundations for industrial plants have been prefabricated and floated in place and supported on screeded gravel beds or on piles driven through mantles and permanently connected to the structure.

77. Moreover immersed tunnel techniques have been considered in connection with prefabrication of concrete offshore drill platforms and for construction of underwater parking and shopping centers along the waterfront of cities with congested or scenic water fronts and high cost land values.

78. Consideration has been given to the possibility of building a depressed highway outside an existing harbour front as a semi-immersed tunnel or open trough designed to be floated in place at high tide and lowered with the ebbtide onto a prepared bund. Land can be reclaimed on both sides of the depressed highway before during and after its construction to suit requirements for land.

79. The concept was recently suggested for a traffic bypass outside the waterfront at Central and Wanchai in Hong Kong as an alternative to an elevated road which would have impacted the scenic view over the harbour and increased the noise level from traffic in front of high class hotels and other important buildings.

80. It is however to be expected that the implementation of any new concept in civil engineering be it a semi-immersed roadway or other novel applications of immersed tunnel techniques will require adequate lead time to study implications the structure may have on existing underground utilities and drainage systems and to devise viable solutions to problems which may exist.

81. Apart from minimal environmental impact the main advantage of immersed tunnel techniques is that the bulk of construction of a project built in accordance with the concept can be carried out in a convenient area away from the permanent site of the works, with a minimum of interference with traffic on existing streets or other activities in the surroundings, in a relatively short time and at competitive cost with other structures.

82. The trend is undoubtedly towards further refinements and developments in immersed tunnel techniques on a project basis and application of the technology to a variety of civil engineering works besides immersed tunnels, which in time may generate the need and economic justification for permanently established casting basins operated like shipyards.

1 foot = 0.305m, 1000 cusec = 28.32 cubm/sec

References:
1. Per Hall and D.A. Young
   Four Lane rectangular tunnel
   placed by trench method
   Civil Engineering ASCE July 1958
2. Armand Couture
   The Louis Hippolyte Lafontaine Tunnel
   The Journal of the Engineering Institute of Canada
   January 1965
3. F.H. Sutcliffe
   A till cofferdam in the St. Lawrence river
   Canadian Geotechnical journal VolII no.3 August 1965
4. F.J. Hansen - Hong Kong MTRC harbour tunnel
   Hong Kong Engineer April 1979
5. P. Hall and F.J. Hansen
   Hong Kong Mass Transit Railway
   design and construction of the immersed tunnel tube
   Pree. Instn Civ Engrs Part 1, 1980.68 Nov

14. Immersed tunnel techniques in a typical delta country

H. J. C. OUD, Rijkswaterstaat, Locks and Weirs Division, The Netherlands

Synopsis

The Ministry of Public Works of the Netherlands carried out an extensive programme of building road and rail traffic infrastructure. The Netherlands have become one of the most densely populated countries in the world. Because the situation of this country in a typical delta area, a lot of tunnels have been constructed since the last century.

Introduction

The flat country known as the Netherlands is situated where the Maas and the Rhine debouch in the North Sea. A typical delta area, with fertile soils and extensive natural water infrastructure. Since such an area is ideally suited for agriculture, stock-raising, trade and industry, the Netherlands have become one of the most densely populated countries in the world. If the water infrastructure already provided by Nature was the first to be extended, road and rail traffic also increased and required a good land infrastructure.
The intersections of both infrastructures quickly increased in number and became ever more important.
Originally the fordable places in the rivers were utilised. Later ferry connections were instituted at many places and bridges were built. Tunnels have been constructed under busy fairways since the last century.

The following criteria are used when the optimum junction point is being selected:
- the physical features of the area, such as the depth and breedth of the fairway;

Immersed tunnel techniques. Thomas Telford, London, 1989.   161

PLANNING

- the total costs of the junction; these are the costs of construction and the capitalised costs of operation, maintenance and repairs;
- the costs for the user, such as waiting time for shipping when a bridge is closed or for road traffic when a bridge is open;
- safety, such as fenders on bridge pillars for shipping or escape routes for road traffic in tunnels;
- nuisance, such as obstructions to shipping on the site of a tunnel under construction;
- influence on the neighbourhood, such as noise, air pollution, visual disruption and the like.

The mutual balancing of such criteria is not always a simple matter. Some may be expressed in terms of money, others can only be measured and yet others can only be assigned a qualitative evaluation.
Systems for reducing the various criteria to a single denominator are available so that proper balancing can be effected.

At important junctions of roads and waterways in the Netherlands, operational maintenance costs, users' costs and safety are heavily weighted criteria in addition to those of construction. It is of major importance for both rail/road and water-borne traffic to be able to cross each other safely and without delay.
A tunnel often appears to be the most suitabel solution.

In the Netherlands Delta area, the upper layers consist of marine and fluvial deposists. In general, tunnels under very busy routes in this area are not constructed at too great a depth (10 to 20 m under water level).
This is why immersed tunnels are preferred in the Netherlands.
Sixteen tunnels have already been completed in accordance with this method of construction, and work is in progress, or will soon be begun, on six more.

The design of immersed tunnels

The immersed tunnel consists of access structures on the banks with the closed section of the tunnel under the intervening fairway. The construction of this closed section is the essence of the immersed tunnel: it is prefabricated in a nearby construction dock in one or more units; after completion these units are floated to their final site and sunk. The access structures are constructed on the site on the banks.
The geometric design for a tunnel, as set out in cross-section and in horizontal and vertical alignments, is decided by internal and external preconditions.

The following should be regarded as external preconditions: the breadth and depth of the fairway, the connecting infrastructure, the intensity and nature of traffic in the tunnel.
The internal preconditions are the profile needed to provide headroom for traffic, the slopes permitted and the vertical and horizontal lines.
The tunnel should be situated as close as possible under the fairway so as to limit the length of the tunnel and hence the costs.
Transportation and sinking exercise a major influence on the cross-sectional dimensions of an immersed tunnel.

While the tunnel units are being transported, a freeboard of 5 to 10 cm is usually maintained. After sinking, the tunnel should exercise a downwards load of 5 to 10 kN/m^2 on the subsurface.

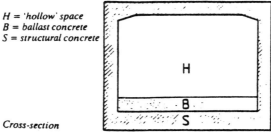

H = 'hollow' space
B = ballast concrete
S = structural concrete

Cross-section

Assuming in rectangular cross-sectional terms, that the area of the required "hollow" space is H m^2, that of the structural concrete S m^2 and that of the ballast B m^2, we arrive at the following two equations.

At the transport stage:
Weight = 0.99 maximum water displacement, or
$2.46\ S + 3.0 = 0.99\ (B + H + S)$ .... (1)

In the final phase:
Weight = 1.075 water displacement, or
$2.42\ S + 2.25\ B = 1.075\ (B + H + S)$ .... (2)

Solution of the equations gives the following results (in m^2):
$S = 0.728\ H - 1.154$ .... (3)
$B = 0.081\ H + 1.319$ .... (4)

A number of mean values have been used in the equations. This means that in specific cases more accurate data must be used.

In the equations, it has been assumed that:
1) the specific weight of reinforced concrete is 2.46;

PLANNING

2) the specific weight of ballast concrete is 2.25;
3) the specific weight of water is 1.0;
4) the immersion equipment and bulkheads have a weight of 30 kN/m length.

Construction of immersed tunnels

A tunnel to be built by the "immersed tunnelling" method is made up of "sub assemblies": de tunnel units, with a length of 70-120 metres. These units each consist of elements, approx. 20 metres long, between which expansion joints have been designed.
The use of these joints prevents the generation of large tensile stresses due to shrinkage after concreting and uneven settlements in the subsoil after immersion.
They allow for minor rotations, due to settlements, and a small longitudinal displacement, due to shrinkage.
Of course these expansion joints must be watertight.
Many solutions have been under consideration. It appeared that the best solution could be reached by using an epoxy resin for the small units and a rubber-metal waterstop with injection tubes for the large elements.
The joints between the tunnel elements are made watertight by means of compressible rubber profiles (Gina profile).
A second rubber seal is usually applied to be on the safe side (Omega profile).
The construction method of the concrete structure itself usualy involves pouring first the floor and, after this has hardened, the walls and the roof.

The differences in temperature and shrinkage of the floor and the wall concrete can lead to vertical cracks penetrating the entire thickness of the wall.

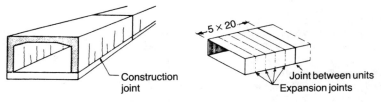

Cracks in the tunnel walls caused by tempe-   Scheme of tunnel
rature influences

To assure the watertightness of the concrete structure these cracks must be prevented.
There are many remedies against crackings.
Not one single method is effective enough, so that a combination of measures is necessery.
The most important methods are:
1. suitable composition of concrete
    a. selection of a suitable cement
    b. decreasing the cement content

      c. use of coarse aggregates
2. reducing the temperature -increase in hardening concrete
   a. reducing initial temperature of mortar by cooling
   b. cooling of newly poured concrete in the walls.

As a rule the tunnelelements are prefabricated in a dregded dock separated from the water by a dike. During prefabrication, a well-point system keeps the working area dry. Upon completion, the dock is flooded and, after the dike has been dregded away, the tunnel units are towed to the place where they are to be sunk. The construction site does not have to be near the tunnel.

Building docks or dry docks as building sites for prefabricated tunnel units have been dug for various tunnels in the west of the Netherlands and are re-used for construction of new tunnels. The effect upon the environment has been practically nil, as a result of the very favourable situation of these building docks and the good condition of the soil. The three best-known exemples are as follows.

1. The Barendrecht dry dock,
   in which the tunnel units of the Heinenoord Tunnel, the pipeline tunnel beneath the Hollandsch Diep and the Oude Maas, the Drecht Tunnel and the Kil Tunnel were built. The dimensions of the dock are about 400 x 120 m. The dock bed is about 10 m below N.A.L. (New Amsterdam Level).
   Some 26 deep-well pumps are disposed around this dock to keep it dry during the building of various tunnels.

2. The Madroel dock, in Rotterdam,
   in which the tunnel units of the Benelux Tunnel and the Botlek Tunnel were built.
   The dock dimensions are about 400 x 100 m and the depth is 10 m below N.A.L. This dock is also kept dry by dewatering.

3. The Amsterdam Noord dry dock
   The tunnel units of the IJ Tunnel, the Coen Tunnel and the Hem railway Tunnel were built in this dock. Originally it consisted of two smaller docks which were combined into one large dock for the Hem Railway tunnel units, by removing the bulkhead between them.

If no suitable building dock is to be found in the neighbourhood, the access to the tunnel may somtimes be used as a building dock (Margriet Tunnel and Zeeburger Tunnel). This has the advantage that the transport route for the tunnel units is very short, but entails the disadvantages of higher constructional costs and longer construction time.

PLANNING

Floating transport of the tunnel elements

In the building docks the tunnel units are provided with temporary watertight bulkheads at both ends allowing them to float during transport.
Other provisions for the submersion are already being installed during concreting of the units in the construction stage. These include facilities for temporary foundations, matching with other units and sandstreaming the units, access to units in various stages, systems for water-ballasting, communication, energy, surveying, anchoring, suspension, etc.

As soon as the construction of the units is finished the ballast tanks in the units are filled with water to prevent them from floating once the building dock is filled with water.

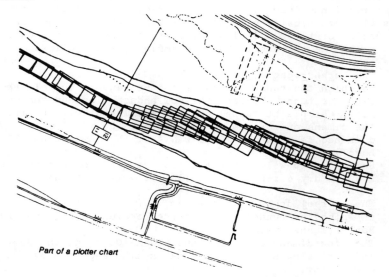

Part of a plotter chart

Before the tow-out of the units laboratory tests are made in order to determine the forces exerted on the tunnel elements during the flowing.
These forces are given by the formula $F = \frac{1}{2} A.r.V^2.R$.
A = the exposed section of the unit perpendicular to the current
r = specific volume of the water
V = relative velocity
R = friction coëfficient

Laboratory tests on models enable us to discover the drag and anchoring forces exterted at difficult points during transport: at the building dock, tributaries, bends and the sink trench. Tidal forces may also play a major part in the time schedule drawn up for the whole operation.

The final number of tugs and the power required to tow and
steer the tunnel units are partly determined by laboratory
tests. The type of tug, the policy towards emergency
situations and the requirements of the river authorities
also bring their influence to bear.
All these requirements result in a fleet of tugs two or
three times more powerful than that suggested by tests and
calculations.
For economic reasons the axis of the transport will not
coincide with the central line of the river. The total
amount of dredging needed along the transport route may be
reduced by using the deep part of the river.
The route of each tunnel unit is checked with echo soundings
before each unit is transported. Shallow areas are dredged
to the required depth and obstacles removed.
The width of the transport channel is determined by the
width of the tunnel unit and its degree of movement under
the influence of inertia, over-corrections, ect.
The transport channel has to be wider at special points such
as bends and tributaries.
A provisional schedule based on average tidal situations
will be drawn up using cost optimalisation.
The difficult points along the route will be passed at
favourable tide conditions. At the same time the schedule
needs to be sufficiently flexible to absorb any setbacks
during transport.
All relevant details are updated in a central control room
during transport, so that the programma may be adjusted if
necessary.

The correct positioning of the tunnel unit is continuously
monitored by a computerised navigational system.
Shipping is limited or held up during transportation.

Immersing the tunnel units

The sinking operation begins when the tunnel unit has been
transported to its destination.
The equipment required during sinking is determined by the
forces operating on the tunnel unit.
These forces are determined by laboratory tests, as for
transport.
They are mainly dependent on:
- the current (river and tide);
- the dimensions and weight of the tunnel unit;
- the cross-section of the river;
- the specific gravity of the water.

Since the sinking operation, most of which takes place under
water, has to pass of quickly and efficiently, simplicity of
action is aimet at. Such simplicity may be discerned in all
the systems for sinking that have been developed.

PLANNING

After the system which will ultimately put it in place has taken over the tunnel unit from the tugs, the sinking operation may be divided into a number of phases:

Vertical Positioning.
The tunnel unit which is suspended from a number of pontoons, or from a jack-up platform, will sink because of a built-in "overweight" (small cross-section) or because of the fact that te unit has been filled with ballast, usually water.

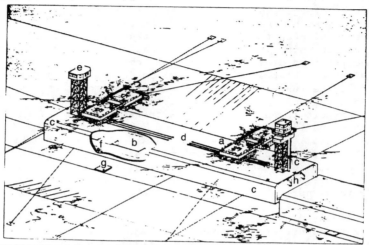

This system shows the tunnel units suspended from the pontoons (a). The ballast tanks (b) are filled with water. For the horizontal movements of the tunnel unit, six moored cables (c) run to winches on both ends of the tunnel unit, and by way of a system of pulley-blocks and pulley/sheaves (d) these cables are connected with the winches at the top of the towers (e). The tunnel unit is temporarily placed on jacks (f) on concrete slabs (g) placed in the trench and on supports (h) on the previous unit.

This overweight will have to be accurately calculated by measuring the gravity of the water. During the construction of the unit the tolerances in wall thickness and the establishment of the specific gravity of the concrete used will have to be minutely followed.
The vertical positioning of the element takes place in a number of steps, becoming increasingly smaller until the element has been positioned onto its supports. This positioning as a rule is done by a construction in which the tunnel unit rests on the previously immersed unit by temporary supports with jacks allowing height corrections. The jacks are positioned at the free end of the unit and supported by piles mounted on slabs placed in the trench.

Horizontal Positioning.
When applying this method, cables are secured at the sides and the ends of the element. These cables run to fixed points at the bottom of the river or at the pontoons or jack-up platform. Winches placed on the pontoons or the jack-up platform allow for a very accurate horizontal movement of the tunnel unit.

Here too the manoeuvring is done in steps, becoming smaller as the previous unit or the land section are approached. The strength of the cables is determined by calculations in which the factors mentioned before - such as flow force, dimensions and so on - are taken into account. Should, however, the forces on the anchorage system become such under influence of these factors, that the system becomes inoprable, we can resort to special measures.

As an example we take the immersion of the central tunnel unit of the Kil Tunnel which was immersed in a sheltered part of the trench and then transported underneath the current to the place where it had to be finally positioned. Once the tunnel unit has been manoeuvred towards the previous one and placed on a support system on this unit and a separate support system at the free end, the last horizontal movement will take place for the final connection.
As has already been said, the GINA profile is usually used in the joints between the tunnel units.

This Gina profile of a rubber ring with a soft rubber tip at the fore-end. The large mass of the Gina profile is hard rubber, but the hardness and the profile may vary depending on the cross section and the depth at which the unit has to be immersed.

By drawing up the unit against the previous unit by means of a special jack construction or cable, the first compression is obtained by the soft rubber tip of the Gina profile. It is useless to try and tighten it now because the water in the chamber thus created will not allow this.
Subsequently the water is pumped out of this chamber, but the water pressure remains at the free outer end.

The pressure in the joint ceases to exist, but the hydrostatic pressure is so great that the unit is pressed further against the previous unit and the whole Gina profile is compressed.
After the final tunnel unit has been immersed, a space of approximately 1 m. width exists between the last two units. This space is fixed by the placing of a number of wedges in the joint. Without these wedges the two units would be moving towards each other as a result of the pressure in the Gina profiles in the previous joints. This is why the joints are opened.
Subsequently, watertight bulkheads are connected onto the sealing joint. The space thus created is pumped dry and the tunnel connection between the last two immersed units can be completed.

PLANNING

Foundations for immersed tunnels

Tunnels are relatively light structures. The weight of a tunnel in itself is between 5 and 10 kN/m^2.
With the additional burden of roadway and traffic the maximum loading on the foundation does not generally exceed 30 kN/m^2.
This also applies to a tunnel under a fairway.
The small settlings which occur present no problem to the tunnel in that the expension joints between the units permit a certain rotation, so that the tunnel lies on the bottom like the links of a chain.
The first immersed tunnel in the Netherlands was built almost 50 years ago. The foundations of the tunnel on a bed of sand that had meanwhile been laid by underjetting, was the first of their type in the world.

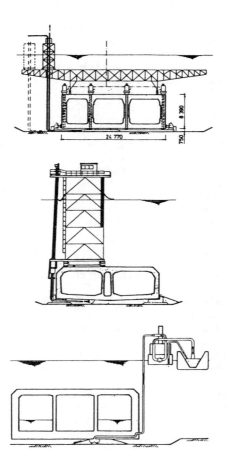

In this underjetting method, sand was introduced evenly under the tunnel by a movable underjetting installation after the tunnel had been placed on temporary foundation.

The method has been further developed in the course of time. A sand injection method is in current use.
Exit openings through which sand can be injected are made at regular intervals in the floor of the tunnel.
In this method there are no obstacles on the tunnel roof, as are needed for underjetting equipment, so that it is highly suitable for use in busy fairways.
In the Netherlands, it is most usual for tunnels to be placed on sand foundations, though other methods are used in special cases.
A special pile foundation was developed for the underground railway tunnel under the Nieuwe Maas in Rotterdam.

As it would be impossible to drive the piles to exactly the required level, a special type of pile with an adjustable head was developed. The piles were constructed in the following way:
First a steel tube,, outer diameter 0.62 m, with a cast iron toe was driven into the soil.
At the bottom of the tube 0.5 m^3 grout was cast on the cast iron toe. Now a prefabricated concrete pile was lowered into the steel tube pushing aside the grout thus forming a good connection between the pile and the cast iron toe.
The steel was withdrawn afterwards and as a result the enlarged pile toe contributes considerably to the bearing capacity of the pile.
The prefabricated piles were provided with an adjustable pile head consisting of a separate concrete part connected to the rest of the pile by a nylon sleeve. The tunnel units were immersed and temporarily placed on an alignment beam, fixed to four special piles.

After placement of the tunnel unit on the temporary structure and ensuring an accurate position, the adjustable pile heads were pressed up against the tunnel bottom by means of a cement grout injection into the sleeve, carried out by divers.
Slight deviations of the pile from the vertical were taken up by the rotation capacity of the pile head. A felt layer on the head of the pile provided for uniform load transfer.

Another example of a tunnel founded on piles is found in the road tunnel under the River IJ at Amsterdam. The poor soil conditions at this place led to the choice of a pile foundation.
As most suitable system bored piles were chosen with a diameter of 1.08 m and a maximum pile toe depth of 90 m below O.D. The piles were produced by using a jack-up platform.

PLANNING

Each tunnel unit is supported by capping beams which interconnect a group of 8 or 10 piles. The capping beam was constructed in a reinforced concrete diving bell suspended from a floating double pontoon anchored to hollow concrete blocks on the bottom of the river.

After the tunnel unit had been immersed to rest on the capping beams by three temporary footings, the definite foundation of the tunnel was realized by grouting 24 rubber slab footings to make contact with the capping beam. The footings did rest on a sliding support made up of chromium-plated steel plates, with teflon as sliding medium. This was introduced in order to prevent transfer of horizontal forces to the piles due to movements of the tunnel as a result of temperature changes.

The last example of pile foundations is that under the Zeeburger Tunnel, to the east of Amsterdam. The foundations here are poor because of a sedimented-up sand quarry nearby. Tubular steel piles of 508 mm diameter, reaching to a depth of 47.5 m M.A.L. were introduced under the exterior wall of the tunnel. In addition to the maximum pressure in use of 200 kN when the tunnel is in use, this connection has also to be able to bear a transverse stress of 200 kN. This transverse stress is caused by shipping or filling the sink trench.
The following solutation has been found: the tubular piles are fitted with lids at the top.
Where the piles are in contact with the floor of the tunnel units, box structures have been incorporated in the floors. After a tunnel unit has been sunk onto its auxiliary supports, the boxes are allowed to drop onto the lids of the tubular piles. In this way both pressure forces and transverse forces can be transferred.

These transverse stresses may be caused by filling the sink trench in which the tunnel units are situated with ballast. To transfer these stresses, the bearing surfaces of the box and lid are provided with ribe which interlock when the box is resting on the lid. In order to recognise in time any executive problems, experiments were carried out with the linkage by blindfold divers in a pool.
The spaces between the lowered boxes and the tunnel floor are closed on the outside with rubber blocks.
Grout is injected into the space formed. A good linkage ensuring the required transfer of forces is obtained after hardening.

This paper has skimmed over the most important aspects of "immersed tube tunnelling".
Because of the construction of immersed tube tunnels, the Netherlands possess extensive know-how and experience in the field.

Construction on four immersed tunnels will begin in the near
future, which will lead to the improvement of existing
techniques and the invention of new ones, increased quality
and reduced costs.
The specific know-how of immersed tube tunnelling is not
only applicable for the conventional inland tunnels, but
also for other hydraulic projects in-shore as well as offshore and for sea strait crossings.

15. Privately financed immersed tube tunnel projects

Ir V. L. MOLENAAR, Consulting Engineer, Breda, The Netherlands

SYNOPSIS: Based upon experience with many immersed tube tunnels of which several are privately financed, the author gives a number of aspects related with a privately financed project. These aspects concern the legal and financial structure with the roles of the parties involved as well as technical matters related to design and execution. Time schedule is of major importance to limit financing costs. A good cover of risks with policies especially prepared for the structure is of major importance. Within the scope of this paper only some aspects can be mentioned briefly.

INTRODUCTION

1. Due to the worldwide concern about the increase of public spending resulting in political unacceptable budget deficits, there is a political desire to decrease the governmental budgets. Therefore governments are looking for privatisation of governmental institutions, government owned companies and other activities previously being considered as being exclusively governmental.

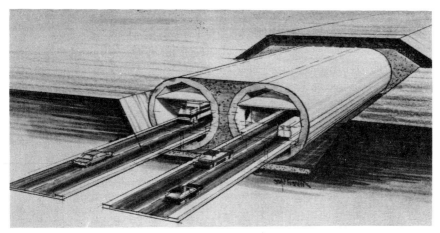

Fig. 1 Sketch of an immersed composit sea strait crossing for cars. It had been developed for the Channel tunnel.

Immersed tunnel techniques. Thomas Telford, London, 1989.

PLANNING

The infrastructure is such a public concern. Roads, bridges, waterways and tunnels are most often financed with public money, allthough exceptions are known of privately financed toll roads, toll bridges and toll tunnels.
Due to the above described privatisation the governments are seeking for more possibilities to privatise projects. Tunnels are ideal for a private approach.
These are well defined projects, ideal for imposing toll due to the often unique connections they form.

LEGAL STRUCTURES
2. The Legal structure of a privately financed infrastructure project seems to be complicated when we have a look at fig. 2. showing all parties possibly involved in such a project, their relations, the agreements and the related flow of money. In fact this figure can be considered as a checklist. In reality we see, that functions are being combined. However this is not yet standard as should be the case with a long established practice. Many alternatives are possible and known. In the following we will not compare legal stuctures of one specific project with another. A general indication of possibilities is more appropriate. However those involved in the various projects will recognise the related structures.

Minimum Governmental involvement
3. The role of the government and authorities is very much dependent upon national practices. Any how the public requirements will have to be establish by the authorities. These requirements will have to be included in an invitation to promotors or concessionaires or in the specifications for the project. Should the government really want to limit its role and responsibility, the only remaining thing to do is to point out the party that will realise and operate the project. Subsequently the government controls wether above mentioned public requirements are met. These public requirements may concern financial, technical as well as social aspects.

In this specific case the contractors very often take the initiative to make a proposal. Their main concern is to execute the project. The creation of an appropriate legal structure is a tool. The way this structure is created is often more important than the tendersum to obtain the project. In order to succeed and be attractive the contractors must give sufficient guarantees to either the government or its representatives and/or to the lenders. These guarantees must be covered by the balance of the company, the system of bonds and possibly an envelope for cost overrun to be included in the financing arrangements for the project. Of course the feasibility of these projects must allow for this envelope.

Full governmental involvement
4. A solution entirely contrary to the one described above could be, that the goverment will establish a private special purpose company with the state as 100 % share holder. This special purpose company can award a contract for the design of the tunnel or make the design themselves and award and supervises the construction. This can be financed through a loan agreement. The required own capital of the company will be in relation to the guarantees. Once a tunnel is in operation the shares of the operating company can be sold to the public. The hight of the price of the shares of course is depending upon the hight of the revenues. This structure does not so much influence the common practice of governmental projects.

Scheme of a structure
5. Fig. 2 shows a complicated legal and financial structure. The parties identified are mentioned in the circles:
- the public sponsor represents the public interest and is generally goverment related
- the tunnel company is the special purpose company created for this specific project
- the operating company may be a separate company for the operation of the tunnel
- the trustee may be created to guarantee a split of the incoming toll payments according to the previously made contracts
- the lenders in this scheme are split in a relatively short term funding for the construction phase and first years of operation of for instance banks to be replaced by a long term financing of for instance pension funds, etc.

The agreements as far as identified are given in the rectangles of this figure including:
- construction agreement
- consultancy agreement
- funding/loan agreements
- a separate cash deficiency agreement
- a trust agreement
- an operation and maintenance agreement

The ovals show the flow of capital including
- invoices
- bank debt funding and services
- long term debt funding and services
- operating costs
- toll payments
- etc.

PLANNING

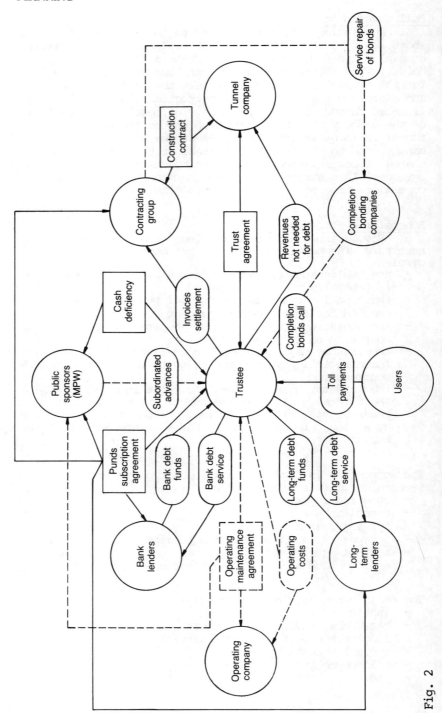

Fig. 2

Contractors and consultants role
6. Apart from technical aspects as described in following chapters the legal and financial structure may have other implications on the contractor and consultant. These implications differ per project. Very often the contractor tries to create a structure himself. As mentioned before, this might be a perfect tool to obtain a contract for a large infrastructure project. Due to the advises required for legal, fiscal and technical matters and the long preparation period this is a costly matter.
The contractor will have to be very selective and needs to have a proper feeling about his chances. The contractor especially when he initiates the special purpose company will need a large balance in order to be able to properly guarantee.

The consultant in most cases will not be able to initiate these activities. The costs related to it will be high. He will join with the special purpose company for the design of the project. The other parties involved such as the lending banks will ask for additional advise from independant specialist consultants. Their know how will be required for a proper risk analysis.

TIME SCHEDULE
7. Previously the construction time of a large infrastructure project consisting of roads, tunnels, bridges, etc. was often influenced by the availability of money on the budget. With governmental projects often there was a lack of an incentive to accelarate the construction time. With privately financed projects this is much different. An important part of the costs, is the cost of financing. Therefore each project will be under the pressure of a short construction time. Construction time is related to the design as well as the construction of a tunnel. In most cases these aspects will have to be considered jointly.

No general rules can be given for reducing the construction time. This can be influenced by the design, use of prefabricated concrete and steel segments, the construction yard facilities and steel versus concrete cross sections. The number of activities need to be limited. An example is the omission of a watertight layer with its required protection and using cooled concrete with high quality cement instead.

In some cases the construction time is influenced by the time required to built a construction yard. This can be rather time consuming. With the developments in the offshore industry the use of large immersible pontoons is common practice. On a number of occasions these pontoons have been used as a construction facility for the prefabricated steel or concrete units.

PLANNING

In order to avoid delay during construction, risks effecting the time schedule, will have to be excluded as much as possible. Practice has learned, that this influences especially the landbased parts. The immersed part has despite its nature never seriously influenced the time schedule of a tunnel. This is contrary to a bored tunnel where the difficulty in estimating the capacity of equipment and in prediction of soil conditions and soil behaviour has lead to serious delays in construction time. It is difficult to accelerate the construction time of a bored tunnel once delay has occured. One is entirely dependant of the capacity of one machine.

Fig. 3 Shows the common way of immersing tunnel elements. The horizontal movement is done through winches placed in towers on top of the tunnel element and through a system of blocks and shieves connected with pre hammered anchors. The vertical movements are controled by winches on floating pontoons and a negative buoyance through waterballast tanks inside the tunnel elements. This system has proven to be reliable in well protected areas but relatively slow due to the time required for the preparation of the tunnel units.

DESIGN
8. In case the tunnel is designed, built and operated by a private enterprise and the whole set up has to be considered on its economic feasibility, each aspect of the project will be investigated on its contribution to the profitability. This not necessarily means, that tunnels financed by government are too expensive, on the contrary, but a private enterprise will not as easily invest in an

extra lane or track or other aspects anticipated on possible future developments. On several occasions this resulted in reduced cross sections compared to the cross sections considered previously

A private company may wish to cut in other costs as well. These costs for instance may concern the electro mechanical installation. It is therefore essential for the government to define and establish requirements related to the project prior to inviting the companies for making proposals. For optimal functioning of the private approach the establishment of these requirement must only be limited to the aspects concerning public interest.

Steel versus concrete
9. The privately financed projects may lead to changing views on the steel versus the concrete projects. Under the specific time pressure especially and in the severe competition as well as the increasing exchange of the international know how, existing prejudices may disappear. In authors' view, steel as well as concrete are both technical good solutions. Steel as well as concrete tunnels have a history of over fifty years. Of both types at least thirty of fourty tunnels have been built. Steel and concrete are proven techniques and perfectly watertight.

The choice between concrete and steel will no more been governed only by the steel versus concrete lobby, existing prejudices and comparable less technical aspects. Other aspects especially related to the construction time will be considered. There the availability of construction sites may have an impact as well. Over the past five years the author has been involved in a number of projects were steel and concrete have been compared with each other. The outcome of these comparison didn't show a clear winner or looser.

EXECUTION
10. The execution of the tunnels is as indicated before closely related to the design, choice of steel versus concrete, availability of casting yards, docks or shipyards all under the pressure of the time schedule. The financier will be prepared to invest in a decreased risk on the time schedule. This investment will concern for instance an increase of construction capacity by using more forms, the forced hardening of the concrete, the use of more shipyards or docks with a steel tunnel, etc.

Another example is shown on fig 4. and fig 5. In these cases the client should be willing to invest not only in equipment, that allows a very fast immersing sequence under rather exposed conditions. The client also was willing to invest in a spare piece of equipment in case the other was damaged. This again was done to decrease the risk of an increase in construction time.

PLANNING

Fig. 4 Shows a Jack Up used for a circular tunnel.
Circular tunnels of limited cross section for train
connections, pipelines and outfalls are ideal for a short
time schedule. The system shown allowed for immersing one
tunnel element per day. That specific tunnel was 2 km long
and had to be built and equipped within one year.

Fig. 5 Shows a piece of immersing equipment which has been
developed for the construction of an immersed tunnel through
rather exposed sea straits such as the Channel. Through a
special D.P. system this semi submersible combined jack up
can very precisely and very quickly immerse one tunnel unit.

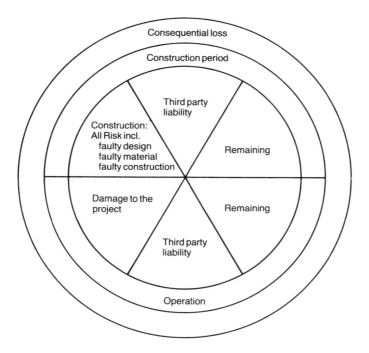

Fig. 6 This scheme shows insurance scheme covering construction as well as operation period. Should the insurance market offer this solution in one and the same policy, it should be the best product for the special purpose company.

PLANNING

Consultant and contractor must realise, that the financing party in principle is not very open for using unproven techniques. Should new techniques been considered this will only be accepted when these financing parties are being convinced of the reality of it. Decrease of risk compared to the past is a convincing argument.

Under the pressure of timeschedule, quality control might be required. Especially with these type of marine structures the client might require quality control as is used in offshore industries. It will give the client additional confidence in the project. We must realise that operating costs are closely related to the quality of the construction.

INSURANCE
11. The insurance of the construction of tunnels was in most cases an independant matter for client, consultant and contractor. The client being a governmental institution often did not insure its projects. With the many projects the governments are executing this should not be wise. The sum of the premiums should surpass the total of damage revenues. The consultant will insure his liability only and the contractor should want to cover his construction risks only. The effect of this approach is a limited cover with different policies.

It is different with a privately financed project which will have to be considered as an independent project. It is for the benefit of the client to insure the project on a policy covering as much as possible including faulty design, faulty material, faulty workmanship, delay resulting from these causes and the operations of the tunnel (fig. 6). The cover must be given under one policy thus avoiding a delay resulting from discussions between insurance companies, what policy or company will bear the cost of an eventual damage.

CONCLUSION
12. Within the scope of this paper we have attempted to mention a wide variety of aspects related to the new trent of privately financed tunnel projects. Only some major aspects wer dealt with. We therefore had to mention the legal and financial structure and its impact on the parties involved and the insurance scheme.
Technical aspects related to design and execution vary per project. We therefore are of the opinion that a succesfull attempt for a privately financed project requires the maximum available know how from all over the world. Within the scope of this congress we therefore want to emphasise that the International Tunnelling Association has only recently started an internation working roup on Immersed and Floating Tunnels. This working group will investigate existing techniques and new developments.

16. Comparative merits of steel and concrete forms of tunnel

D. R. CULVERWELL, MA, FICE, FCIArb, Consulting Engineer, UK

SYNOPSIS. The characteristics of the two main forms of immersed-tube tunnel, namely the steel shell and the concrete tube, are compared and illustrated by reference to different tunnels of each type. The distinctive practices of the USA and Western Europe in this regard are noted and it is suggested that a more common approach might be followed.

INTRODUCTION
1. There are two general forms of immersed-tube tunnels and, for the purposes of this paper, they are referred to as the "steel shell" and the "concrete tube". Neither term is entirely appropriate because there are a number of variants within each form; however, they are convenient and provide sufficient distinction.
2. The immediate differences between the two forms lie in the methods of construction of the units. Bearing in mind the qualification about variants, one may describe the methods as follows:
Steel-Shell : a stiffened steel shell is fabricated on land, with its ends sealed by temporary bulkheads, and is launched down a slipway into the water. While it is afloat, structural concrete is placed internally and sufficient ballast concrete is placed externally, in compartments provided for the purpose, until the unit is at a state of near-neutral buoyancy. It is then towed to the site and sunk into position in the trench, more ballast concrete being added as necessary.
Concrete Tube: a large basin is excavated on land adjacent to the tunnel site and is de-watered and left dry. The units are fabricated in the basin, according to the normal methods used for reinforced concrete, and, when complete, the ends are sealed by temporary bulkheads. The basin is flooded and a passage formed between it and the waterway. The units are then floated out of the basin, towed to the site and sunk into position in a manner similar to that for the steel shell form.
3. In the case of the steel shell, the main structure is usually circular in shape; it may comprise a single structure or two together in a binocular layout, though one of the early tunnels of this type comprised four such structures. Several Japanese tunnels of this type have been rectangular in shape, but this is unusual. In the case of the concrete tube, the main

structure is usually rectangular, with a number of separate compartments, though circular shapes, usually singular but sometimes binocular, have also been used.

4. Of the 67 immersed-tube road and rail tunnels completed in the world up to the end of 1986, 33 are of the steel shell form and 34 of the concrete tube form. They are listed in Table 2 at the end of this paper (ref.1), with their cross-sections and other details, and, in Table 1 below, a summary is given by country, form and size of road tunnel.

Table 1. Summary by country, form and size of road tunnel

Form/Lanes	USA	Japan	West.Europe	Elsewhere
Steel	21	11	0	1
Concrete	2	3	21	8
2-lane	13	2	0	1
4-lane	2	3	6	4
6/8-lane	1	2	10	2

5. Features of note in this table are the high number of 2-lane steel shell road tunnels in the USA (13) and the small number of road tunnels less than 4-lane elsewhere in the world (3). In total numbers of immersed-tubes, the USA leads, with Holland and Japan equal second and other individual countries well behind. As can be seen from Table 2, no steel shell form of tunnel has been built in Western Europe and only two concrete tubes have been built in the USA, both circular and 2-lane.

6. The steel shell form is generally thought, other things being equal, to be more expensive than the concrete tube form. The difference is likely to be small, however, and the immersed-tube itself only represents a part of the total cost of the tunnel. The steel shell form is likely to offer a shorter construction period and this may well offset the difference in capital cost.

7. Traditionally, screeded bed foundations are associated with the steel shell and pumped or jetted sand with the concrete tube. Such associations are not absolute and the form of foundation will generally not affect the choice between one form and the other.

8. About a quarter of all road and rail immersed-tubes are now over thirty years old, with the first steel shell dating back to 1910 and the first concrete tube to 1928. Both forms seem to have a satisfactory record for safety and durability.

9. As regards their application, neither form has advantage as regards natural circumstances, such as water depth, length of crossing and bed conditions. In these respects, their characteristics are similar. Where they are different is in such matters as configuration of cross-section, construction period and the construction facilities required. These are referred to in the following sections.

CROSS-SECTION

10. The maximum loadings to which the steel shell is normally subject occur during construction. Of these, the longitudinal

Table 2. Immersed-tube road and rail tunnels

N°	YEAR	NAME	PURPOSE	LOCATION	TUBE LENGTH	CROSS-SECTION	LANES TRACKS	FORM
1	1910	DETROIT RIVER	RAILWAY	MICHIGAN, U.S.A / ONTARIO, CANADA	800m		2 x 2	S
2	1914	HARLEM RIVER	RAILWAY	NEW YORK	329m		4 x 1	S
3	1927	FREIDRICHSHAFEN	PEDESTRIAN FOOTWAY	BERLIN	120m		—	R
4	1928	POSEY	ROAD	OAKLAND, CALIFORNIA U.S.A	742m		2	R
5	1930	DETROIT-WINDSOR	ROAD	MICHIGAN, U.S.A / ONTARIO, CANADA	670m		2	S
6	1940	BANKHEAD	ROAD	MOBILE, ALABAMA U.S.A	610m		2	S
7	1941	MAAS	ROAD	ROTTERDAM NETHERLANDS	587m		2 x 2	R
8	1942	STATE STREET	RAILWAY	CHICAGO, ILLINOIS U.S.A	61m		2 x 1	S
9	1944	AJI RIVER	ROAD	OSAKA JAPAN	49m		2 x 1	S
10	1950	WASHBURN	ROAD	PASADENA, TEXAS U.S.A	457m		2	S
11	1952	ELIZABETH RIVER (1)	ROAD	PORTSMOUTH, VIRGINIA U.S.A	638m		2	S
12	1953	BAYTOWN	ROAD	BAYTOWN, TEXAS U.S.A	780m		2	S
13	1957	BALTIMORE	ROAD	BALTIMORE, MARYLAND U.S.A	1920m		2 x 2	S
14	1957	HAMPTON ROADS (1)	ROAD	VIRGINIA, U.S.A	2091m		2	S
15	1958	HAVANA	ROAD	CUBA	520m		2 x 2	R
16	1959	DEAS ISLAND	ROAD	VANCOUVER CANADA	629m		2 x 2	R
17	1961	RENDSBURG	ROAD	KEIL WEST GERMANY	140m		2 x 2	R
18	1962	WEBSTER STREET	ROAD	OAKLAND, CALIFORNIA U.S.A	732m		2	R
19	1962	ELIZABETH RIVER (2)	ROAD	PORTSMOUTH, VIRGINIA U.S.A	1056m		2	S
20	1964	CHESAPEAKE BAY (a) THIMBLE SHOAL TUNNEL (b) BALTIMORE CHANNEL TUNNEL	ROAD	VIRGINA, U.S.A	(a) 1750m (b) 1661m		2	S
21	1964	LILJEHOLMSVIKEN	RAILWAY *	STOCKHOLM	123m		2	R
22	1964	HANEDA	ROAD	TOKYO	56m		2 x 2	S
23	1964	HANEDA	MONORAIL	TOKYO	56m		2	S

PLANNING

N°	YEAR	NAME	PURPOSE	LOCATION	TUBE LENGTH	CROSS-SECTION	LANES/TRACKS	FORM
24	1966	COEN	ROAD	AMSTERDAM	540m		2 x 2	R
25	1967	BENELUX	ROAD	ROTTERDAM NETHERLANDS	745m		2 x 2	R
26	1967	LAFONTAINE	ROAD	MONTREAL, CANADA	768m		2 x 3	R
27	1967	VIEUX-PORT	ROAD	MARSEILLES FRANCE	273m		2 x 2	R
28	1967	TINGSTAD	ROAD	GOTENBURG SWEDEN	452m		2 x 3	R
29	1968	ROTTERDAM METRO	RAILWAY *	ROTTERDAM NETHERLANDS	1040m		2 x 1	R
30	1969	IJ	ROAD	AMSTERDAM	790m		2 x 2	R
31	1969	SCHELDT E3 (J.F.K. TUNNEL)	ROAD/ RAILWAY	ANTWERP, BELGIUM	510m		2 x 3 ROAD 2 TR	R
32	1969	HEINENOORD	ROAD	BARENDRECHT NETHERLANDS	614m		2 x 3	R
33	1969	LIMFJORD	ROAD	ARLBORG DENMARK	510m		2 x 3	R
34	1969	PARANA (HERNANDIAS)	ROAD	SANTA FE ARGENTINA	2356m		2	R
35	1969	DOJIMA RIVER	RAILWAY *	OSAKA JAPAN	72m		2 x 1	R
36	1969	DOHTONBORI RIVER	RAILWAY *	OSAKA JAPAN	25m		2 x 1	S
37	1969	TAMA RIVER	RAILWAY	TOKYO	480m		2 x 1	S
38	1970	KEIHIN CHANNEL	RAILWAY	TOKYO	328m		2 x 1	S
39	1970	BAY AREA RAPID TRANSIT	RAILWAY *	SAN FRANCISCO CALIFORNIA U.S.A	5825m		2 x 1	S
40	1971	CHARLES RIVER	RAILWAY	BOSTON, MASS.	146m		2 × 1	S
41	1972	CROSS-HARBOUR TUNNEL	ROAD	HONG KONG	1602m		2 x 2	S
42	1973	EAST 63rd ST. TUNNEL	RAILWAY (PART*)	NEW YORK	2 x 229m		4 x 1	S
43	1973	I 10	ROAD	MOBILE, ALABAMA U S A	747m		2 x 2	S
44	1973	KINUURA HARBOUR	ROAD	HANDA, JAPAN	480m		2	S
45	1974	OHGISHIMA	ROAD	KAWASAKI, JAPAN	664m		2 x 2	S
46	1975	ELBE	ROAD	HAMBURG, GERMANY	1057m		3 x 2	R

188

PAPER 16. CULVERWELL

Nº	YEAR	NAME	PURPOSE	LOCATION	TUBE LENGTH	CROSS-SECTION	LANES/TRACKS	FORM
47	1975	VLAKE	ROAD	ZEELAND NETHERLANDS	250m		2 x 3	R
48	1975	SUMIDA RIVER	RAILWAY *	TOKYO	201m		2 x 1	S
49	1976	HAMPTON ROADS (2)	ROAD	VIRGINIA, U.S.A	2229m		2	S
50	1976	PARIS METRO	RAILWAY *	PARIS	128m		2	R
51	1976	TOKYO PORT	ROAD	TOKYO	1035m		2 x 3	R
52	1977	DRECHT	ROAD	DORDRECHT NETHERLANDS	347m		4 x 2	R
53	1978	PRINSES MARGRIETT	ROAD	SNEEK NETHERLANDS	77m		2 x 3	R
54	1978	KIL	ROAD	DORDRECHT NETHERLANDS	330m		2 x 3	R
55	1979	WASHINGTON CHANNEL	RAILWAY *	WASHINGTON D C U S A	311m		2 x 1	S
56	1979	KAWASAKI	ROAD	KAWASAKI, JAPAN	840m		2 x 3	S
57	1979	HONG KONG MASS TRANSIT RAILWAY	RAILWAY *	HONG KONG	1400m		2 x 1	R
58	1980	HEMSPOOR	RAILWAY	AMSTERDAM	1475m		3 x 1	R
59	1980	BOTLEK	ROAD	ROTTERDAM NETHERLANDS	508m		2 x 3	R
60	1980	DAIBA	RAILWAY	TOKYO	670m		2 x 1	S
61	1980	DAINIKORO	ROAD	TOKYO	744m		2×2	R
62	1982	RUPEL	ROAD	BOOM, BELGIUM	336m		2×3	R
63	1984	SPIJKENISSE	RAILWAY	ROTTERDAM	530m		2×1	R
64	1984	COOLHAVEN	RAILWAY	ROTTERDAM	412m		2×1	R
65	1984	KEOHSIUNG HARBOUR	ROAD	TAIWAN	1042m		2×2	R
66	1985	FORT McHENRY	ROAD	BALTIMORE	1638m		4×2	S
67	1986	ELIZABETH RIVER (3)	ROAD	VIRGINIA	762m		1×2	S

NOTES
1 Year is that of completion
2* Denotes part of underground railway system
3 Form of tunnel denotes thus:
   S—Steel shell
   R—Reinforced or pre-stressed concrete box

189

loadings, that occur during launching and when the complete unit is afloat, can be fairly readily accommodated. The transverse loadings that occur during concreting are more onerous and can best be accommodated by using a circular shape. As can be seen from Table 2, by far the majority of steel shell tunnels are of this shape. The exceptions are several rectangular box shape tunnels in Japan and these required rather heavier steelwork than normal.

11. By contrast, the concrete tube form allows a high degree of flexibility in the choice of cross-section. This feature gives it great advantage in certain applications and can indeed, in some, be over-riding. If one studies Table 2, one can see that concrete tube tunnels have ranged from the single 2-lane circular shape of the Posey (4) to the simple dual 3-lane twin rectangular shape of the Kil (54) and to the more complex multi-compartment rectangular shape of the Scheldt (31) which carries road, rail and cycles.

12. Circular steel shell road tunnels have so far been limited to 2 lanes per tube, the largest diameter so far being that of the Fort McHenry (66) which is 10.4m internal diameter. This provides a carriageway 7.92m between kerbs, with a vertical clearance of 4.88m. This is less than the 5.1m clearance normal in the United Kingdom, though greater than that in many continental tunnels. To accommodate 3 lanes would require an internal diameter of about 13m. This would increase the stiffening required and for normal applications the surplus space would also be a disadvantage. For 3-lane carriageways, a rectangular concrete tube is generally the more suitable form. Beyond this size, difficulties arise due to hydraulic loadings and buoyancy considerations.

13. Although all but the most recent steel-shell tunnels in the USA have been of circular shape, the Japanese seem to have been more flexible in their approach and their first tunnel, the Aji (9), though very short, was a rectangular steel shell. This was followed by the two Hanada tunnels (22 & 23), which were of similar shape. More recently, they have used a circular shape for their rail tunnels, such as the compact binocular form of the Tama River (37).

14. While the steel shell type cannot offer the flexibility of shape and layout that is possible with the concrete tube, it may nevertheless offer more scope in these respects than is generally accorded to it. Japanese practice has illustrated this and also the recent third Elizabeth River (67) in the USA, where a compromise horse-shoe shape was used. Another development that might be pursued would be the use of an elliptic shape for a 3-lane tunnel, so as to reduce the surplus areas below the carriageway and above the traffic envelope.

15. For tunnels in deep water, where the hydraulic pressures are high, the circular shape clearly has advantage. An example of this was the proposed EuroRoute scheme in 1985 for the crossing of the Channel, which included a 20km length of immersed-tube beneath the shipping lanes in mid-Channel. The requirement was for a dual 2-lane crossing with hard shoulders, and thus a carriageway width of 9.80m. The design proposed by the United

Kingdom consulting engineers was a circular steel shell of binocular form, similar to the Hong Kong Cross-Harbour Tunnel (41) but with a larger and continuous central services duct and an internal diameter of 12.90m and ring thickness of 0.80m. This design was able to accommodate the high water pressures of 60m or so and also offered considerable resistance to possible sunken ship loadings, which were a matter of some concern.

16. An alternative French design was also of interest. This was for a lozenge shape unit, with a flat top and bottom and semi-circular sides, of overall dimensions about 24m x 13m. The centre part contained two rectangular traffic compartments, one on top of the other, and the semi-circular spaces each side served as air ducts. Unlike other steel shell designs, there was both an internal and external skin, with web plates between and stiffeners, similar in form to the double skin construction of a tanker hull. The intervening space was filled with concrete which acted compositely with the steel. Some of the shear forces were rather high and the design probably needed more study but it was nevertheless an interesting proposal and one that seems likely to find application elsewhere.

17. To summarize: the concrete tube type readily lends itself to cross-sections of different layouts and dimensions and has a clear and natural advantage in this respect. The steel shell type is naturally suited to a circular shape and, on this account, is less flexible in its application; there is probably, however, greater scope for using alternative shapes and layouts for this form than practice generally indicates.

18. An instance where USA practice did divert from the conventional circular shape for a steel shell was East 63rd Street (42). This cross-section is about 12m square, with four internal compartments, each containing one rail track. The shell and the stiffeners were of normal dimensions but internal vertical and horizontal bracing was also provided to give additional strength prior to concreting.

VENTILATION OF ROAD TUNNELS

19. An important factor, which may to some extent qualify the comments made above regarding road tunnels, is ventilation and the provisions needed in this regard for air ducts. The matter becomes somewhat complicated, as it depends on the size and length of the tunnel under consideration.

20. For a 2-lane bi-directional tunnel of length between, say, 500 and 1000m, it will normally be necessary to use a semi-transverse ventilation system, though it may be noted that, in the past, fully transverse systems were used for some tunnels near the higher end of this range. For both systems, dedicated space for air ducts is required. The circular cross-section naturally lends itself to this requirement by virtue of the segmental spaces that are available above and below the space needed to accommodate the traffic. Further, the areas and profiles of these spaces are reasonably suited to air quantities required.

21. This convenience which the circular shape offers in regard to space and layout accords with the structural advantage

it offers for a 2-lane tunnel of the steel shell form. These two factors running together render this form attractive for 2-lane tunnels of this length. A number of examples can be found among the USA tunnels, e.g. Detroit-Windsor, Bankhead, Washburn, Baytown and Elizabeth River (1 & 2).

22. For similar tunnels between 1000 and 2000m long, a fully transverse system will normally be necessary. Again, the circular shape is convenient, as the space beneath the carriageway can serve as the supply duct and that above the ceiling as the exhaust duct. The duct spaces are less suited to the greater air volumes required but not unduly so. Once more, the two factors run together and render the form attractive. Examples are Hampton Roads (1 & 2) and the two Chesapeake Bay Tunnels.

23. Where a 2 x 2-lane tunnel might be justified in the range of 500-1000m length, with uni-directional traffic flow normal in each compartment, other considerations apply which favour the rectangular concrete tube. For such a tunnel, a longitudinal system of ventilation may be used, thus obviating the need for special air ducts. Accordingly, the segmental spaces afforded by the circular shape become redundant and a simple rectangular cross-section can be adopted, containing only the two traffic compartments. It follows from this that no great extra cost is normally involved in making such a tunnel 4-lane, instead of 2-lane, and this should be borne in mind in the planning of such works. Examples of rectangular concrete tube tunnels of this type are Coen and Benelux.

24. For a 2 x 2-lane tunnel between 1000 and 2000m long, opinions differ as to whether a semi-transverse system may be used or whether a fully transverse is necessary. For the first Hong Kong Cross-Harbour Tunnel (41), a special form of semi-transverse system was used, with provision to prevent inadequate air distribution in the event of traffic jams and provision for conversion into a reduced transverse form in the event of fire. This has shown itself to work satisfactorily. In the USA, however, it is more likely that a fully transverse system would be used for a tunnel of this length, as was so for Baltimore (13). As the length increases, the duct spaces become less suited to the air volumes, especially so with transverse systems.

25. In this regard, the rectangular shape, and thus the concrete tube, has the advantage that the air ducts can be sized to suit the air flows and so achieve the optimum balance between capital and operating costs. In practice, however, this balance is only telling in broad terms. In the first place, arrangements for funding are likely to favour lower initial capital cost and, in the second place, the uncertainties in estimating future interest rates, electricity costs, engine emission rates and traffic volumes and composition render accurate assessment impossible. An incidental factor that may favour the rectangular shape, in this regard, is the need to accommodate pipes and cables, as these may cause excessive drag where air duct space is already less than might be desired.

26. As a generalisation, it is probably true to say that in the upper part of this range of 1000 to 2000m length, the rectangular concrete tube will have advantage over the circular

steel shell. Above 2000m the advantage will normally be decisive.

27. These comments on ventilation all refer to road tunnels. This feature is of very much less significance with rail tunnels and is unlikely to affect the choice between one form and the other.

WATERPROOFING

28. A salient feature of difference between the steel shell and concrete tube forms lies in the means of waterproofing. This is a matter of great importance, not only because of the harmful effects of seepage on the concrete and reinforcement of the structure and of dampness on the internal finishes, fixtures and fittings, but also because of the difficulty and inconvenience involved in remedying defects after the tunnel is completed. Access to the exterior of the structure will be very difficult and practical working conditions perhaps impossible. Measures taken internally may be of limited effect and may seriously disrupt use of the tunnel, especially a rail tunnel.

With the steel shell form, the waterproofing is an inherent and fundamental feature, being provided by the steel shell itself. The fabrication of the shell is normally done using purpose-made equipment and under near-factory conditions. Much of the welding can be done by automatic machines and effective testing of all welds is possible. Testing of the complete shell for waterproofness, prior to launching, can be done simply, using a soap solution and compressed air. As regards the joints between units, these include lapping plates, welded in position so that the steel shell is effectively continuous over the full length of the immersed-tube.

29. As the units are buried well beneath the bed of the waterway, the likelihood of significant corrosion is minimal and this has not proved to be a feature of concern in practice. However, it may be prudent and expedient to protect the shell by a system of cathodic protection or by coating it with shotcrete.

30. With the concrete tube, the waterproofing is not an inherent or fundamental feature of the structure but is ancillary. It is normally provided by steel plate beneath the base of the unit and extending part-way up the sides and by a bituminous or similar membrane extending up the remainder of the sides and over the top. Whatever form the membrane may take, it or its joints have to be formed in-situ, as have the joints between it and the steel plate. An alternative is to use steel plate throughout but this is expensive and its fixing and the joints between sheets may pose difficulty.

31. By using care in the design and forming of the membrane, a high standard of waterproofing can be achieved, as has been demonstrated in many tunnels. Nevertheless, the method has the basic disadvantages that the work has to be done in site conditions that may not be ideal and that the only means of testing or proving during construction is by inspecting the interior of the units after the basin has been flooded. By this stage, remedial measures, for other than very minor defects, will be time consuming and may be difficult.

PLANNING

32. An alternative means of waterproofing, which has been used successfully in Holland though not, I believe, elsewhere, is to use concrete that is effectively impermeable and take special measures to seal joints within the concrete and prevent cracking due to shrinkage or other causes. To be successful, this requires considerable care and control and is still open to the objection that immediate testing or proving is not possible. Associated with this method is an arrangement whereby the units, in their final state, comprise a series of contiguous elements, about 15m long, with shear keys and water bars between them but otherwise separate. The series is free to flex and move slightly, so avoiding longitudinal tensile stresses between the elements.

33. In regard to waterproofing, it is also necessary to consider the structures involved. The circular steel shell is basically a simple structure. Ignoring the presence of the internal shell stiffeners and longitudinal reinforcement, it is possible to envisage this as comprising a series of individual annular concrete rings placed end-to-end throughout the full length of the shell. The loadings due to earth and water pressures are then taken mainly by the rings in compression and any longitudinal bending, as might be induced by vertical loadings, is resisted by the shell, which, being ductile, can do so without harm. Longitudinal thermal effects are accommodated by compression or separation of the rings and, as necessary, by ductile movement of the shell. The point of importance in all this is that the waterproof membrane, i.e. the steel shell, remains secure and intact throughout.

34. Evidence of the behaviour of a steel shell in practice, where the tunnel had suffered longitudinal deformation, is provided in regard to the Baytown (12) and Second Hampton Roads Tunnels (49) in a paper published in 1979 (ref.2). Settlements of some 300mm, with appreciable angular distortions at joints, had occurred but without significant damage.

35. By contrast, the rectangular concrete tube is a complex structure. The earth and water pressures give rise to considerable loadings in bending and in shear and thus to tensile stress in the concrete. Such stresses may also arise under longitudinal bending. Longitudinal movement due to shrinkage, creep and thermal effects will also tend to induce cracking, except in so far as it may be resisted by reinforcement or stressing or the provision of movement joints. The longitudinal movement will also be affected by friction forces due to earth pressure, though the incidence and magnitude of these may be uncertain.

36. It is against this background that a suitable form of waterproofing has to be devised and against which its likely performance has to be assessed. By comparison with the steel shell not only is the problem more difficult but also the consequences of leakage may be more severe.

PROGRAMME

37. Assuming that the physical requirements of any particular tunnel were suited to both forms, the main advantage of the

steel shell is likely to be a shorter overall construction period. The reasons are, first, that no time need be spent in building a casting basin for the units and, second, that the units will come available in series, one by one, for sinking and placing, rather than in batches of, say, five at a time.

38. To illustrate this, one might consider a road tunnel with 10 x 100m long units. With the steel shell method, one might assume, after allowing for mobilisation and shop drawings, that the first shell was launched after 8 months and, thereafter, that successive shells were launched at monthly intervals. Allowing 3 months for concreting the first shell and fitting it out, it would be ready for sinking and placing after 11 months or, allowing one month float, after 12 months. By this time, it should be possible to build and commission the sinking rig, complete sufficient of the in-situ tunnel, or ventilation building, against which the first unit would abut, and dredge enough of the trench. This being so, sinking of the units could commence and, assuming an average monthly cycle, they would all be placed by the end of month 22. Allowing five months for finishes and equipment and a fairly generous three months for contingencies, the tunnel should be complete after 30 months.

39. With the concrete tube method, a period has first to be allowed for constructing the casting basin. This will clearly vary according to circumstances but is unlikely to be less than 12 months. Assuming two batches of five units each and a 6-month period for constructing each batch, the first units would be ready after 18 months. Allowing a suitable period for removing the units, re-sealing the basin and pumping it out, the second batch could be ready after 26 months. The last unit could then be placed by the end of month 31, giving completion after 39 months, or 9 months later than for the steel shell method. An alternative would be to use a larger basin and a single batch of ten units, but the outcome would be similar.

40. A possible expedient is to start the casting basin in advance of the main tunnel contract. This will save time but may pre-empt the number and length of the units the successful contractor might wish to use and is likely to involve divided responsibilities.

41. As a practical example of the saving in time, the estimated period of construction for the original rectangular concrete tube design for the Hong Kong Cross-Harbour Tunnel (41) was 54 months. Space being short, the casting basin was to be six miles away and at a difficult site. The period actually taken for the construction of the steel-shell alternative, which involved 15 units, each about 100m long, was 35 months.

42. The schemes proposed for the rail tunnel across the Storebaelt in Denmark are also of interest in this regard. The construction period required was effectively 39 months and there were to be about 39 units, each about 144m long. Tenders were invited for both steel shell and concrete tube forms. In the case of each tenderer, the steel shell tender was the lower. Part of the reason for this may have been the relatively fast construction period for a tunnel of this length, since the concrete alternative was envisaged to require three separate

PLANNING

casting basins.

43. The opposite situation seems to have applied with the Conway Tunnel. This involves 6 x 118m long units and a construction period of about 48 months was postulated. Tenders were invited for both the steel shell and concrete tube form but none were received for the former. Maybe there were difficulties with shipyards, but it also seems likely that the relatively long period did not give opportunity to exploit the time advantages of the steel shell form.

44. This issue of time brings into question the tendering procedures that are used. Where both forms of tunnel would be suitable, to prescribe the construction period may favour one rather than the other and so pre-empt the choice. If the procedures were more flexible, so as to allow tenderers more choice in the period of construction, then the value of time could be brought better into account. This is likely to happen with privately financed projects, including those where the contractor is responsible for funding, as well as for designing and constructing the works.

CONSTRUCTION FACILITIES

45. Each form of tunnel requires its own type of construction facilities: for the steel shell these are normally a suitable shipyard and, for a concrete tube, suitable land adjacent to the waterway for a casting basin. Each may pose problems and these may have important bearing on the choice.

46. An order for steel shells may be less attractive for a shipyard than might be thought. It may mean turning away long-term shipping customers and it will, in any case, probably only engage a limited part of the yard's resources. The work is unusual and the tolerances required are generally closer than those for ships. Moreover, planning such work is difficult as not only are civil engineering and shipping projects prone to cancellation or delay but the tenderer who enquires and expects a commitment may not win the contract.

47. These problems can usually be overcome by spreading the enquiries over a number of yards. This is brought out by experience in the USA where it is not unusual for steel shells to be fabricated in yards far distant from the site. A recent example is the Third Elizabeth River (67) where the yard was 2000 miles away in Texas. Within Western Europe, this would be as though a shipyard on the Clyde were to fabricate shells for a tunnel near Barcelona or a shipyard in Italy were to do so for a tunnel in the south of England. This would mean countries foregoing preference for their own shipyards but perhaps we may see this after 1992.

48. Where commitment by a local yard is not possible then an alternative to using a distant yard is to set up an ad hoc fabrication yard at or near the site. This was done for the Hong Kong Cross-Harbour Tunnel (41) and the raw plate and other steelwork was shipped from the United Kingdom. Side launching was used, down simple slips, the first unit being launched after 12 months, with a rate of one a month thereafter. The area of the yard was about 4 hectares.

49. The problems of a casting basin are different. The area required is quite large, perhaps 6 hectares or more, and such space may not be available adjacent to the site. If this is so, the units will have to be towed from elsewhere and this may require deepending of the waterway. The basin is itself a major piece of work, not without its own hazards, and, in a built-up area, there may be risk of damage to adjacent structures through lowering of the ground water.

50. There may also be administrative problems. While it might be expedient, on commercial grounds, to allow tenderers to find their own basin sites, planning or environmental objections may make it difficult for them to do so in the time available. For this reason it is often necessary for the promoter to make advance arrangements for a site. He may thereby pre-empt the choice, though this risk might be avoided or reduced by charging the contractor for use of the basin site.

51. A useful expedient, which will reduce the space required and also the cost, is to combine the casting basin with the excavation for the open approach. A notable example of this was the La Fontaine Tunnel in Montreal (26) and a current example is the Hong Kong Eastern Harbour Crossing.

SERVICE TUNNELS

52. There are probably about twenty immersed-tube services tunnels in the world, though no list of these seems to have been published. Most are of the concrete tube form, the steel shell offering little or no advantage for a tunnel of this size and being inherently more expensive.

53. Several recent concrete tube services tunnels have been constructed by pre-casting the units as a series of separate elements, about 6m long, and then laying these end-to-end and stressing them together. By casting the elements vertically, under controlled conditions, rapid production and a high standard of concrete can be achieved. Further, a casting basin may be unnecessary as the units can be assembled on a ship-lift or a submersible barge or in a small dry-dock. Such methods can give construction periods comparable with those for a steel shell, or better.

54. This is an attractive method of construction for services tunnels required to carry, for example, cables, pipes, conveyors, effluent or cooling water and one may expect to see it being used more widely in the future. It could also be used for rail tunnels and indeed was the basis for one of the tenders for Storebaelt.

CONCLUSIONS

55. There seems little doubt that the market for 2-lane road tunnels in the USA has done much to encourage and consolidate the use thereof the circular steel shell. Underlying this market has been the readiness to raise finance by bonds, to charge tolls and meet demand by stages, using phased construction. By contrast, Western Europe, perhaps starting later in the expansion of its highway system, has found it necessary to build larger 4 or 6-lane tunnels at the outset and this has

favoured the rectangular concrete tube.

56. Nevertheless, it is difficult to believe that the circumstances are so different, as between Western Europe and the USA, as to justify the almost total allegiance that each seems to show to its particular form of tunnel and it seems that tradition and other such influences must be playing a part. In view of the style of each method, I have often thought that they reflect the different attitude and approach sometimes attributed to the New World as opposed to the Old. The steel shell form is straightforward and lends itself to fast and robust construction methods. It may not always be the most economic but it is proven and understood and has been widely used. The concrete tube form, by comparison, is sophisticated and complex, both in its design and methods of construction. It may sometimes be more economic but one feels that, in a race, the Americans would have finished their tunnel and be on with the next, while Europe was still exploring the niceties of their method. Perhaps one should credit the Japanese with being perceptive and pragmatic, as they seem to use both forms.

57. As with most generalisations, these comments oversimplify and need qualification. Even so, I feel sure that there is scope to reduce the barriers that seem, at present, to exist. It is likely that the concrete tube form would have useful application in the USA for 4, 6 and 8-lane tunnels and for combined road and rail tunnels. It is also likely that the steel shell form would have useful application in Western Europe for rail tunnels and for 2 and 4-lane road tunnels. For such suggestions to be tested and pursued will require not only willingness on the part of promoters and those involved in the industry but also a conscious effort to ensure that the planning of a project is such as would allow the respective features of each form to be properly exploited.

REFERENCES
1. D.R. Culverwell. World List of Immersed-Tubes. Tunnels and Tunnelling, March 1988. (Table published by permission of the Editor)
2. Schmidt B. and Grantz W.C. Settlement of Immersed-Tunnels. ASCE September 1979.

Discussion

MR D.R. CULVERWELL, Consulting Engineer
Reference is made in paragraph 4 of Paper 8 to the Owner Risk Assessment which was carried out by the employer to assist him in the comparison between the immersed tube and bored forms of tunnel. This comparison was important and posed some difficulty because of the disparate nature and magnitude of the respective risks inherent in each form. The tender price of the bored tunnel was expected to be lower (and was so) but the consequences of things going wrong during construction, mainly through adverse ground conditions, were likely to be more severe and, moreover, the opportunity to recover lost time would be small. By contrast, the tender price for the immersed tube was expected to be higher (and was so) and the likelihood of things going wrong was expected to be greater; however, the consequences of things going wrong were likely to be less severe and the chances of recovering lost time were greater.

It was concluded that methods of statistical risk analysis were unlikely to be of much value and that a better method would be one based on the judgements of a small group of engineers who, between them, had wide knowledge and experience in bored tunnels, immersed tubes and local conditions. Over a period of about three months, the group went through all the various risks of significance they could think of, and assessed, for each, the likely over-runs in cost and time. The next stage was to assess the incidence to be assumed for these risks, the extent to which they might be concurrent or consecutive and the opportunities for recovery. This had to be done by judgement, and each member of the group was required to complete a questionnaire in which he gave his overall estimates for over-run in cost and time for each form of tunnel and also certain subjective data about his appreciation of risk. From the answers were derived figures which summarized the overall views of the group. The results, which remain confidential, helped considerably in the comparison between the two forms. In the end, the choice was finely balanced and it is probably true to say that a proper decision could rightly have gone one way or the other.

The assessment had a useful incidental benefit in that, by

Immersed tunnel techniques. Thomas Telford, London, 1990.

formulating the risks, SBF and the consulting engineers were in a better position to anticipate them and to take steps to reduce their likely incidence and effects.

In paragraph 13, reference is made to sunken ship loading. Where the profile of the immersed tube is such that it protrudes above the bed of the waterway then such loading may be critical where this is used by large vessels and particularly by laden ore-carriers. In the case of the proposed immersed tube across the English Channel, these loadings were important. Would the Authors comment on their importance at Storebaelt, both in the longitudinal and transverse senses, with regard to the part of the tunnel which was to be founded on an embankment beneath the deepest part of the waterway. Also, what protective measures, if any, would they suggest, such as raised berms alongside, and what comments have they in a probability approach to such risks?

MR H.A. KAMP NIELSEN, Comar Engineers A/S, Virum, Denmark
With reference to Paper 8, two design teams were working on this tunnel design, and Comar Engineers contributed to the design of the concrete scheme. For this reason, the difference in quantities between the steel and concrete schemes was attributable to causes other than those inherent in steel/concrete. These were

(a) a smaller free profile for the steel scheme
(b) an outer ballast on the steel scheme might have made refloating difficult
(c) more attention was paid to durability in the concrete scheme.

In conclusion, the difference in quantities between steel and concrete was not of the same order in the case of the Conwy Tunnel, which was designed by one team.

MR H.D. OSBORN, HAKA UK Ltd
With reference to Paper 10, a slide which Mr Neilson showed indicated what I thought was a trailer suction dredger working on the dredging of the trench for the Sydney Harbour Tunnel.

Could Mr Neilson explain the dredging methods that were employed, and the problems encountered, particularly since he mentioned that sandstone was indicated in the geophysical conditions at the site?

MR D.R. CULVERWELL, Consulting Engineer
Although Paper 12 describes how the tunnel now being constructed came to be adopted, it omits reference to its

DISCUSSION

origins. The scheme for the tunnel was conceived by Freeman Fox & Partners and was put forward, in some detail, at the public inquiry in 1975 on behalf of the Aberconwy District Council who were rightly very concerned about the official Welsh Office bridge scheme adjacent to the Castle. The tunnel scheme was strongly opposed by the Welsh Office, who were advised by Travers Morgan & Partners, but the Inspector saw its merits and it was eventually adopted, very largely as proposed. Freeman Fox took no further part in the project but it would have been fitting for the Authors to have acknowledged their role in conceiving the scheme.

The feasibility study for this crossing began in 1969 and, if the works are complete by 1991, as is planned, it will have taken 22 years to complete the successive stages of inquiry, design, tender and construction. The need for the crossing was clear before the start of the study and there is something very wrong with a procedure that is so slow and so wasteful in human resources.

The Paper notes that tenders were invited for both the steel shell and concrete tube forms of tunnel, but that none were received for the former. I wonder, however, whether the tender procedure allowed a fair comparison between the two forms. As I explain in Paper 16, this is generally only possible where some flexibility is allowed in the construction period and where tenderers know that the value of time will be taken into account in comparing the tenders made for each form. Was this done here and, if not, might this not account for the absence of steel shell tenders?

MR M.W. MORRIS, Acer Consultants (Far East) Ltd
I should like to refer to the effects of earth leakage currents in mass transit railway tunnels. This is a particular problem of dc traction systems, and considerable measures were necessary in the Hong Kong Mass Transit Railway System (including the immersed tube crossing). These included: provision of rubber insulating pads under the rail; plastic clip insulation of plinth reinforcement; and welding of the track bed reinforcement into an earth drainage mat. Longitudinal bonding of reinforcement and prestress cables was carried out to allow monitoring of earth leakage currents and to provide for an impressed current system to be applied later.

It is worth noting that the problem applies not only to tunnels carrying dc traction systems but also to those adjacent to existing systems. The Sydney Harbour Tunnel passes close to the existing suburban railway system on opposite sides of the Harbour, and tests have shown that a voltage gradient exists between these two points. A monitoring system is therefore being incorporated in the tunnel to allow potential current drainage to be monitored and controlled.

PLANNING

MR D.R. CULVERWELL, Consulting Engineer
Reference is made in paragraphs 45-55 of Paper 13 to the Hong Kong Cross-Harbour Tunnel, completed in 1972. A six-lane rectangular concrete box form of immersed tube, with semi-transverse ventilation, had been proposed by the Joint Engineers in their 1961 report, and the four-lane tunnel emerged as a compromise in 1965. The design was developed further by the Joint Engineers (Fig.7), with Per Hall Associates providing some advice, notably on handling and sinking techniques. The steel shell that was eventually adopted, which is similar to the Baltimore Tunnel, was designed, in the first place, by Parsons Brinckerhoff; the design was then developed by them and the Joint Engineers to the form shown in Fig.8 of Paper 13.

The tunnel is notable in being an early example of a composite finance-design-build-and-operate project. The original shareholders of the promoting company provided the £50 000 needed for the original studies in 1959-61, and later increased this substantially to cover further studies and the preparation of detail designs and tender documents. They saw no real return on their investment for over 13 years. The original intention was to seek finance by direct loans but this did not prove feasible and it was later decided to go for contractor finance. Costains foresaw the export credit advantages of the steel shell in this context and offered it as an alternative to the concrete box. The steel shell was estimated to be more expensive than the concrete box but it probably offered the only avenue of finance at that time. It also allowed a faster three-year construction period, thus reducing interest charges and producing revenue sooner. The actual construction period was 35 months.

The Tunnel Company's organization was very simple, by comparison with that suggested by Fig.2 in Paper 15, although most of the activities indicated were carried out in one way or another. The Board of Directors included shrewd and experienced Hong Kong businessmen, and the consulting engineers were involved in administrative matters to a greater degree than normal, but most of the administrative work was done by the Tunnel Company's very able manager with minimal assistance. The project was a considerable success, covering its costs within about four years, and it currently carries some 110 000 vehicles per day.

As is often the case with something new, much of the procedure followed through this project was devised, step by step, as the need arose, and those involved, including Government, were perhaps unaware of the complications that could be introduced. While the Cross-Harbour Tunnel organization may have been exceptional in its simplicity, I do question whether projects like this, including perhaps the current Euro Tunnel, really do need large and elaborate organizations, and indeed whether they can succeed in this form. Smaller and simpler organizations, in which there is better scope for the entrepreneur and for close collaboration between promoter and engineer, each using their skill and ingenuity, would seem more suitable for such projects.

MR W.S. CHARLES-JONES, Laing Civil Engineering
With reference to Paper 15, in the UK certainly, and possibly elsewhere, the Government would prefer that, whenever possible, infrastructure projects are built with private capital rather than with public finance. There will, therefore, be an increase in such projects in future and we need to adapt to deal with them.

The proposals for these projects seem normally to be initiated by the construction companies. Rarely will there be more than a conditional promise of external finance, at best, until the proposal is accepted by the Authorities, statutory procedures are completed and the project is free to go ahead. Before this, the construction group will have to set up at least the skeleton organization of the single purpose company which will be responsible for the entire project (build, own, operate and transfer) as well as the design and construct company which is more familiar to them.

It follows that the first directors of the single purpose project company will be members of the construction group, who will have to negotiate the construction price with their 'cousins' in the same company who manage the design and construct company, to produce a viable total cost proposal for the whole project. This incestuous relationship is not always easy.

The objectives of the project company which will be in existence for the whole period of the project - i.e. to satisfy the requirements of the lenders - are different from those of the design and construct company which will aim to make a profit during construction, a small part of the total project duration. There is a potential conflict here, as the construction group will seldom have sufficient internal finance available to make more than a token contribution, to indicate commitment, to the total finance of the project. Consequently, the overall control of the project will lie with the financiers. For example, the Project Company may require rapid completion of construction to get the revenues flowing as early as possible; the Construction Company might prefer a longer programme to obtain more economical use of materials and plant.

We occasionally hear of construction companies complaining of the actions of the Project Company: the monster they have created but no longer control. What is required is understanding and give and take between the parties, not warfare.

This is a very simple example of the complexities of the private finance project. Many other factors - e.g. the interim requirements of long-term, short-term, or mezzanine finance - can influence the construction plan.

PLANNING

What is required are financiers who understand construction and what can be constructed, so that they can properly evaluate the project risk and not kill it with pessimistic contingency provisions, and engineers who can understand financial objectives; not only what is economic but what is financeable.

These skills are not usually part of the engineer's tool kit: engineers tend to be cost-wise but financially innocent.

For the time being we must do the best we can with the skills we have accumulated from our experience. We need to think how we should plan the career development path of engineers so that they gain these financial skills. Otherwise, engineers will either lose all control of the projects they initiate, or, worse still, fail to get useful projects started at all, because of the inability to convince the financiers of the viability of those projects.

MR M.W. MORRIS, Acer Consultants (Far East) Ltd

In discussing the relative merits of steel and concrete tunnels, it is important to remember that if a tunnel unit has a cross-section which gives it a displacement of 1.0 tonne, then 1.05 tonnes (typically) of steel, concrete or ballast are required to make it stay firmly in place with an adequate factor of safety. In our Sydney Harbour Tunnel study (Paper 11), we tried to show that in Sydney this 1.05 tonnes was most economically provided in a rectangular concrete tunnel, using local resources and materials. In the USA and elsewhere, the situation is clearly different but the end result is the same: i.e. the most economical tunnel is constructed.

MR G.J.T. VAN HECK, Interbeton Inc., Williamsburg, Virginia

In the case of the Hampton Roads Tunnel, now in execution in Virginia, USA, 12 of the 15 units are now in place.

This project is being executed by a joint venture comprising Morrison Knudsen and Interbeton (HBG), an American/Dutch combination where typically American designed steel tubes were completed in a typically Dutch way.

The steel tubes are produced in the shipyard of Bethlehem Steel in Baltimore, where they are assembled and finished with about 50% of the total concrete in the dry dock of the same shipyard, before floating out of the dock. The dry dock is available because shipbuilding activities in the USA are currently depressed.

After additional ballast concrete to about 75% is added in Baltimore, the tubes are towed to Newport News, about 200 miles away. During the time of bidding, offers for steel tube construction were received from as far away as Brazil.

This example demonstrates that numerous factors influence the final way the tubes are constructed, even when bid as typically American steel tubes.

DISCUSSION

MR NEILSON, Paper 10
With reference to the question from Mr Osborn, the geology of the harbour comprises

(a) sandstone bed rock extending at tunnel trench level up to 200 m from the shore line and up to 10 m deep at the shore line
(b) alluvial sands and silts up to 20 m deep generally on the north side of the harbour
(c) clays, peats up to 30 m deep generally on the south side of the harbour.

The dredging methods to excavate the IMT trench (generally 10 m deep) include

(a) the use of a trailer suction dredge to remove sands and softer clays
(b) the use of a cutter suction dredge to remove rock
(c) the use of a grab dredge to remove clays and other OTR materials close to the shore line where the trailer suction dredge could not operate.

To date, the cutter suction dredge has, without difficulty, successfully completed the excavation of all rock at the southern interface. The trailer suction dredge has excavated satisfactorily in sands but was unsuccessful in excavating in the harder clays.

The environmental requirement that the harbour be maintained 'clean' has dictated that the trailer suction dredge should operate without overflow and hence pay loads, particularly in silts and clays, were low. A boom and curtain has been used around the work area of the grab dredge to contain silty water. All material is being dumped at sea at an approved dumping ground.

MR MORTON, Paper 11
In reply to Mr Morris's first question, in the design of the new railway crossing in Hong Kong, further steps have been taken to prevent return current from leaking to the tunnel reinforcement. In addition to the steps mentioned by Mr Morris, the track bed and ballast concrete are isolated from the tunnel by a membrane. Track bed reinforcement is connected to the return earth fault.

With regard to Mr Morris's second question, it is worth noting that the exercise carried out in Sydney to review the comparative cost of steel and concrete tunnels in Australia was done for a client of whom one partner had experience in the construction of concrete units and the other had extensive experience in steel fabrication. The review was, in my view, carried out quite impartially, and several interesting steel sections, designed to minimize dredging, evolved from it. Nevertheless, concrete did prove more

PLANNING

economic in that area of Australia.

Mr Culverwell has raised a very interesting point regarding simplicity of build and operate management systems. However, when the constructor is the designer as well the initial owner and operator of a facility which is later to be transferred to public ownership, then a series of commercial and technical safeguards are demanded. The ultimate beneficiary and the security agents acting for the lending agencies require assurance of workmanship, structural integrity and durability. The constructor as the operator needs commercial safeguards, which inevitably means separating his role as the owner from that of the constructor. Consequently, complex management structures arise simply to ensure that the interests of all parties are met and that conflicts of intent do not arise.

MR DAVIES, Paper 12
Mr Culverwell makes several points ranging from the origins of the tunnel scheme through to the tendering procedure.

During the original feasibility study, a tunnel scheme known as Route 7 was developed as one of several options considered at that time. This involved a 1200 m long immersed tunnel near Deganwy Narrows, together with approach roads which passed over the Causeway on a major overline bridge. The scheme proved both costly and intrusive, and was rejected by the Secretary of State who favoured the original bridge.

At the Public Inquiry, Freeman Fox and Partners, acting on behalf of Aberconwy Borough Council, proposed a variant of Route 7 which more or less followed the Green Route alignment. It involved a shorter tunnel (at 990 m) and was also less damaging in that it passed beneath the Causeway in an underpass. The scheme incorporated many features now included in the adopted scheme, although there were also many differences.

Paper 12 deals with developments following the Inquiry, hence the absence of any acknowledgement of the contribution by Freeman Fox and Partners or of others who gave important evidence at that time. Details of the original studies can be found in the paper by Lober and Wilson: 'A55 North Wales Coast Road, St Asaph to Aber: route location' (Proc. Instn Civ. Engrs, Part 1, 1988, 84, Oct., 895-937).

The timescale for the project is really 13 years rather than 21 years, since new procedures were required once a decision was taken not to proceed with the bridge. Nevertheless, I agree with the overall sentiment. Measures must be found to reduce the planning periods needed for major schemes of this type.

With regard to the tender procedure, tenderers were free to submit alternative tenders with shorter construction periods for either form of construction, and had they done so any savings would have been taken into account in the tender comparison. For this scheme, however, the overall period is constrained by the construction of the cut and cover lengths and by the earthworks operations needed for the approaches. Any savings in constructing the units would not have produced a saving overall.

17. Shell composite construction for shallow draft immersed tube tunnels

M. J. TOMLINSON, BSc(Eng), and A. TOMLINSON, BSc(Eng), Tomlinson and Partners, M. Ll CHAPMAN, BSc, and A. D. JEFFERSON, BSc, MSc, Sir Alexander Gibb and Partners, and H. D. WRIGHT, BEng, University of Wales College of Cardiff

SYNOPSIS. An alternative immersed tube tunnel scheme is presented which uses medium draft units that can be constructed remote from site. The structure conceived, uses a novel dual skin composite construction and has a non-uniform rounded cross-section. The paper suggests a design method for the system and also describes the non-linear finite element program developed to model its behaviour.

INTRODUCTION
1. There are two basic forms of immersed tube tunnel which have been commonly constructed to date. In Europe a rectangular concrete cross section has been used exclusively, whilst in the USA cylindrical stiffened steel shells, with a concrete lining have been preferred.
2. The shape and constructional form of both types of tunnel has remained relatively unchanged since their conception. Both forms have advantages in certain circumstances, but both have major limitations which ensures that neither form finds much favour with the advocates of the other, and possibly prevents more general use of immersed tube tunnels worldwide.
3. In the past twenty years the advent of computer aided design has led to a much fuller understanding of structural behaviour, and combined with improvements in material and structural technology has resulted in major advances in the design of marine structures. The most notable example of this would be the development of the offshore industry.
4. A development programme was undertaken by the authors to investigate the characteristics of existing tunnel forms and to assess if improvements, whether economic or technical, could be made by applying modern technology to the design and installation of immersed tube tunnels.
5. It was recognised at the outset that in certain circumstances improvements over the currently used techniques would be difficult to achieve, for example, the deep draft, rectangular concrete cross section is ideally suited to the special environmental conditions found in Holland.
6. It was also recognised that these circumstances are unlikely to exist at many other potential tunnel locations,

DESIGN

where the use of an immersed tube tunnel to form a crossing may be inappropriate using present techniques, either on technical or financial grounds.

MAIN CHARACTERISTICS OF RECTANGULAR CONCRETE AND STEEL CYLINDRICAL IMMERSED TUBE TUNNELS

7. The initial stage of our investigation aimed to identify any advantages of the currently used techniques which we wished to retain, and any major disadvantages which we wished to avoid. The main characteristics were perceived as :

8. (a) The rectangular cross section provides a much more economic use of internal volume than does the cylindrical shell, which has a large void above the roadway. This potentially wasted space is important since internal voids have to be compensated for by additional ballast, so not only is the tunnel cross section enlarged, but must be enlarged even further to provide space for ballast.

9. (b) A further serious disadvantage of the cylindrical cross section is that for a given road level it will obviously require a deeper dredged trench than a rectangular shape, which could greatly influence the cost of a project. This is particularly true in very hard ground, where dredging becomes more difficult, or very soft ground, where the dredged trench requires shallow slopes.

10. (c) Another disadvantage of the cylindrical cross section is that for maximum road gradients, the lower level of the roadway within the cylindrical shell will make the overall length of the tunnel much longer than if a rectangular cross section were to be used.

11. (d) The rectangular form resists applied forces (predominately hydrostatic) primarily in bending. This is much less structurally efficient than the shell action generated in the cylindrical shape, and results in a structure with thick walls and pronounced haunches at the corners, and inevitably a deep draft.

12. This becomes particularly important as water depths and thus the hydrostatic pressures increase. This, in practice, would limit the rectangular concrete cross section to relatively shallow crossings.

13. (e) The distinction between strength and ballast has become blurred in concrete tunnels, the heavy structural members providing not only strength but also most of the negative buoyancy of tunnel element. This prevents consideration being given to alternative ballast materials, which may be cheaper or more efficient than concrete.

14. (f) Recent European experience indicates that the rectangular concrete cross section is generally a more cost effective solution than the steel cylindrical shell. This is primarily because of the high cost of fabrication of the shell, in particular the large amounts of the stiffeners required. The major function of this stiffening is to provide strength during constructional operations, such as launching, and its structural contribution during the permanent phase is much less effic-

ient than provided by concrete.

15. (g) Conventional concrete units have a large draft and as a result lengthy estuarine or seatows are often impractical, since not only would dredging long approach channels be prohibitively expensive, but also the concrete unit would be unlikely to be strong enough to withstand the rigours of a long seatow, especially in North European waters. In addition the virtually neutrally buoyant unit would be extremely difficult to control. This means that a casting basin must be provided adjacent to the tunnel line. In Holland, tunnel builders have been fortunate enough to re-use one casting basin for several tunnels. This would be unlikely to be practicable elsewhere.

16. The provision of a non permanent casting basin adds considerably to the constructional costs, contributing perhaps 10% to the project budget. In urban areas the large acreage required simply may not be available or may be prohibitively expensive to acquire. In addition, the dewatering of the ground may cause damage to structures adjacent to the site. Even where suitable sites exist, a casting basin may be visually or environmentally undesirable.

17. A not unimportant factor in these days of private finance is that construction of tunnel units cannot begin until the casting basin is complete, (which may take as much as eighteen months) and that tunnel units generally have to be constructed simultaneously, since the casting basin has to be flooded to float out the units. Both these factors can add considerably to a project's length and cost.

18. (h) Cylindrical stiffened steel shell tunnel units, are generally constructed in a fabrication or shipyard facility, and launched into the water. This means that units can be constructed sequentially, in a controlled environment, with a high degree of automation. Fabrication can begin immediately a contract is let, allowing installation to begin much earlier in the programme than would the concrete alternative (associated works on the site permitting). The units after launch are generally towed to a fitting out quay, where ballast and the concrete lining are added, but if necessary, the unballasted units, which have a very shallow draft can be towed considerable distances to the tunnel site before fitting out and installation. This realises several major benefits since not only could a fabrication facility be located in a convenient area where materials and labour are perhaps more freely available than at the tunnel site, but once established the facility could, if required, be used for several tunnels. In addition, the effect of the constructional works on the area surrounding the tunnel site is dramatically reduced since not only is no casting basin required, but fewer materials would have to be brought directly to the site, reducing traffic congestion, noise and other environmental problems, which would all help to minimise any local objections to the tunnel works.

DESIGN

DESIGN PARAMETERS

19. Our appraisal lead us to several main design criteria :
(a) That the cross section should be basically rectangular, providing maximum efficiency in the use of internal volume, combined with minimum dredging requirements. Overall tunnel length would also be minimised if maximum vertical road gradients are to be used.
(b) That if necessary the units could be constructed remote from the tunnel site, and brought in as required, whilst minimising dredging requirements in approach channels. By implication therefore a relatively shallow draft would be required.
(c) That a purpose built casting basin should be avoided, unless a future use can be found, as a marina perhaps.
(d) That sequential construction of units should be possible, and begin as early in the programme as required.
(e) That the unit should be robust enough to withstand a seatow.
(f) That constructional costs must not exceed those currently expected.
(g) That strength and ballast functions should be separated.

20. The tunnel design discussed in the remainder of this paper is a result of gradual evolution and compromise over an extended period. To assist the design process, a linked financial and engineering computer program was constructed. This was used to assess the financial effects of changes in technical parameters, for example the effect on dredging costs resulting from an increase in wall thickness and hence in draft.

21. It was recognised that neither existing solution could be adapted to satisfy all of our criteria, especially those of high strength combined with shallow draft, and so a totally different solution was developed. This entailed the use of double composite construction, which consists of two skins of steel separated by a concrete core, the whole acting compositely Connection between the structural elements being achieved through headed studs welded to the steel plate.

22. In order to understand the behaviour of the system, and to produce a reliable design procedure, a series of model tests have been carried out. The subsequent sections of this paper describe the behaviour of the system, as observed in these tests, and then present appropriate design and analysis methods.

THE BEHAVIOUR OF DUAL SKIN COMPOSITE ELEMENTS

23. The dual skin composite tunnel as just described can be formed using steel plate as permanent forms for the concrete. The connection between the concrete and steel can be made using welded stud shear connectors which are popular in composite beam construction. Figure 1 is a photograph of a test specimen prior to casting which shows the various components and form of construction.

24. Dual skin composite elements could be assumed to behave in a similar way to reinforced concrete elements. There are, however, several aspects of dual skin composite behaviour that differ markedly from conventional R.C. construction.

FIG. 1 Curved Model test before casting

25. The main differences may be described using the bending of a simply supported double composite beam as an example. Consider the possible failure modes of the beam as shown in fig. 2.

26. It is preferable that the tension steel yields in gradual ductile failure. This means that the compression steel should be heavy enough to ensure that the section is not over-reinforced. In the case of the simply supported beam over-reinforcement could occur. However in a tunnel section where stress reversal can occur each face needs to be reinforced with approximately equal thicknesses of steel.

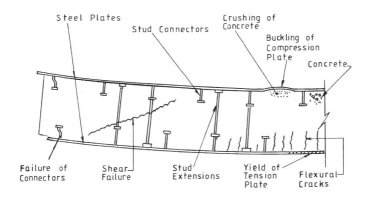

FIG. 2 Failure modes of a dual skin composite beam

DESIGN

27. However the steel plates are only connected to the concrete with shear connectors at discrete centres. The thin steel plate in compression can buckle if the centres of the studs are high. Consequently the centres of the shear connectors must be limited to avoid buckling.

28. The purpose of the shear studs is, however, to resist shear forces in the beam. The main transverse shear forces within the concrete are resisted by links in a R.C. beam. In the double skin composite beam it is the shear studs that must perform this function. A proportion of the studs must, therefore, extend through the depth of the beam and be anchored into the compression area. In fig. 1 these long studs can clearly be seen. These are formed by "piggy-back" welding studs on top of one another.

29. The other shear forces are slip shear between the steel plate and the concrete. These are similar to the anchorage and bond stresses between bars and concrete in R.C. construction. The stud shear connectors must, therefore, carry these forces which may, of course, occur at both the tension and compression plate interfaces. In the compression region the studs are well confined by the concrete and the connection is strong and stiff but in the tension zone the concrete is assumed cracked and the stiffness and ultimate strength of the connectors are reduced.

30. The example of the simply supported double skin composite beam element can only go so far in explaining the complex action of the highly indeterminate tunnel structure.

31. The tunnel walls are subject to external pressure loading which generates axially compressive forces in addition to the bending action already explained. For a pure cylinder the bending forces become very small and the section is subject to circumferential compression. The behaviour of dual skin composite elements under this form of loading is again similar to R.C. construction. The steel skins assist the concrete in carrying the compressive forces, the short shear studs act as links to prevent the steel plate buckling and the longer studs are required to resist transverse shear.

32. Although a more circular section would prove structurally efficient, it would need to be large to accommodate the traffic envelope. A balance between a circular section taking mainly circumferential compression and a rectangular section taking combined moments and axial forces is required.

33. The possibility of stress reversal within the tunnel is likely during transportation of the segments. This stress reversal leads to a further failure mode when the inner skin of the curved sections of the wall are subject to tension. This is shown in figure 3, a photograph of a test on a curved portion of the tunnel wall. The negative moment around the curved portion is simulated by a hydraulic jack forcing the walls outwards. The area of tension in the elbow of the corner has caused pull out forces on the connectors which in this case has led to failure.

34. The behaviour of the tunnel has so far only been consid-

FIG. 3   Curved Model test

ered in two dimensions. The behaviour of the tunnel segments in three dimensions is important both during transporting and in the final service state. During sea transportation from the dry dock to site the segment acts as a ship and is subject to considerable longitudinal bending from wave action. Even once bedded on the sea bed in its final state variable soil conditions can give rise to substantial moments in the longitudinal direction.

35. Most R.C. structures are designed at the ultimate state and their behaviour at service loads is often approximated. The stiffness of a reinforced concrete structure is normally high and can be estimated by assuming that the concrete and reinforcement act fully together. It is also often assumed in the analysis of reinforced structures that the concrete remains homogeneous and cracking in the tensile zones is ignored.

36. In the dual skin composite structure the movement between the steel plates and the concrete does affect the stiffness to a greater degree, This is particularly pronounced in the tensile areas of the section where the stiffness of the connectors are reduced by the presence of cracks in the concrete. These cracks are often discrete at each stud position as the studs act as crack inducers. The inclusion of mesh reinforcement around the connectors as a means of distributing the cracks appears to be effective.

37. The complex action of dual skin composite construction gives rise to complications in the design of the tunnel. As the slip between concrete and steel affects the stiffness of the section this will have implications for the stress distribution. The analysis of the stress distribution must therefore take into account the non-linear load slip relationship of the connectors and the shear stress patterns in the concrete.

DESIGN

DESIGN AND ANALYSIS

38. The design of a structural system normally requires a mathematical model of that system together with a set of design rules which define the safe limits of structural behaviour.

39. In order to develop the appropriate design rules and mathematical models for dual skin composite structures, a substantial amount of experimental data was needed. Many tests have been carried out at University of Wales College of Cardiff during the past three years to help provide this data and this programme continues under an on-going research contract.

40. The models that have been developed for dual skin composite structures can be divided into two categories. The first are relatively simple models that are intended for use in the design of conventional structural components such as beams and columns and in initial concept design of structures. The second are comprehensive and accurate computer models which account for all the main factors controlling the structural behaviour. These computer models have also been used in the derivation of the design rules and in the understanding of the behaviour of the system.

41. In the design of standard reinforced concrete and composite structures, it is not normally necessary to produce a complex mathematical model, since simple methods provide results adequate for design. However in the present case, in which the structure has a non-uniform shape and for which the structural behaviour is relatively complex, it is necessary to have a method of accurately predicting the behaviour of the structure. The computer programs developed for this analysis use the finite element method. The details of the elements and material models available in these programs are described later in this section.

42. The procedure adopted for the design of a tube cross-section will now be described.

43. Assuming that a traffic, ventilation and access envelope has been defined, a cross-sectional shape is sketched. The shape is chosen to give the maximum amount of 'shell action' while avoiding too much 'dead' space. A few approximate calculations are made as to the section's buoyancy and the shape is refined. The draft of the tube will be relatively low and the design will aim to produce a section with the largest draft permitted by the site conditions. The buoyancy of the section is then checked accurately. Throughout the design calculations the 'buoyancy' of the cross-section will be checked a number of times and because the shape of the section is non-uniform it is useful to have a computer program to calculate the buoyancy properties. A program has been developed in the present work to perform these calculations which makes use of Green's theorem to convert line integrals into area integrals.

44. When the shape satisfies the installation and permanent buoyancy criteria, the first main structural analysis of the section is performed with an elastic plane frame program. This analysis is carried out for all relevant loading cases. The results from these analyses are then used for the design cal-

culations. To assist with these calculations, a computer program was developed which reads the resultants from the plane-frame output and then, for each section, calculates the crack depth, the concrete and steel stresses and the shear flows at the plate-concrete interfaces. This, together with the transverse shear resultants, allows all the components of the section to be designed. These include plate sizes, concrete grade, mesh reinforcement, stud connectors for slip shear and elongated connectors for transverse shear.

45. It should be noted that the plane frame analysis provides only an approximate model of the section's behaviour and the design calculations performed with its results can not be regarded as definitive but rather as the next stage in the refinement.

46. Following on from the plane frame analysis and the above design exercise, the section is analysed with the non-linear finite element program. Only the most onerous load cases, as identified by the plane frame program, are analysed. From the results of the finite element program, the design is further refined. Once the design satisfies all the relevant requirements a parametric study is performed to investigate the affects of various factors. These include, long term creep, loss of plate thickness due to corrosion, failure of a certain percentage of studs and the loss of support under a section of the base of the tube.

47. In addition to the main structural considerations a number of secondary but no less important, factors need to be checked. These include;
(a) plate buckling between studs when the plate is in compression
(b) limits of stud diameter to plate thickness
(c) plate pulling away when tension occurs in inwardly curved plates
(d) fatigue affects from cyclic loading
(e) fire resistance
(f) and others

48. The behaviour in the longitudinal direction has also to be checked, particularly during transportation and installation. If simplified analyses identify a problem in the longitudinal direction, it may be necessary to perform a three dimensional finite element analysis of a section of the tunnel.

49. The cross-section of a tunnel designed with the above procedure is illustrated in fig. 4.

50. In the next section the finite element programs developed for the analyses will be described.

FINITE ELEMENT PROGRAMS

51. The essential aspects which govern the behaviour of a dual skin composite section have been described in detail earlier in the paper. The computer analysis needs to model all these aspects and thus should contain :
(a) A concrete model which allows for crushing and cracking,
(b) A connection model which simulates the behaviour of the studs

DESIGN

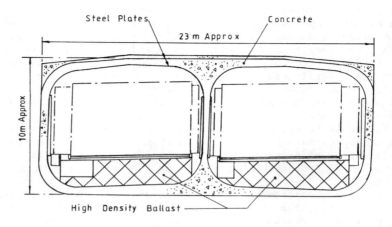

FIG. 4 Cross-section of Immersed Tube Tunnel

(c) A steel model for both the plates and the reinforcement.
52. The elements and material models used to model the above components in the two-dimensional program are outlined below. These elements and material models have direct equivalents in the three-dimensional program. Full details of the models are given by Jefferson in Reference 1.
53. The concrete is modelled by 8-noded quadrilateral elements. The stress-strain or constitutive model allows for both cracking and crushing of the concrete. This material model uses non-linear hardening plasticity theory to model the crushing and a uni-directional softening for the cracking.
54. The steel plates are modelled by 'line' elements which are capable of transmitting only axial forces. The material model for the steel is a standard elastic-plastic model.
55. A 'contact' element is employed to simulate the slip interface between the steel and the concrete. The in-plane stud behaviour is modelled by a new non-linear plasticity model which accounts for the gradual yield of studs and also allows for the reduction in stiffness when the concrete at the interface cracks.
56. Embedded lines of stiffness are added to the quadrilateral elements to model the reinforcement.
57. The above element and material models are combined to simulate the behaviour of the dual skin composite system. A typical arrangement of elements is shown in Figure 5.
58. Finally, in this section, the finite element mesh (Fig.6) and predicted crack plot (Fig. 7) are shown from an analysis of the test illustrated in Fig. 1. The agreement between the analytical and experimental results, from this and other tests, has been consistently good (see Reference 1).

CONCLUSIONS
59. A dual skin composite immersed tube tunnel has been presented as an alternative to more conventional reinforced concrete and steel shell systems. The system developed would

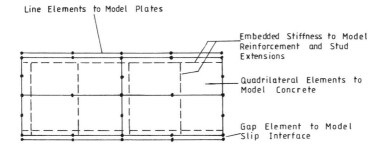

FIG. 5 Typical arrangement of elements

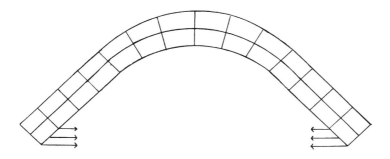

FIG. 6 F.E. Mesh for curved model test

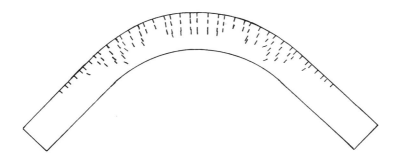

FIG. 7   Predicted crack pattern

give a unit with a relatively shallow draft which could be constructed in an existing dry dock or shipyard and then be towed to site. Dense ballast would be used to provide the additional weight needed to achieve the required negative buoyancy.

60. The structural behaviour of the system, as observed in a number of experimental tests, has been described and the design

implications of certain aspects of this behaviour have then been discussed.

61. A design procedure for a tunnel unit has been outlined. The procedure uses the results from an elastic plane frame program to produce a first design. This is then checked and refined with the aid of a non-linear finite element program. The elements and material models available in the finite element program are also described.

62. The above dual skin composite system has been used in the complete design of an immersed tube tunnel although a tunnel using this method has yet to be built.

63. The authors believe that the system would be viable in certain situations.

REFERENCES

1. JEFFERSON, A.D. "Finite element analysis of composite structures". Ph.D. Thesis, University College, Cardiff, 1989.

18. Concrete immersed tunnels: the design process

L. C. F. INGERSLEV, BA, MA, Parsons Brinckerhoff Quade and Douglas, Inc., Boston

SYNOPSIS. A guide is given to aspects of preliminary design of concrete immersed tunnels up to the start of final design. Recommendations should ensure sufficient strength, serviceability, and stability, and that structures should survive overload, albeit possibly damaged, without flooding. An outline of construction aspects which may affect design is included. While this paper is written with highway tunnels primarily in mind, the content applies equally to rail, utility, or any other kind of tunnel.

DESIGN ASSUMPTIONS

1. It is essential to itemize assumptions made during the design stage, as the design must be amended if assumptions change during construction. Such assumptions might relate to:

   (a) Quality control during fabrication.
   (b) Supervision of construction to ensure compliance with specifications, methods, and drawings.
   (c) The magnitude and control of unavoidable imperfections (e.g. dimensional tolerances).
   (d) The qualifications and skill of personnel.

2. The design life of the project is chosen by the client to suit the purpose of the tunnel. The design life of road, rail or utility tunnels usually lies in the range of 50 to 100 years. It should be assumed that the design extreme and exceptional loadings will occur at least once during the expected service life. The design fire resistance should comply with local codes, should correspond to this life, and should probably not be less than 4 hours.

3. Road or rail tunnels with more than one tube should be designed so that in the event that one tube is shut down, traffic can be carried in another tube. Multi-lane tubes may need to be capable of operating in a bi-directional mode complete with appropriate cross-over areas, possibly considerably changing ventilation requirements.

4. In addition to structural requirements, inundation of the tunnel by floods caused by the combined effects of surge, tide and waves must be prevented. In common with the similar hazards

of flooding over dykes, levees etc., the height and shape of surrounding walls and the elevation of access road surfaces must be chosen such that entry of water is prevented; a water level with the probability of being exceeded 0.001 times in any one year is generally used.

5. It is essential to use a common datum level throughout the project for both marine and land works. Tide tables, marine charts, local authorities, utility companies, topographic maps etc. often each have their own datum levels which should be converted to a common base.

SIZING OF TUNNEL SECTION

6. Requirements of the client, existing land utilization, projected traffic data and make-up, and design speeds will usually dictate the location of the proposed alignment, connections to existing roads, profiles of the tunnel, and the number of traffic lanes required. Alternative routes should be considered. Long or steep uphill grades may result in a need for climbing lanes for heavy vehicles. Climbing lanes should be avoided within the immersed tunnel elements themselves because this may complicate manufacture considerably and thereby drive up costs unacceptably. As water quantities are small, cross slopes of the roadway can be reduced to a nominal 1% within tunnels. Navigational clearances to the seabed, varying channel widths, position, and geotechnical profiles need to be considered together with costs when designing the horizontal alignment and vertical profiles.

7. Once the number of traffic lanes and clearance envelope are determined, together with any additional space needed for lighting and signs, decisions regarding the width and provision of emergency walkways can be made. Provided for the use of maintenance personnel and for vehicle occupants in case of breakdown or other emergency, such walkways should not be narrower than 0.6 m and should be set back from the kerb sufficiently for vehicle overhangs not to endanger pedestrians. The use of un-mountable kerbs can further protect users. When there is more than one lane in a tunnel, walkways on each side of the roadways should be considered. Emergency lanes are not usually provided as this may lead to a great increase in the cost of the tunnel.

8. With the width and height of internal spaces, preliminary calculations can be made regarding methods of ventilation. Ventilation methods need to satisfy local requirements for air quality under conditions of traffic (usually stalled) or fire: Longitudinal ventilation has been used for traffic tunnels of up to 1 to 2 km long, semi-transverse ventilation as tunnel lengths increase, graduating to fully transverse ventilation as air quality dictates (the primary method used in USA). Transverse ventilation considerably increases tunnel costs as it requires the provision of separate ventilation ducts and substantial fan buildings. Even when longitudinal ventilation is used, an additional duct is often provided for utilities. Longitudinal ventilation may be provided by jet fans within the

traffic tunnel, requiring an increase in cross section and noise attenuation, or by fans in ventilation buildings, as used for transit tunnels in Hong Kong. Ventilation building locations must be selected, preferably on land where they cannot be hit by shipping. The end faces of the buildings may also make a suitable interface between immersed tunnel elements and on-shore techniques. In addition to emergency access into the adjacent tube usually provided at say 50 m intervals, air ducts could also be considered for an alternative path of escape.

9. Structural dimensions of the tunnel can be estimated when air, traffic, electrical, mechanical, and utility duct requirements have been evaluated. Concrete can be cast into any shape, so economics can influence whether rectangular or circular sections are used, steel immersed tubes tending to be circular. Shallow profiles usually dictate rectangular sections, side vented if required. Where more than two lanes per tube are needed, concrete rectangular tunnels are usually the most economical choice. Haunches may help to alleviate shear problems at the edges of larger spans. In making an initial sizing of the cross-section of the immersed tube, the overall weight must be such that the section can meet requirements for floating. Tolerances to allow for construction variations and possible misalignment of elements must be incorporated without encroaching on roadway clearances. Allowance must be made for the addition of more concrete or permanent ballast in order to meet requirements not to float under final conditions. Having made this estimate of size, the approximate overall structural height is then known.

10. The depth to the structure below the clearance envelope required for the shipping channel and elsewhere in the waterway will vary according to the amount of over-dredge expected, (typically about 1 m, depending upon the method used for dredging), and the amount of protective backfill required over the tunnel. Protective backfill is often about 2 m thick, though the thickness may depend, for example, on the ability to distribute loads from sunken ships or other foreseen loads, to the tunnel structure. Anchor loads are usually small compared with structural capacity, but may cause local surface damage. The type and grading of backfill layers are selected so that the tunnel structure is not damaged by them and so that material does not get washed away under anticipated currents. The effects of hydraulic intrusion of a tunnel into an existing waterway regime may require study.

11. The process of sizing the tunnel may need to be repeated several times during the optimization of the structural cross-section before tunnel dimensions are finalized.

CONSTRUCTION OF TUNNEL ELEMENTS

12. Tunnels can sometimes be built in situ across waterways using temporary cofferdams and constructing the tunnel in sections as on land, but by far the most common method uses precast elements.

DESIGN

13. The location of a facility for fabrication of precast tunnel elements needs to be chosen. Because of the great weight of the tunnel elements out of water and therefore the great difficulty of launching them, construction typically takes place in a graving dock or dry dock which can easily be flooded when elements need to be removed. If the site for the dock is far from the tunnel location, towing costs may be significant and take considerable time, even though the shape of the ends of a element may be improved for towing by attaching a temporary bow and stern. If only a few tunnel elements are required because the water crossing is relatively short, and if the program of construction will permit it, it may be possible to use the shallows or immediate landfall area of the tunnel (which have to be excavated anyway) as the dry dock, then float the tunnel along the alignment into position. This can avoid environmental impact problems of constructing the dock in a location which otherwise would not need to be touched (ref. 2). In the United States, it is often the difficulty and expense of meeting environmental impact problems of dry dock location that result in comparatively light steel shell elements being constructed in a fabrication yard on land and launched. The design of the dry dock may be governed by:

(a) A compromise between size and the number of reuses. The construction schedule may influence the choice.
(b) Choice of location and interior layout of the dry dock. The location of construction joints both around the section and longitudinally, determining the length of elements, will need to be taken into account both in the dock layout and in the design analysis.
(c) Results of the geotechnical profile in the dock area and of soils tests.
(d) The stability of the side slopes forming the exterior boundary of the dry dock and the stability of the soil layers below the dock and the exterior walls.
(e) Dock floor depth adequate to float out elements.
(f) The methods of dewatering the dock or of making cut-off walls. Flow tests may be required to confirm dewatering procedures.
(g) The design and stability of the gate for removing elements, and the possibility of reusing it. The gate may either be a structural element or merely removal of a dyke.
(h) Accessibility of the dock and the ease of element fabrication.

14. Ease of construction can be achieved by a high degree of mechanization and line fabrication. Prefabricated forms and work shelters (heated in winter) running on rails, shop welding of steel bottom plate, if used, and of prefabrication of

reinforcement cages and prestressing cables can all be used to advantage. Tunnel element lengths of 100 m are common, though 230 m lengths have been used (ref. 2). Casting sequences, methods of curing to avoid cracking, and concrete technology all have to be determined at this time.

CONSTRUCTION ASPECTS.
15. Although not considered in detail here, construction aspects which need to be considered even at the early design stage include:

(a) Topography, geology, hydrology, and meteorology.
(b) Schedule for construction as well as for design. Constraints of construction sequence will need to be incorporated.
(c) Choice of location with adequate draft for outfitting of elements, if not completed in the dry dock.
(d) Temporary bulkhead installation with access to the interior both afloat and after placing.
(e) Towing method and route.
(f) Survey methods during sinking and placing, and within the tunnel.
(g) Navigation, fairway requirements, and the method of sinking (including plant, equipment, and temporary ballast) and of joining tunnel elements. Together, these may limit practical tunnel element lengths.
(h) Foundation method and stability thereof for tunnel elements, both immediately after sinking and finally, e.g. temporary supports, and screeded gravel foundation, sand flow or sand jetting. (Piled foundations are proposed in Boston, USA, adjacent to an existing subway tunnel which is to be crossed using immersed tunnels).
(i) Methods, and cost of excavation and disposal of the precast tunnel trench materials which might include gravels, sands, silts, clays or even rock and boulders. Sometimes a longer shallower tunnel on a different alignment may be cheaper than a shorter deeper tunnel, especially if rock excavation is involved, since removal of rock below water is particularly expensive.
(j) Source, type, and method of placing, of backfill around the precast tunnel after placing. Special backfill may be required for seismic conditions.
(k) Protection against anchors and over dredging.
(l) Adjacent rock mounds to carry sunken ships. A history of ship grounding may indicate a need for special precautions.
(m) Construction methods and sequence for the immersed tunnel approach structures. This would also include excavation support and methods for de-

DESIGN

watering, cut-and-cover tunnels, approaches, and ventilation buildings.
(n) Method of waterproofing both the exterior faces of the immersed tunnel. A waterproofing layer may also be needed below the roadway surfacing.
(o) Drainage of the interior, including sumps at low points of the alignment and at portals.
(p) Electrical, mechanical, operational and maintenance aspects, often on the critical path and can delay opening of tunnels.

STABILITY OF ELEMENTS

16. In checking tunnel elements for stability while floating, due attention should be paid to effects of variations in structural dimensions including results of thermal and hydrostatic effects. Items to consider include:

(a) Sufficient freeboard, say 0.3 m for marine operations, so that the element is relatively unaffected even when a wave runs over the top.
(b) A positive metacentric height (static stability) of say 0.20 m can avoid the effects of an angle of loll.
(c) Designing for a 1 % positive buoyancy margin to guard against sinking due to variations in dimensions and densities of both tunnel materials and of the surrounding water.
(d) Cross curves of stability should show a factor of safety of 1.4 of the area under the righting moment curve against the heeling moment curve.

17. After placing tunnel elements on the seabed, a suitable factor of safety against flotation ignoring assistance from adjacent elements should not usually be less than 1.04 and may need to exceed 1.08 to combat tidal lag effects. Temporary ballast, stable even under the most adverse conditions, may need to be placed either inside or on the tunnel roof. It should be noted that in areas such as river estuaries where water of different salinities may exist, several distinct layers of water may form. Layers with a higher salinity will lie below layers with a lower salinity, and the speed and direction of flow of each layer may differ, tidal data usually referring only to the uppermost layer.

18. Completed structures are recommended to have a factor of safety exceeding 1.2 against flotation including any backfill above the plan area, and a factor of safety never less than 1.02 against flotation when backfill, road surfacing and all removable items are excluded.

DURABILITY, WATERPROOFING AND CORROSION PROTECTION

19. Corrosion of the reinforcing steel can be controlled by providing concrete cover between the reinforcing steel and the atmosphere or oxygen laden water. Rates of corrosion are

considerably reduced where structures are buried and replenishment of the corroding medium is prevented (e.g. no circulation of water).

20. Cover to reinforcement and prestressing ducts should be maintained appropriate to exposure, with the assumption that waterproofing membranes leak. Leaks may also occur where walls or compression zones are thin, or when membrane tensile stresses exist. Air ducts and traffic tubes may also be exposed to humid and salty conditions and/or wetting and drying.

21. External waterproofing has been omitted in several Dutch tunnels, but if waterproofing membranes are used, protection against damage and backfill should be provided.

22. Joints between tunnel elements are generally sealed with a rubber gasket and a secondary independent flexible seal is common practice. Each seal should be capable of resisting the external hydrostatic pressure and should allow for expected future movements. Protection should also be provided against damage to the seals from within the tunnel.

23. Thicknesses of concrete sections are large enough to require precautions to be taken against cracking due to heat of hydration. Cooling of concrete may be required: Ice or chilled water may be a constituent of the mix, and cast concrete may be cooled by the use of buried tubes. The use of insulated forms may also be required. Partial replacement of cement using pulverized fuel ash (P.F.A.) can help to reduce heat of hydration and may also increase strength and durability.

24. Concrete should be durable under the conditions of exposure, with due regard paid to good workmanship with little free water in the mix (low water/cement ratio). External form vibrators may detrimentally increase the water content locally in the layer of concrete touching the forms. A chloride-free concrete of low permeability is desireable. P.F.A. and especially microsilica help to fill the interstices between the cement particles and may considerably increase durability, strength, and waterproofing. Sulphate resisting cement and low C_3A content may be appropriate under certain conditions of exposure. Special precautions may be required in areas exposed to wetting and drying, especially so in regions where ambient temperatures may exceed 32 °C in salty environments so that protective coating of the concrete may be necessary. Where marine borers can be expected, the choice of aggregate may be especially important.

25. Whilst cathodic protection systems for concrete structures are still under research and development, it is generally preferable to use sacrificial anode systems rather than impressed current systems. Where externally exposed steelwork (e.g. baseplate etc.) requires protection, electrical isolation from steel reinforcement and prestressing should be provided.

26. Continuity of the running rails, electrical isolation and external grounding are desireable where rails are used as electrical ground by transit systems. Leakage currents can be a cause of severe deterioration.

DESIGN

APPLIED LOADS

27. Loads should be applied in accordance with relevant codes. Where structural codes do not consider the particular conditions that apply to immersed tunnels, the following classifications are proposed:

  (a) Dead loads, to include all long term loads and mean water level.
  (b) Live loads, creep, shrinkage, prestress, temperature, backfill (the effects of which can vary with temperature etc.), erosion of the seabed, siltation, traffic, and variations in water level, current, storm loads and earthquakes, each with an annual probability of being exceeded of 0.2 or greater.
  (c) Exceptional loads, including loss of support (subsidence) below the tunnel or to one side, and storms and extreme water levels with a probability of being exceeded once during the design life.
  (d) Extreme loads, including where appropriate: sunken ships, ship collision, water-filled tunnel, explosion (e.g. vehicular), fire, the design earthquake predicted for the location, and the resulting movement of soils. Some of these loads may be affected by categories of dangerous goods permitted through the tunnel.

28. Load combinations should be selected with regard to simultaneous probability: Extreme earthquakes may be assumed to occur without storm loads.

DESIGN LOADING CONDITIONS

29. At least the following conditions should be considered during analysis and design:

  (i) Normal operating conditions, with the tunnel operation unaffected by environmental loads.
  (ii) Abnormal conditions, demonstrating that either a tunnel with live loads can remain operational under the after-effects of extreme or exceptional loads (but not flooded) or settlement, or that with operations ceased and closed to traffic, the tunnel can survive some loss of support beneath or to one side.
  (iii) Extreme actions with the tunnel closed, the abnormal conditions above combined with one of the extreme loads.
  (iv) Construction conditions, including temporary structures (e.g. sheet piles) and loads due to handling, transporting, and placing, combined with environmental loads appropriate to the season, duration of use, and location. Abnormal and extreme conditions may be inappropriate.

METHODS OF ANALYSIS

30. Where tunnels are designed using working stress analysis, ultimate load design checks should be carried out. Increased allowable stresses may be used when considering exceptional and extreme loads. It is important in choosing appropriate factors that a consistent design code be used. For example, where load factors are chosen in accordance with a particular design code, it is necessary to choose both partial factors for material strength (or strength reduction factors, phi), and the allowable stress levels and crack control criteria from the same code. For slender structural elements, second order deformation effects due to construction tolerances should be calculated.

Ultimate load analysis

31. The behaviour of structures under ultimate load analysis (load factor, or strength design) should be investigated; The theory used should be appropriate to the response to loads of the structural material and the structure. Effects to be examined include:

(a) Adequate safety against failure of structures and their components.
(b) Static equilibrium of structures as a whole.
(c) Buckling.
(d) Fatigue.
(e) Deformations, particularly at those locations where plastic or creep deformations could transform the structure into a mechanism.

32. Each load in a combination should be factored. For immersed tube tunnels, load factors (appropriate load combination multipliers) are proposed in Table 1 below. Where alternate factors are given, use the most adverse combinations.

Working strength analysis

33. For working strength analysis, linear elastic methods of analysis are usually appropriate. Model testing and analysis may be very suitable for some conditions, particularly for examining installation procedures. Except under load combination (i), overstress is usually permitted. The behaviour of structures under working loads (serviceability) should be checked for compliance with criteria including:

(a) Cracking.
(b) Deformations.
(c) Vibration.
(d) Durability, including corrosion of steel and deterioration of concrete.
(e) Water tightness.

DESIGN

Table 1: Load Combination Multipliers.

Loading	Load case: (i)	(ii)	(iii)	(iv)
Dead load & road ballast	1.4/0.9	1.2/0.9	1.05	1.1
Traffic load:				
Normal	1.6/0	1.2/0	–	–
Exceptional vehicle	1.4/0	1.2/0	–	–
Centrifugal	1.6/0	1.3/0	–	–
Braking load:				
Normal	1.6/0	1.3/0	–	–
Exceptional vehicle	1.4/0	1.2/0	–	–
Wind and waves	–	1.2	1.05	1.2
Current	1.4	1.2	1.05	1.2
Water pressure allowing for tidal variation	1.4/0.9	1.2/0.9	1.05	1.2
Additional water pressure due to surge	–	1.2/0.9	1.05	1.2
Backfill pressure:				
Horizontal	0.9/1.6	0.9/1.4	1.05	1.3/0
Vertical	1.6/0.9	1.4/0.9	1.05	1.3/0
Creep & shrinkage	1.3	1.3	1.05	1.1/0.9
Temperature effects & prestressing	1.3	1.3	1.05	1.1
Sunken ship	–	1.2/0	1.05	–
Earthquake	–	1.2	1.05	–
Other extreme loads	–	–	1.05	–
Imposed loads during construction	–	–	–	1.3

34. Where frequently reoccurring cyclic loads cause significant stresses, fatigue strength should be considered. Detailed analyses should consider both high cycle/low amplitude and low cycle/high amplitude fatigue.

FOUNDATION ANALYSIS

35. Structures should be designed to accommodate expected movements due to deformation of foundations without limiting normal operations. Soil pressures should take account of the soil surface profile as well as the geometry of the structure. Geotechnical considerations should also include effects due to seepage, erosion, a change from drained to undrained conditions and cuts in soft clays. Differential settlements can be expected at interfaces between types of construction, at locations where subaqueous tunnels extend into the shoreline, and during construction.

ANALYSIS OF EARTHQUAKE EFFECTS

36. Both structural and geotechnical aspects enter into earthquake analysis. Maximum ground acceleration and period for design should be selected with care. Local records may give guidance on the selection of suitable seismic waves and characteristics for dynamic analysis.

37. Conditions which should be considered during analysis (see also ref. 1) should include:

(a) Risk analysis, assessment of faulting, soil liquefaction potential, consequences thereof, and suggested methods of treatment during construction.
(b) Dynamic structural response analysis and assessment of displacements using soil/structure interaction.
(c) Slides triggered by earthquake activity.
(d) Sufficient ductility to absorb imposed deformations.

38. Dynamic amplification of earthquake and other cyclic loads can result in increased load effects. Structures need to be ductile in the overload range, particularity under seismic loads. Methods of analysis should allow both for foundation-structure interaction and for the effective mass of surrounding water.

39. Primary members forming the tunnel, including its joints, should be designed to be ductile so that seismic energy is dissipated by yielding. Structural integrity must be maintained against collapse so that catastrophic inundation of the tunnel does not occur.

40. Effects of cyclic seismic loads should be considered with dynamic properties of soils using either a reduced undrained soil strength or increased pore pressures, obtained from appropriate tests. An effective drainage system can reduce pore pressures. As the density and shear strength of clays may change as the result of repeated earthquakes, the least expected value should be used.

CRACKING IN CONCRETE

41. It is important that cracking in structural concrete members is limited so that durability of the reinforcement is assured. Cracking can be regulated by limiting tensile steel stresses, the basis for crack width formulae in codes. Actual crack widths will be less than calculated for thick slabs. For prestressed concrete, no tensile stresses should be considered in the direction of prestress under normal loads. In the direction normal to the prestress, tensile strains at the level of the prestressing steel should be limited. Thin-walled sections susceptible to instability should not be permitted to crack under working loads.

42. Factors which directly affect the quantity of reinforcing steel in the concrete include the selection of permissible surface widths of cracks due to imposed loading and measures taken to reduce cracking of concrete during construction. Both of these must take due account of the type of waterproofing, if any, and the condition of the surrounding water and soils.

43. Research has indicated that crack widths up to 0.3 mm may affect only marginally the time for corrosion of the

DESIGN

reinforcement until spalling occurs, considerably in excess of 50 years. The onset of spalling would correspond to the end of the useful life of a structure. Since the useful life of structures is affected by the amount of exposure to air and water, this can be extended by the protection of a waterproof membrane or other impervious layer. (See also corrosion protection below).

44. It is essential that crack width limit checks be considered together with the calculation method appropriate to the same document. Design to FIP Recommendations for the Design and Construction of Concrete Sea Structures clause R3.2.3.2 (the same as BS CP 110 clause 2.2.3.2) using a crack width of 0.004 times the cover would be considered sound and appropriate. For American codes, corrosion control requirements are effectively satisfied assuming a cover of 50 mm, disregarding the excess when checking appropriate values of z to ACI 318.

45. Where cracking and consequential hydrostatic pressure in the cracks can significantly change structural loading and behavior, reinforcement should be provided across the cracks in these locations, anchored in compressive zones.

46. Differential temperature effects may lead to severe cracking in areas of structural restraint. This may include results of temperature differences across walls and slabs. Creep strain induced by temperature loadings may also be significant during the life of some members.

JOINT ANALYSIS

47. If shear strengths of the joints are less than those of the adjacent tubes, shear checks of all load conditions will need to be made. Joints should be ductile (refer to earthquake effects above) in addition to accommodating longitudinal movements. Where the potential of ground liquefaction exists, movement of the tunnel can be minimized by permitting plastic rotation only at joints and ensuring a high plastic resistance. Tension ties may be required to limit movement so that joints do not leak or break open. Temporary joints in elements can relieve construction stage moments. In casting in situ closure joints, spacers may be required to prevent the joint closing during casting.

48. Tunnel Bulkheads - shear capability at end closure junctions should take into account the influence of normal forces and bending moments on the shear capacity of the section. The effects of higher or lower internal air pressure in sealed tunnel elements due to variations in temperature etc. due to compression, expansion and immersion should be considered.

FINISHES

49. Due to the difficulty of maintaining and replacing road pavements in tunnels, durability is especially important. Light colored surfaces (concrete or stone chips) will reduce lighting requirements initially, but may darken with time.

50. In selecting finishes for walls and ceiling, the final finish should not permit build up of water vapour behind it. When tiles are used, care must be taken to ensure long term adherence of the tiles, particularly in ceilings. The use of prefabricated ceiling and wall panels avoids this problem and allows for rapid installation. Drainage should be provided assuming that water leakage can occur. Interior design should be arranged for ease of cleaning by machine.

OPEN RAMPS
51. Open ramps forming the approach to immersed tunnels generally extend below existing ground water table. Conventional gravity ramps are often designed with integral retaining walls on each side and to be heavy enough to prevent heave due to pore water pressure on the underside. This can lead to gravity ramps with slab thicknesses in excess of 6 meters as used on the Fort McHenry Tunnel (Baltimore, USA) and up to 9 meters as is being considered for the Third Harbour Tunnel (Boston, USA).

52. Relief of pore water pressure is commonly used below buildings when permanent lowering of ground water levels is permitted. This approach was used for the Guldborgsund Tunnel, Denmark, (ref. 2), where the ramps lie between dykes extending into the sea, obviating the need for retaining walls. Analyses should take into account the new pore water profile in the design of the side slopes. Although there is a higher risk of flooding due to portal pump failure, this may be acceptable in view of the large cost savings.

53. Where soil conditions or nearby existing structures do not permit lowering of the ground water table, pore water pressure can be resisted using vertical permanent ground anchors, as used below two tunnels in Florida, USA, and proposed in lieu of gravity slabs for Boston, USA. This method is in common use for buildings and dry docks. Although waterproofing details will require more detailed analysis and although anchors need to be corrosion protected to meet the design life, considerable cost savings over gravity ramps can be achieved. Normal working loads are generally well below design loads, but additional anchors can be provided if the need for a greater margin of safety is required. Kaohsiung Cross-Harbour Tunnel, Taiwan, uses tension piles to resist pore water pressure as well to provide support under conditions of liquefaction during earthquakes.

REFERENCES
1. Recommended lateral force requirements and commentary. Seismology Committee of the Structural Engineers Association of California (SEAOC).
2. SØRENSEN E.A. Tunnel under Guldborgsund. Strait crossings, Stavanger, Norway 1986.

19. A computerized concrete hardening control system and its application in tunnel construction

W. C. HORDEN, MSc, E. MAATJES, MSc, and A. C. J. BERLAGE, MSc, Hollandsche Beton- en Waterbouw bv, Gouda, The Netherlands

SYNOPSIS. Within the Hollandsche Beton Groep nv a computerized concrete hardening control system has been developed. The purpose is a better control of the hardening process of concrete and the related quality and cost aspects. The system covers both technological and costing features and is available on personal computer. The main fields of application are:
. selection of optimum execution methods;
. adjustment to changing circumstances during execution;
. quality improvement and quality control;
The system is used succesfully for the design, work preparation and construction of e.g. concrete tunnels.

INTRODUCTION
1. The excellent construction material concrete has one property which can be troublesome at times: it has to harden. We have to cast it in formwork and subsequently wait until it is strong enough to remove the formwork. Moreover, the development of this hardening process is dependent on a large number of parameters (concrete mix, geometry of structure, type of formwork, ambient conditions etc.) and the heat development may cause problems if deformations are restrained.
2. It is obvious that a better control of the hardening process can improve quality and reduce costs of the concrete works performed. Some illustrating examples are:
- selection of optimum execution methods, covering casting sequence, section lengths, cycle times, etc.
 In general the use of less formwork with shorter cycles and more repetition can reduce the costs, due to the lower investment in formwork and due to improved efficiency (the so-called learning effect). However, this industrial approach requires an increased control of the hardening process to realize these short cycles and to avoid disturbances in the planning;
- adjustment to changing circumstances during execution. This includes e.g. changes of weather, loss of time and holidays. Such circumstances may require additional measures to steer the hardening process;

DESIGN

- control of heat and temperature development in mass concrete to avoid cracks.

3. Within the Hollandsche Beton Groep nv a computerized concrete hardening control system has been developed to enable this better control. The system combines the 'state of the art' knowledge and techniques in the fields of concrete technology and execution methods. The main advantage is, that a wide range of parameters can be evaluated in a short time to arrive at solutions which are both cost effective and feasible with respect to technology. The system is available on personal computer. Successful implementation throughout the organisation is reached via in-house training programs.

4. This system is a powerful tool in the design, work preparation and construction of concrete tunnels: the relatively great lengths demand an effective optimization of the execution methods and the mass dimensions necessitate the control of the temperature development (ref. 1). In fact this application was one of the main reasons for developing the system.

5. The paper comprises a short general description of the system, some theoretical backgrounds of the calculation models and examples of application in concrete tunnel construction.

DESCRIPTION AND USE OF THE SYSTEM

6. A general scheme of the system is given in figure 1.

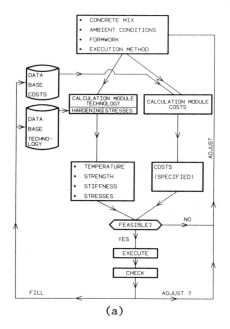
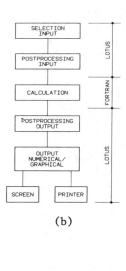

Figure 1. General scheme of the system and structure of modules: (a) general scheme; (b) structure of modules

The module technology calculates the development of:
- temperature;
- degree of hydration;
- strength and stiffness;
- stresses due to restrained deformations

as functions of time.

The module costs calculates the costs for a specified execution method, taking into account the repetition of the formwork and learning effects.

The databases contain a growing number of data from tests and executed projects, which can be selected as input for the calculation modules.

7. The (sub)modules can be used separately; the general procedure is given in figure 1. The blocks with calculations are programmed in Fortran. All other blocks are Lotus worksheets. These worksheets are fully menu driven. The result is a user orientated package, which enables quick calculations.

8. The use of the system in the project preparation phase can be elucidated as follows:

(a) select a set of input data (i.e. concrete mix, environmental conditions, formwork, execution method);
(b) investigate this solution with the technology module. Possible criteria are sufficient strength at planned striking time of formwork and stresses within allowable limits.
(c) if the solution is not feasible, adjust one or more input data and try again.
(d) if the solution is feasible calculate costs with the cost module.
(e) in this way a number of feasible solutions with corresponding costs are obtained.
(f) based on minimum costs and sensitivity a final solution to be used during execution is selected.

9. During execution the selected solution is checked as part of the quality program: field measurements are compared with predicted values and the data obtained are incorporated in the databases. Additional measures required by changing circumstances can be evaluated in the way described under paragraph 8.

10. It is acknowledged that some large 3-dimensional finite element programs are available for the calculation of temperature and stress developments in hardening concrete. These give accurate results but are very time and money consuming and consequently they are not suited for the purposes described here. In practice their use is limited to the final check for critical solutions. The somewhat simplified models used in this system provide quick and cheap results with satisfying accuracy for the majority of practical cases.

DESIGN

THEORETICAL BACKGROUNDS
Temperature development

11. The basic part of the module technology is the calculation of the temperature development in the hardening concrete. This temperature development is a special case of non-stationary heat flow, which can be described with the Fourier equation (ref. 2).
The program provides a solution for 1-dimensional problems based on the differential method: the structure is divided in elements and the calculation is performed in time steps. A 2-dimensional version is under development. The solution technique as such is rather straightforward and will not be elaborated here.

12. The specific problem in the case of hardening concrete is the realistic modelling of the hydration heat development, which is a function of both the actual temperature and the hardening state. The reaction speed will increase at higher temperatures and decrease as more cement has hardened (a comparison with a fire is always very clarifying in this respect). The modelling of this process is discussed further. In principle the approach suggested by Van Breugel (ref. 2) is followed; the influence of the temperature on the heat development is modified (ref. 3).

13. Each concrete mix has specific heat development characteristics depending on cement type, cement content, water cement ratio etc. A theoretical description, based on mix composition, of these characteristics is not available. Instead the heat development characteristics are derived from an adiabatic temperature development of the considered mix. The term adiabatic means that there is no heat exchange with the environment, so the complete heat production development becomes apparent in the temperature development. The adiabatic temperature development can be obtained from a temperature measurement in a well insulated cube of sufficient dimensions or from special adiabatic test methods, figure 2.

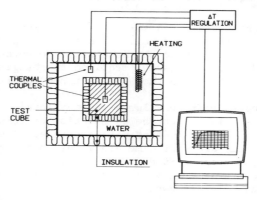

Figure 2. Adiabatic test set-up.

The adiabatic temperature development is approximated with an exponential function, figure 3.

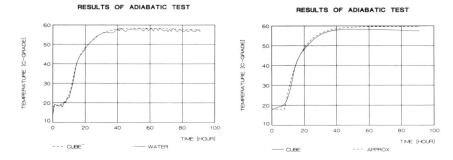

Figure 3. Approximation of adiabatic temperature development

Different heat development characteristics of concrete mixes will become apparent in different function parameters. Since the development of temperature and heat production are fully affined under adiabatic conditions (assuming constant specific heat of the concrete), the adiabatic heat production can be described with the same function. It is convenient, although not entirely correct, to define the degree of hydration Hg(t) as:

$$Hg(t) = \frac{Q(t)}{Q_{max}} \quad (1)$$

with
$Q(t)$ = heat production until time (t) (J)
Q_{max} = maximum theoretical heat production (J)

This implies also full affinity between temperature development and degree of hydration under adiabatic conditions. Relation (1) is however also valid under non-adiabatic conditions.

14. The hydration heat development per time step as referred to under 12 can now be calculated according to the following procedure:
  (a) at the beginning of the timestep the actual temperature and degree of hydration are known for each element;
  (b) the adiabatic heat production for the actual degree of hydration $\Delta Q_{adiab}(Hg)$, is determined as described under 13;
  (c) because of heat dissipation during the process the actual temperature will be lower than in adiabatic conditions. The heat production has to be corrected for this difference; this is done according to:

DESIGN

$$\Delta Q_{real} (Hg, T_{real}) = \Delta Q_{adiab} (Hg, T_{adiab}) * g \qquad (\frac{T_{real}-T_{adiab}}{10}) \qquad (2)$$

with
    g = temperature sensitivity factor of the mix considered; normally a factor 2 is used.

This formula is also used to adapt adiabatic curves for different initial temperatures.

(d) With known temperatures and degrees of hydration at the beginning of the time step and known heat productions during the time step, the heat dissipation is calculated, resulting in the starting conditions for the next time step.

15. Further features of the temperature program are:
- varying ambient temperature is possible;
- removal of formwork during process is possible;
- cooling during the process with cooling pipes embedded in the concrete can be simulated by "taking away" part of the heat production;
- thermal properties of concrete are assumed constant during the process.

Mechanical properties

16. The development of the mechanical properties during the hardening process can be related to the development of the degree of hydration, which is the parameter describing the hardening state of the concrete and which is available as result of the temperature calculation. These relations are derived from a series of strength tests on cubes of the considered mix. These are stored at constant temperature (20°C) and tests are performed at different times, e.g. 1, 3, 7, 14 and 28 days. The development of the degree of hydration follows from a "temperature" calculation for this isothermic condition. Combination of the data obtained yields the desired relation, figure 4.

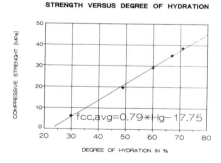

Figure 4. Relation compressive strength vs degree of hydration.

Usually a linear relationship between compressive/tensile strength and degree of hydration is very satisfying. The Youngs modulus is related to the compressive strength via a standard formula. The obtained relations are input for the temperature program and thus the (predicted) development of mechanical properties become available as calculation results.

Calculation of stresses
17. The development of stresses during the hardening process can be calculated for a section consisting of maximum 3 rectangular blocks which can represent for instance a tunnel section, see figure 5. In the middle block (the wall) the effect of cooling can be incorporated. As basis for the stress calculations the developments of temperature, strength and stiffness for points 1 to 4 are calculated with the temperature program. The stress calculation uses the average values over the thickness of the blocks. The influence of gradients over the thickness can be calculated separately.

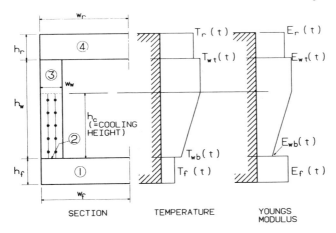

Figure 5. Model for stress calculations.

18. Further input consists of:
(a) the geometry of the section;
(b) the cooling height of the middle block; over the cooling height a linear variation of properties is assumed.
(c) the planned casting sequence; the blocks can be cast at different times;
(d) reduction factors for the calculated Young's modulus to take account of relaxation effects.

19. The program determines the development of stresses caused by (internal) restraints within the section under the assumption that plane sections remain plane. The influence of subsoil stiffness and length of the structure can be accounted for by specifying additional restraint factors.

DESIGN

The risk of cracking is judged by comparing the developments of tensile stresses and tensile strength. In future the stressmodel will be more refined with respect to relaxation and external restraints.

Calculation of costs
20. The module cost consists of simple but effective spreadsheet programs which are tailored for different types of structures. A further development of this module with the related database will be performed in 1989.

Verification of models
21. The described models have been verified with in situ measurements on several projects (ref. 4). The temperature calculation can be verified simply by performing in situ temperature measurements in the hardening concrete. Verification of the stress calculation is more complicated because the stresses are caused by restrained deformations; both stresses and restrained deformations cannot be measured directly. In situ verification is restricted to checks on the time of occurrence of cracks. Results of verifications performed uptil now are considered satisfactory.

APPLICATIONS IN TUNNEL DESIGN AND CONSTRUCTION
22. The use of the system will be illustrated with following examples:
 (a) striking time of formwork for tunnel roof;
 (b) cooling in tunnel wall;
 (c) parameter study for concrete tunnels.

Striking time of formwork
23. We consider a tunnel roof with a thickness of 1,0 m. Suppose that removal of the formwork is allowed if an average strength of 25 MPa is achieved and that we aim to remove the formwork 4 days after casting. A more or less standard mix with 320 kg Dutch blast furnace cement per m^3 is used.
24. The first solution investigated comprises:
- steel formwork below the roof;
- no protective covering on top;
- average ambient conditions, temperature $10 \pm 4^\circ$ C; wind velocity 4 m/sec. above and 1 m/sec. below the roof;
- initial mix temperature 15°C.

Some selected graphical calculation results are presented in figure 6, which shows the development of temperature and strength in time for typical points in the section; the average values over the section are given in figure 7. An average compression strength of 25 MPa is reached after 78 hours and thus removal of formwork according to the planned schedule is possible under these conditions. One should notice however, that the development of strength at top and bottom of the slab is relatively slow.

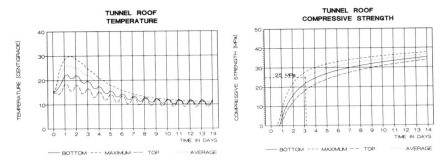

Figure 6. Striking time of formwork. Steel formwork, average conditions

25. Next the given solution is investigated for (Dutch) winterconditions:
- temperature 2 ± 4° C., wind velocity 4 m/sec. above and 1 m/sec. below the roof.
- initial mix temperature 10°C (preheated mix).

The average results for these conditions are shown in figure 7. Now the required strength of 25 MPa is obtained only after 150 hours and the development of strength at top and bottom is retarded further. It is concluded that additional measures are necessary to realize the planned schedule under winter conditions; these may consist of
- use of faster cement and/or accelerating additives;
- further preheating of mix;
- insulation at topside or at top- and bottomside
- external heating

or a feasible combination of these.

The programs developed enable quick evaluation of these solutions. As an example the results for the case with insulation at top and bottom are shown in figure 7. This appears to be an effective solution as an average strength of 25 MPa is already reached after 72 hours.

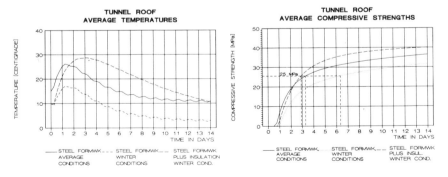

Figure 7. Striking time of formwork. Summary results.

DESIGN

Cooling of tunnel wall

26. This application is taken from the work preparation activities for the Willemspoortunnel in Rotterdam which is now under construction. The elements for the immersed part are fabricated in a dry dock. Figure 8 gives a typical section for these elements. The outer walls and roof are cast in one pour on the already hardened base slab.
It is known that within this execution method, measures may be necessary to limit the temperature difference between the wall and the base slab. If not, cracks may occur in the wall due to the deformation restraint imposed by the base slab.

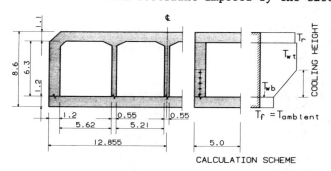

Figure 8. Cooling of tunnel wall. Typical section.

27. Cooling of the wall with cooling pipes embedded in the concrete has proven to be an effective solution. The cooling intensity is maximum at the construction joint with the floor and decreases upward in the wall. Calculations to evaluate the necessity and the required rate and time of cooling were very time consuming in the past, which prohibited optimization. The computerized hardening control system offers improved calculation possibilities. Some results for the mentioned case are discussed hereafter.

28. The presented calculations are based on the following conditions:
- wooden formwork, removed after 4 days.
- ambient temperature $15 \pm 3°C$, wind velocity 2 m/sec.
- initial mix temperature $15°C$.

The mix used is about the same as in the first example. The calculated developments of (average) temperature, tensile strength and Young's modulus for roof and wall (cooled and uncooled part) are input to the stress calculations. As first approximation no temperature rise is allowed at the construction joint. To determine the influence of the cooling height on the tensile stress level this parameter was varied. Temperature and mechanical properties are assumed to vary linearly over the cooled part of the wall. To account for relaxation the Young's modulus is reduced to 50% during the first 2 days and 75% during the following 3 days; this reduction is based on test results.

Figure 9 gives the calculated stress development and the resulting stress distribution after 25 days in the wall for a cooling height of 0 m (which means no cooling) and 6 m.

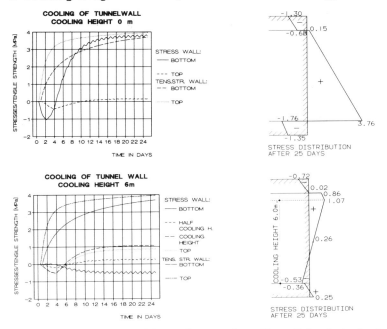

Figure 9. Cooling of tunnel wall. Calculated stresses for cooling height of 0 and 6 m.

The resulting maximum tensile stress for intermediate values of the cooling height can be read from table 1.

Table 1. Maximum tensile stress vs cooling height

Cooling height (m)	Maximum stress (MPa)
0	3.8
1	3.5
2	3.1
3	2.6
4	2.1
5	1.6
6	1.0

To minimize the risk of cracks the allowable tensile strength was taken as 1.0 MPa in this case, which resulted in a cooling height of 6.0 m.

29. Because it is difficult to limit the temperature rise at the construction joint to zero an additional investigation

DESIGN

was carried out, figure 10. Limitation of the temperature rise to zero gives lowest stress at the construction joint, but somewhat higher stress in the upper part. A temperature rise of about 5°C at the construction joint gives the best overall results for this case; this value was aimed at in practice.

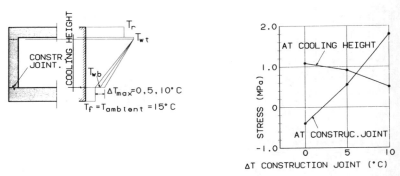

Figure 10. Cooling of tunnel wall. Maximum tensile stress vs temperature rise at construction joint (cooling height 6 m).

Parameter study for concrete tunnel

30. An extensive parameter study for concrete tunnels has been carried out with the system (ref. 3). The main purpose was to determine the cost effects of the numerous variables. The considered cross-section and scheme for calculations are given in figure 11. The length was taken as 800 m.

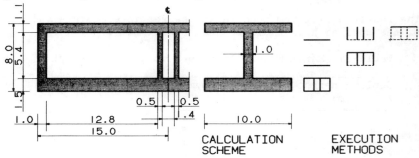

Figure 11. Parameter study concrete tunnels. Considered section.

The investigated variables included:
 (a) casting sequence, figure 11
   - section cast in 3 separate parts: roof after wall after base
   - section cast in 2 parts: roof and wall after slab
   - complete section cast in one part

246

(b) length of section to be cast: 7.5, 10, 12.5 and 15 m
(c) cycle time: 2, 3, 4, 5 and 10 days
(d) type of concrete mix
(e) type of formwork: steel, wood
(f) additional insulation
(g) cooling and preheating
(h) ambient conditions: average, summer, winter.

31. A matrix of feasible combinations of items (a), (b) and (c) of above list was determined and for these combinations both technological and cost calculations, as explained earlier, were carried out. The technological criteria were sufficient strength at desired striking time, stresses below allowable values and limited temperature gradients. The cost calculations covered all relevant cost items: concrete (including cooling, preheating), reinforcing steel, formwork, insulation and joints. The learning effect was incorporated dependent on the number of repetitions.

32. Some relative cost results are presented in figures 12 and 13.
Figure 12 shows the influence of the section length on the total costs and on the separate cost items for the execution method "wall and roof after base". The smaller section lengths (7.5 m à 10 m) provide the cheapest solution. The increased costs of the joints are more than compensated by the savings resulting from less formwork and from the learning effect.

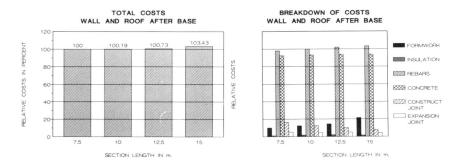

Figure 12. Parameter study concrete tunnels. Influence of section length on cost-level.

The influence of the execution method can be read from figure 13. For the investigated conditions the execution method "wall and roof after base" turns out to be the best solution. Casting in 3 parts is more expensive due to the necessary cooling in the roof (included in the concrete costs). Higher formwork and insulation costs increase the costs for casting of the full section.

DESIGN

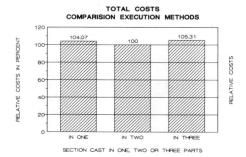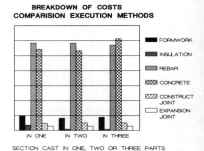

Figure 13. Parameter study concrete tunnels. Influence of execution method on cost-level.

33. Some general conclusions from the study are:
- it is attractive to work with small section lengths
- the learning effect has a relative large influence and should be established well
- the use of insulation to influence the hardening process is often more cost effective than cooling or preheating.

CONCLUSION
34. The computerized concrete hardening control system has been in use for two years and has proven its value in the design, work preparation, execution and quality assurance of concrete works, including tunnels.
At present emphasis is laid on the further development and integration of the cost calculations and the data bases.

REFERENCES
1. VOS Ch.J. Developments and quality safeguards in immersed tunnel technique. Proceedings of Delta Tunnelling Symposium II, Amsterdam, 1987.
2. VAN BREUGEL K. Artificial cooling of hardening concrete. Delft University of Technology, Delft, 1980, report 5-80-9.
3. BERLAGE A.C.J. Verhardingsbeheersing in de tunnelbouw. (Hardening control in tunnel construction). Graduation thesis civil engineering, Delft University of Technology, Delft, 1987.
4. MAATJES E. and BERLAGE A.C.J. Ontwikkeling en toepassing van een geautomatiseerd verhardingsbeheerssysteem. (Development and application of a computerized hardening control system). To be published in Cement, 1989.

20. Electrical and mechanical aspects relating to the civil design of immersed tube tunnels

J. F. L. LOWNDES, BSc(Eng), and C. R. WEEKS, BSc, Mott Hay and Anderson Ltd, Croydon, UK

SYNOPSIS. In this paper, the electrical and mechanical services provided within modern tunnels are reviewed. Particular reference is made to an immersed tube road tunnel in North Wales, an immersed tube cable tunnel in Singapore and, for comparison, a bored road tunnel in Cornwall. The electrical and mechanical requirements generally common to both immersed tube and bored tunnels are outlined and differences are noted. The particular requirements of each of the immersed tunnels are treated in more detail and the implications for the designers and contractors are discussed.

INTRODUCTION
Conwy Road Tunnel, North Wales
1. This tunnel is of immersed tube design comprising a twin bore two lane highway forming part of the A55 North Wales Trunk Road to Holyhead in Anglesey. The tunnel crosses under the River Conwy at the town of that name. The tunnel is under construction and is due to be completed in November 1990. The Welsh Office is sponsoring this tunnel, together with two other land tunnels (both bored) as part of the upgrading of the A55 trunk road.
2. The tunnel is of reinforced concrete construction, with six immersed elements, each approximately 120 metres in length. At each end of the immersed section a length of cut and cover tunnel is being constructed. The total length of the tunnel, portal to portal, will be 1.09 km. A typical cross section of the tunnel is shown in Fig. 1.

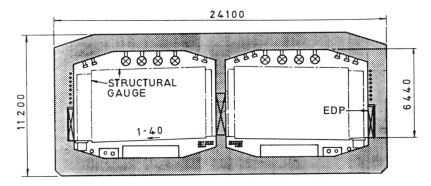

Fig. 1. Typical cross-section of Conwy Tunnel

Immersed tunnel techniques. Thomas Telford, London, 1989.

DESIGN

Pulau Seraya Cable Tunnel, Singapore

3. The Public Utilities Board of Singapore (PUB) have constructed a new power station on the island of Pulau Seraya. As part of the project, but under separate contract, a cable tunnel was built to carry a total of seven 230 kV, 500 MVA oil filled power transmission circuits from Pulau Seraya to Singapore's mainland. The 2.6 km long tunnel, constructed by the immersed tube method, is 6.5 m wide, 3.7 m high and is divided longitudinally into two bores of slightly unequal width, see Fig. 2.

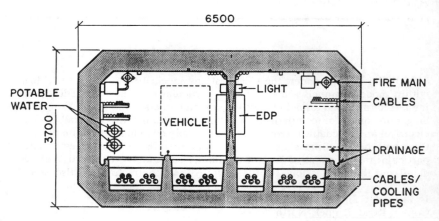

Fig. 2. Typical cross-section of Pulau Seraya cable tunnel

The larger bore also contains two 200 mm potable water pipes and space for battery driven, rail guided maintenance vehicles, see Fig. 3.

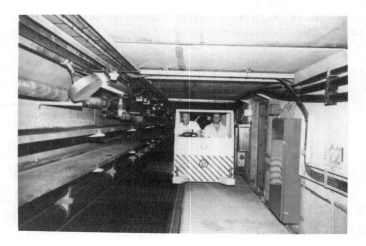

Fig. 3. Interior view of larger bore, Pulau Seraya cable tunnel

The power cables were laid directly into the invert of the tunnel in troughs, with two 90 mm cooling water pipes per circuit laid immediately on the cables. The whole was protected by cement stabilised thermal sand. The majority of the tunnel invert is approximately 25 m below sea level. Terminal buildings at each end of the tunnel contain control, cooling and fire protection equipment, cable jointing basements and oil pressure tank rooms.

4. The original concept for the cable crossing was the laying of six power circuits buried directly in trenches in the seabed. The tenders proved to be considerably greater than the original PUB cost estimates, mainly because the submarine cables themselves are considerably more expensive than their land-laid counterparts and it was a requirement of the Port of Singapore Authority that they be laid in deep trenches in the seabed to allow for possible future dredging. The necessity to lay each circuit in a separate trench for future maintenance and as a precaution against damage by ships' anchors added significantly to the cost.

5. Following receipt of these tenders, the PUB appointed Mott, Hay and Anderson as consultants for the feasibility study and, subsequently, the design and construction supervision of a tunnel alternative. This radical change resulted in a very tight programme.

6. The geology of the area ranged from fresh sandstone and mudstone, outcropping at seabed level, with zones of deep weathering towards the centre of the crossing and soft marine clay close inshore.

7. At feasibility stage it was concluded that construction of a bored tunnel was unlikely to be achieved within the available time. The principal factors being the variability of the geology and the difficulty of the E&M equipping of the tunnel, the majority of which could only take place in a bored tunnel after completion of the whole of the civil works.

8. An immersed tunnel was chosen because it was considered more likely to overcome the difficulties mentioned above.

9. The tunnel consisted of 26 no. 100 m long reinforced concrete elements. Each element was made up from 29 no. precast concrete segments of 3.5 m length which were joined together by grouted joints and post-tensioned longitudinally. The relatively small cross-section of the tunnel enabled the precast segments to be cast in the vertical position without any construction joints between floor, walls and roof, see Fig. 4.

Fig. 4. Vertical casting of tunnel segments, Pulau Seraya cable tunnel

DESIGN

10. Assembly for each element was carried out on either one of two marine lifts which were constructed from the foreshore adjacent to the casting yard, see Figs. 5-6. The marine lift platforms were 120 m long x 10 m wide. They were suspended by steel cables from 12 no. heavy lift hydraulic jacks, each having a capacity of approximately 300 tonnes.

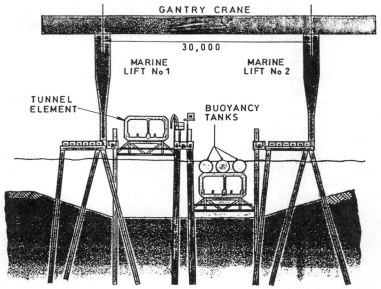

Fig. 5. Marine lifts and gantry crane, Pulau Seraya cable tunnel

Fig. 6. Immersing tunnel element on marine lift, Pulau Seraya cable tunnel

11. The tunnel elements were virtually complete in respect of the civil engineering prior to float out, and thus had negative buoyancy. During float out they were supported by steel buoyancy tanks. For the sinking operation the buoyancy tanks were water ballasted and the whole suspended from two cranes mounted on a lay barge.

Saltash Road Tunnel, Cornwall

12. This is a bored tunnel some 400 m in length, excavated in rock. The rock was supported initially with rock bolts and shotcrete or with steel arch ribs and shotcrete, depending upon the depth of rock cover and the quality of the rock. The final lining is unreinforced insitu concrete. At either end of the tunnel, the portal section was constructed in reinforced concrete using the cut and cover method. Recesses for emergency and electrical equipment have been provided at intervals along either side of the tunnel. The tunnel provides an intown bypass for the busy A38 trunk road in Cornwall. The tunnel carries three lanes of road traffic without any restriction on content. Traffic levels at peak time average 1300 vehicles per hour in each direction. Fig. 7 shows a typical cross section of this high arch profile tunnel.

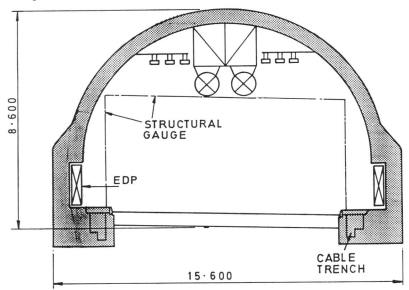

Fig. 7. Typical cross-section of Saltash Tunnel

TUNNEL SERVICES

13. The tunnel services common to the majority of modern tunnels comprise the following:-

            ventilation   lighting
            drainage   communications
            fire protection

In addition, the Pulau Seraya cable tunnel required special cooling equipment.

Ventilation

14. Normally forced ventilation is required for two main reasons:
   a) to maintain a good physiological state so that normal activities can take place within the tunnel.
   b) to remove smoke and fumes from the tunnel bore in event of a fire.

15. Natural ventilation may be possible in the case of a relatively short bored tunnel, such as at Saltash, but in the case of Saltash two-way heavy

DESIGN

Table 1. Air Velocity Required

Mode	Conwy	Pulau Seraya	Saltash
1) Normal operation	2.5 m/s	1.0 m/s	1.0 m/s
2) Maximum heat dispersal	N/A	2.5 m/s	N/A
3) Smoke & combustion gas dispersal	4–6 m/s	5 m/s	4 m/s
4) As 3) but in reverse direction	4–6 m/s	4.5 m/s	4 m/s

Table 2. Fan Requirements

	Conwy	Pulau Seraya	Saltash
Method of ventilation	Jet fans	2-stage axial fans	Large diameter jet fans
Number of fans for normal operation	18 per bore reversible	1 per bore 1-stage low speed	3 fans
Number of fans for maximum heat dispersal	N/A	2 per bore 2-stage low speed	N/A
Number of fans for smoke and combustion gas dispersal	36 per bore reversible	2 per bore 2-stage high speed reversible	6 fans reversible

Table 3. Drainage Requirements

	Conwy	Pulau Seraya	Saltash
Profile of the tunnel	U shape	U shape	down gradient westwards
Is water ingress into the bore an important factor	No	No	No
Disposal of fire water	Yes	Yes	Yes
Disposal of chemicals or hydrocarbon spillage	Yes	No	Yes
Disposal of washing down water	Yes	No	Yes
Are tunnel sumps and pumps required	Yes	Yes	No
Are the pumps external or internal to the tunnel	Internal	Internal	None

traffic generating large quantities of CO, NO and NO_2 necessitated the use of forced ventilation. An added complication exists in that the system has to respond to winds from various quarters which, under certain circumstances, requires the reversal of the direction of forced ventilation. Forced ventilation is normally essential for all but the very shortest sub-aqueous tunnels.

16. Two methods of ventilation are commonly used for vehicular tunnels:
a) longitudinal system using multiple jet fans suspended from the roof, or in the upper corners of the tunnel bore, or
b) semi-transverse system using axial fans located in ventilation buildings

17. Conwy and Saltash tunnels both use the multiple jet fan system. Pulau Seraya cable tunnel uses a longitudinal system with axial fans located in one of the terminal buildings.

18. The number of fans, their size and method of operation is dependent on the environmental requirements in each tunnel. Table 1 gives the parameters called for in each case.

19. The method of meeting the criteria in each case depends on the ventilation design philosophy, which includes producing the optimum solution at the most economic cost. Table 2 gives the fan requirements provided to meet the criteria.

Drainage

20. Drainage requirements in each tunnel vary according to a number of important considerations. These can be listed as follows :
    profile of the tunnel   ingress of water into the bore
    disposal of fire water   disposal of chemical or hydrocarbon spillage
    water table and need for dewatering in the surrounding area(s)
    disposal of washing down water

The above points are listed in Table 3 for comparison.

Fire detection & control

21. In the road tunnels fire mains are provided and have standard fire hydrant outlets connected. Additional facilities are provided in the form of long 25 mm dia hose reels and hand held extinguishers. This enables fire fighting to commence in advance of the arrival of the fire brigade.

22. In the case of the cable tunnel the situation is rather different to the road tunnels. The cable tunnel is normally unmanned, except for infrequent visits by maintenance staff, thus other means of fire fighting is required. This is provided by means of medium density foam generation, actuated by a fire detection system. Extinguishing by means of halon gas release is used in certain areas of the terminal buildings but is not practicable for the long length of tunnel.

23. In some countries fire detection in a road tunnel is assisted by means of ionised smoke and rate of rise heat detectors. This method is not favoured in the United Kingdom. The provision of telephones at frequent intervals enabling the public to call for the fire brigade is considered to be the better means of fire detection . In the case of an unmanned tunnel this has to be done remotely, and in the case of the cable tunnel this has to be performed by ionised smoke and rate of rise heat detectors. The cable tunnel bores are divided into 100 metre long detection zones, corresponding to foam flooding zones.

DESIGN

Lighting

24. Adequate lighting is essential in both road and cable tunnels, each at a suitable level to allow normal activity both by day and by night. The lighting level and type of activity are very different but the basic requirements are no different in that a reliable lighting system is provided to meet all circumstances.

25. The uni-directional road tunnels have additional lighting provided at the entry and frequently the exit from the bore. Where bi-directional working is normal, as at Saltash, then full additional lighting is provided at both portals. This is graded according to the external lighting levels to allow the driver's eyes to adjust to the internal lighting without the driver having to reduce his speed. This 'boost' lighting is adjusted automatically in accordance with output from photometric devices located on the approach to the tunnel. The range of adjustment varies from responding to a bright sunny day through to darkness with a minimum of surrounding illumination. This 'boost' lighting is not required in the cable tunnel as neither contrast or speed of movement are factors.

26. In addition to providing sufficient light levels to allow normal activity, it is also necessary to cater for possible failure of electricity supply. This failure can take two basic forms :
a) loss of main supply to the tunnel
b) loss of local supply to the luminaires

27. The common design approach today is to supply adjacent luminaires, or lamps within a luminaire, from different electrical supplies. Normally, the electrical supply to an area within the tunnel is at 415/250 V, 3 phase, each lamp being connected across a different phase, thus the failure of one phase fuse only extinguishes one lamp in three. Further security is normally obtained by providing two separately derived supplies to an area, thus the failure of one phase on one supply extinguishes only one in six.

28. To cover the situation where there is a total failure of supply to the tunnel, it is now the policy to provide a limited battery fed supply. This can be either derived from within the luminaire or from a source external to the tunnel such as a bank of batteries housed in a terminal building. Normal practice now calls for one lamp in ten to be battery fed. Thus it will be noted from the above that the likelihood of a person, whether as a driver or a maintenance man, being suddenly plunged into total darkness is minimal. This approach has been adopted in both the road tunnels and the cable tunnel.

Communications

29. In any remotely controlled installation communication to the outside world is essential, the most common means being telephones located at frequent intervals along the tunnel bore. Recently more sophisticated methods are being employed, such as leaky feeders or VHF aerials. In some road tunnels the BBC are installing aerial systems which enable the drivers to receive instructions and information via their car radios tuned to popular radio stations. Further, the emergency services are being provided with radio link facilities within the tunnels. The benefits to maintenance staff of radio links are obvious.

Emergency Distribution Panels (EDP's)

30. These various items of emergency and communications equipment can be conveniently grouped and contained within an EDP The content of an

EDP may typically be as follows :
fire hose reel
fire hydrant
fire extinguisher
emergency telephone

local electrical switchgear (fuses) for lighting and fans
communications minor concentration point
traffic control electronics

31. The spacing between each EDP will vary according to the local requirements, but generally is governed by the maximum length of hand held fire hose available, i.e. 25 to 45 metres. Where hand held hoses are not provided, then optimum spacing of electrical distribution points becomes a significant consideration. In the cable tunnel emergency doors between bores were provided conveniently at 100 m intervals with EDPs adjacent to the doors.

32. With the amount of equipment concentrated within an EDP, the enclosure dimensions can be in the order of 2m high x 4m wide x 0.75 m deep although for the cable tunnel the EDP was rather smaller. They are of steel construction, frequently painted red in the past, although a trend towards unpainted stainless steel is taking place as the latter material is much more resistant to aggressive tunnel atmospheres.

Methods of cooling cable tunnels

33. In the earliest purpose built cable tunnels the cables were installed in water filled troughs. The flow of water in the troughs was controlled by weirs in order to maintain a suitable depth and flow rate. This method has the major disadvantage of introducing dampness and humidity to the tunnel and gives rise to major maintenance problems.

34. Where the load carried and the consequent heat losses are not great, cable tunnels are cooled with air passed along the tunnels from ventilation stations at each end of the tunnel. The limiting factor here is the air velocity and temperature rise which can be tolerated in the tunnels.

35. Cables may also be laid in sand filled troughs in the floor of the tunnel. By this means some of the heat is conducted to the surrounding ground and some is convected from the tunnel floor where it can be removed by an air flow. The load carried again limits this method.

36. Where a very high load is carried and losses consequently rise, chilled water pipes are laid in the sand adjacent to the buried cables. This method has a high installation and running cost but can be designed to deal with any load. As described above, this is the method chosen for the Pulau Seraya tunnel.

Cable ducts

37. The tunnel services are fed via cables from the service building. The method of containing these cables varies according to the design requirements of the civil structure. Their location within the tunnel can also varies according to the method of tunnel construction. The two common methods for road tunnels being :
 a) within a cable trench where the cables are clipped to the sides.
 b) in cable ducts which are cast into the structure, which require cable draw pits to be provided every 50 to 100 metres for access purposes.
Cable trays and wall mounted cables were used for the Pulau Seraya cable tunnel.

Luminaire suspension

38. Common practice today is to suspend the luminaries from a grid of

DESIGN

unistrut or a similar multi purpose system. This allows greater flexibility and is mainly independent of any discontinuities in the civil structures. The cables feeding the lighting are contained within steel conduits which are also fixed to the unistrut. The unistrut itself is connected to the tunnel structure by either cast-in fixings or by subsequently drilled or bolted mountings.

Stray current protection

39. Cathodic protection against stray electrical currents may be required in the immersed design as a protection to the reinforcing steel and/or to the steel waterproofing membrane. This would not normally be required in a bored tunnel design.

RELATIONSHIP BETWEEN E&M AND CIVIL DESIGNS

40. Having completed the survey of tunnel services and equipment, it is now appropriate to consider the differences in tunnel construction and the inter-relationships of the E&M and civil designs.

Space requirements

41. Space allocation for E&M equipment in tunnels always tends to be at a premium. It is uneconomic to provide an enlarged structure to accommodate over-generous E&M equipment designs. This is especially true for an immersed tunnel design.

42. In a bored tunnel, having a circular or partially circular cross-section, there is normally a suitable space at roof level to accommodate a suspension grid for lighting, ventilation fans and certain other equipment. Depending on the terrain, a bored tunnel may be unlined rock, in situ concrete lined or of segmental ring construction. It is generally a relatively straightforward matter to provide recesses, sumps or cross passages to accommodate various items of the E&M services.

43. In contrast to the bored tunnel, the provision of recesses, sumps and other special features creating a non-standard cross-section will have a significant impact on the civil designer in respect of buoyancy and structural considerations. The overall height of an immersed tunnel structure needs to be kept to a minimum for an economic design. Any increase in height carries a major cost penalty.

44. The location of cable routes is a major concern to the E&M designer, in a bored tunnel, these are usually contained within the walkway. In the case of an immersed design they may have to be mounted on or buried in the side walls, which creates particular problems for cables entering the EDP's. In the United Kingdom the mounting of exposed cables on side walls of road tunnels is not acceptable. At Conwy, the E&M designer has had to accept some cables in wall ducts. The E&M designer's preference, based on easy access and maximum security, for the below ground option may have to be overruled because of the additional civil costs.

45. The civil engineer in his design may prefer to bury conduits in the wall between the wall mounted equipment and roof mounted equipment. This can create early design problems as decisions have to be made on conduit sizes long before E&M detailed design would normally have commenced. Any subsequent changes can cause significant problems to both parties.

46. A similar problem arises with the necessity of early design of pipe routes, including fire water mains, particularly where these are, for preference, to be cast into the structure of the section. Additional

difficulties arise in designing for couplings and expansion joints for each pipe section.

Standardisation by the Civil Engineer

47. In the immersed tube tunnel the standardisation of sections minimises civil costs. This can have penalties on the E&M side, where for example, multiples of small drainage sumps may require more pumps than optimum to perform the duty. These have to be squeezed into very restricted areas. The total pumping system design has to be completed very early on in the design period so that sump sizing can be completed by the civils. One advantage of the immersed tunnel design is the likelihood of minimal leakage of water into the bore, as compared with bored/segmental ring design.

48. In the case of Conwy, the mid river sumps, one for each bore, have been accommodated within the structure with need for enlargement. Each sump has dimensions of 12000mm x 4000mm x 700mm, with a trough at each end of some 400 mm deep to house the two submersible pumps. No other sumps have been provided within the tunnel, all other sumps are located external for storm water collection and spillage storage.

49. In the case of Pulau Seraya there was no ballast concrete in the tunnel invert and no space within the structural concrete to accommodate the sump. It was necessary, therefore, to modify the tunnel cross-section at the sump location by providing a blister underneath the cable troughs, see Fig. 8. This blister was kept to a minimum possible size. This was assisted by the use of non-submersible positive displacement pumps mounted on the centre wall above the sump. Nevertheless the non-standard cross-section had to be carefully considered by the civil designers from the outset to ensure that its effect was catered for structurally and in respect of the buoyancy of the unit, ballasting of the buoyancy tanks and, not least, that it could be accommodated on the marine lift used for launching the elements.

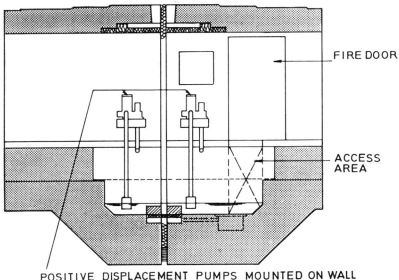

Fig. 8. Longitudinal section at sump, Pulau Seraya cable tunnel

DESIGN

50. The optimum construction length of each immersed tunnel element, from the civil aspect, may not correspond conveniently to a multiple of the optimum spacing of the EDP's. Adjustment of EDP spacing for standardisation of the civil construction would result in increased E&M costs. This needs to be balanced against the saving in civil costs.

INSTALLATION OF TUNNEL SERVICES
Initial fixing

51. This term is used to cover the initial E&M works where non cast-in equipment mountings, unistrut mountings and primary grids are installed and conduits or trunkings are installed ready for wiring. This gives the E&M contractor an early start close behind the civil contractor and before the need to install sensitive equipment. This approach would be used in most immersed tunnels and is common in bored tunnels, once the main excavation and tunnel lining has been completed. Subject to the method of sinking the immersed elements, it could be possible for the E&M contractor to be working beside the civil contractor on dry land. If the intention is to sink the elements with both ends sealed and the interior maintained substantially dry, much of the initial fixing could be done before sinking. A minor problem arises at the joints, where the E&M contractor has to prepare his installation such that closures can easily made between the two elements after sinking.

Final installation

52. Whether the tunnel is bored or immersed, as a general rule cabling and equipment installation cannot satisfactory take place until the tunnel is complete as far as civil engineering is concerned. The partial installation of cabling and equipment within an immersed tunnel element prior to floating out and sinking the element can leave the E&M equipment more vulnerable to accidental damage by water than it would be in its fully installed state. Leaks in the tunnel structure or temporary bulkheads, and in temporary water ballasting equipment, if used, are possible. Furthermore, in road tunnels, there is normally significant civil works in the form of ballast concrete and road formation to be completed after sinking and joining the immersed tunnel elements.

53. In the case of the Pulau Seraya cable tunnel the situation was rather different. As described above, the design and construction programme was exceedingly tight and the tunnel services provision was included under the main civil contract. The ability to briefly dip the tunnel elements beneath the water, prior to floating out and sinking, produced a high degree of confidence that leak damage would not be a problem. Ballast water was introduced into the separate buoyancy tanks, never into the tunnel itself. The method of construction of the tunnel elements, using precast segments, allowed a useful time 'window' for the internal installation of the E&M equipment whilst external civil works were being completed on the marine lift. This enabled the contractor to mount the majority of his E&M equipment on a modular basis within the element prior to sinking. A notable exception was the installation of main tunnel services feeder cables which did not take place until a number of elements had been sunk and joined, since these main cables ideally extend for several elements without cable jointing.

54. The great benefit of this approach at Pulau Seraya was that it enabled the contractor for the main power transmission cables to commence pulling these cables through the full length of the tunnel a mere three weeks after the sinking of the last tunnel element.

DESIGN PROGRAMME

55. To accommodate the civil design requirements it is necessary for the E&M design to be agreed by all parties and finalised much earlier, perhaps up to 2 years ahead of the normal design plan for an equivalent bored tunnel. Having committed himself, the E&M designer will have considerable difficulty in materially changing any of his space requirements once the civil design is well advanced. This cut off occurs well ahead of design finalization and contract award. By contrast, in the bored tunnel design, variations in requirements can within limits be accommodated even during construction, although there may be a cost penalty.

CONCLUSION

56. In conclusion it must be said that the immersed tunnel design concentrates the mind of the E&M designer more fully and much earlier than any other design. The immersed tunnel concept is very much less able to accept changes in E&M requirements, which are often requested by the client at a late design stage and even after construction has commenced. It is necessary for the client to accept that any change of requirements may entail major cost penalties and delay in completion. These constraints do not arise to the same extent in the bored tunnel design, although any changes after contract award would normally still involve the client in additional expense.

ACKNOWLEDGMENTS

The Authors would like to thank the following: Directors of Mott, Hay and Anderson; Department of Transport; Welsh Office; Travers Morgan and Partners and the Public Utilities Board, Singapore for permission to publish the various parts of this paper, and R.D. Tomlin of Canterbury Consultancy Services for his assistance in its preparation.

21. Impact of the development of tunnel service systems on the planning and design of the immersed tunnel for the Sydney Harbour crossing

A. G. BENDELIUS, PE, Parsons Brinckerhoff Quade and Douglas, Inc., Atlanta, and K. R. MEEKS, Parsons Brinckerhoff International

SYNOPSIS. This paper outlines the development of tunnel service systems including ventilation, lighting, power, traffic surveillance, drainage and fire protection and the supporting necessary facilities to provide sufficient dimensional information to the immersed tunnel (IMT) designer in a timely fashion to meet the design and construction programme for the IMT of the Sydney Harbour Crossing.

INTRODUCTION
1. The Sydney Harbour Tunnel is a proposed subaqueous vehicle crossing of Sydney Harbour, the first new crossing since the Sydney Harbour Bridge was built in 1932. The south portal of the Sydney Harbour Tunnel, which crosses roughly parallel to the Sydney Harbour Bridge, will be less than 500 metres from another Sydney landmark, the Sydney Opera House (see Figure 1).

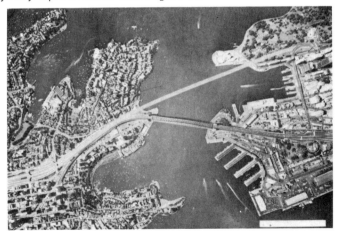

Figure 1 Sydney Harbour

2. In its 2,300 metre length, the proposed crossing incorporates three different tunnel construction methods (IMT, driven tunnels and cut-and-cover tunnels) which result in the use of five different tunnel cross sections and air duct configurations (see Figure 2). It is the tunnel services within the IMT portion, 1,000 metres in length, which will be addressed in this paper.

Immersed tunnel techniques. Thomas Telford, London, 1989.

DESIGN

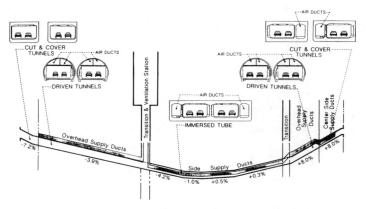

Figure 2 Tunnel Cross Sections

3. The tunnel services, including ventilation, lighting, power, traffic surveillance and control, drainage and fire protection, are key elements in the design of any vehicular tunnel and have a particularly significant impact on the civil and structural design of IMTs. The soundness of these designs will ensure the safe and efficient operation of the tunnel. The services considered in the design of the Sydney Harbour Tunnel are described in the following section.

TUNNEL SERVICE SYSTEMS
Ventilation System
4. The modified semi-transverse ventilation system was devised for the Sydney Harbour Tunnel to meet the criteria established for this tunnel as shown in Table 1. The system has the capacity to limit

Table 1. Design Criteria & Data

Ventilation Criteria
Tunnel Length:	Northbound	2292	metres
	Southbound	2302	metres
Number of Lanes per Carriageway:		2	
Maximum CO Concentration:			
Design Condition		125	ppm
Congested Traffic Condition		150	ppm
Avg. Exposure Limit (15 min.)		100	ppm
Design NO_X Concentration:		5	ppm maximum
Design Particulates Concentration:		1.5	mg/m^3 maximum

Traffic Data

Traffic Vehicle Mix	% Total	% Diesel
Trucks	2.5	2.5
Buses	1.3	1.3
Utilities and Panel Vans	5.8	3.0
Remainder	90.4	1.0

Design Traffic Flow Rates		
Peak Traffic Flow	2000	PCU/hr-lane
Maximum Density (stalled)	140	vehicles/km-lane

adverse atmospheric air contaminants in the vicinity of the portals, control smoke during an emergency, maintain the prescribed level of contaminants within the carriageways and accommodate varying levels of traffic. This ventilation system accomplishes all of this while minimising the visual impact upon the harbour area by locating all ventilation equipment on the north shore either underground or within the existing north pylon structure of the Sydney Harbour Bridge.

5. The underground intake ventilation structure (see Figure 3) housing the required supply ventilation equipment was located on a site situated above the tunnel and adjacent to the north shore of the Harbour. The site is directly across the Harbour from the Sydney Opera House and adjacent to the North Pylon in Bradfield Park. The only surface visual impact will be a series of air intake louvres measuring totally 50m x 8m as seen on Figure 4.

6. The north bridge pylon, located 100 metres from the tunnel, is

Figure 3 Underground Ventilation Structure

Figure 4 Air Intake Louvres

DESIGN

the site selected to house the exhaust ventilation equipment. The use of the pylon allows the vitiated tunnel air to be discharged vertically at a distance of 88 metres above the parkland below. The upward discharge of the vitiated air at the top of the pylon is at a maximum velocity of 20 metres per second. This dispersion of air from the top of the pylon is considered to be sufficient to protect surrounding environs.

7. The potential problem of discharging significant amounts of pollutants through the existing portals was addressed by utilising a single point exhaust system. These single points consist of one large (8m x 9m) exhaust opening at the ceiling level of each carriageway in the vicinity of the Tunnel Ventilation Station (TVS). This system creates longitudinal air flow toward the exhaust opening, thus minimising portal outflow. The contaminated air can then be drawn longitudinally through the carriageway away from the portals toward the exhaust openings and transported to the bridge pylon by exhaust fans.

8. The pylon will be connected to the tunnel by use of an underground exhaust duct tunnel. This duct tunnel includes two 48 m^2 ducts approximately 100 metres in length. The exhaust air will be drawn from the single exhaust opening in each carriageway by varying combinations of the 16 axial flow fans. The 16 exhaust ventilation fans are arranged in groups of eight, each group to serve one of the two carriageways.

9. The supply air system, incorporating 14 axial flow fans, is sub-divided into four zones, with each carriageway section north and south of the TVS being considered a separate ventilation zone. The air ducts in the IMT not only carry the supply air quantity required for that section, but also the supply air required for ventilation of the land tunnels on the south side. The introduction of air to the carriageway within the IMT is achieved by low level supply air outlets, spaced at 7.6 metres, which penetrate the duct sidewalls (see Figure 5).

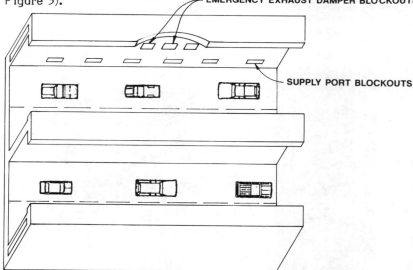

Figure 5 Ventilation Blockouts and Ducts

10. The type of fan selected for this tunnel is the axial flow type. The axial flow fan is suitable to cater to the need to conserve space in the Pylon Ventilation Station (PVS) and the TVS and to match the vertical configuration of the PVS. All fans will be electrically reversible. The supply fans will be equipped with two-speed motors to provide the necessary steps of control for the ventilation system. The fans will have the capacities as shown in Table 2.

Table 2. Installed Ventilation System

System	Zone	Number of Fans in Zone	Total Zone Air Flow (m3/s)	Single Fan Air Flow (m3/s)	Fan Total Pressure (Pa)
Supply	NB-N	3	192	64	1,190
Supply	NB-S	4	220	55	840
Supply	SB-N	3	159	53	1,000
Supply	SB-S	4	352	88	1,370
Exhaust	NB	8	752	94	1,640
Exhaust	SB	8	784	98	1,770

11. This system provides flexibility for effective smoke control in the event of a fire. This is achieved by inclusion of the four ventilation zones, the selective reversible capability of both the supply air intake and exhaust air fans, and the enhanced exhaust capability through the installation of large emergency exhaust dampers connected to the supply air duct. The emergency exhaust dampers are normally kept closed and can be selectively opened remotely in the event of a fire.

12. The primary exhaust ventilation system can generate the longitudinal air velocity required to control smoke and heated air by removing or supplying air via the single exhaust openings located within each carriageway. The supply ventilation system, when operated in reverse mode, can provide enhancement of the primary exhaust system by exhausting smoke locally through remotely-activated emergency exhaust dampers located along the length of each carriageway. This system not only offers the required protection to the stopped motorists but also acts to limit the length of carriageway potentially exposed to heated air.

Electrical Power System

13. The electrical power system necessary to support the tunnel and its ancilliary facilities includes primary power provided by Sydney County Council (SCC) in the form of three 11 KV lines on the north side and three 11 KV lines for the south. This primary service at 415/240 V, three phase and neutral, 50 Hertz, is then transformed at four locations associated with the tunnel as follows:

Tunnel Ventilation Station	2	1500 KVA Transformer Substations
Pylon Ventilation Station	3	1500 KVA Transformer Substations
North Portal	1	750 KVA Transformer Substation
South Portal	1	750 KVA Transformer Substation

DESIGN

14. There are two high voltage feeders from the SCC Substation at Berry Street to the site. These feed the Pylon Substation and the TVS Substation such that if one main is severed, power supply can be maintained. The South Portal is being supplied from the City network which has a highly reliable supply. The North Portal has a Substation with uninterruptible power supply and generator support in the event of power or transformer failure. The two ventilation station substations can still operate with one transformer each down.

15. The power distribution cabling serving the tunnel is installed in conduit embedded in the walkway of each bore. Electrical distribution boards are installed at approximate 120 metre intervals within the IMT. The sub-circuit power cabling is run from the panels and boards in embedded conduit to the individual items of tunnel equipment.

Tunnel Lighting System

16. Tunnel lighting (at 240 V) will be achieved by a combination of high intensity discharge (high pressure sodium) and fluorescent lamps. The interior zone will be illuminated by one continuous row of fluorescent lamps, ceiling mounted. The threshold and transition zones will be illuminated by both continuous fluorescent luminaries and an array of high pressure sodium (HPS) lamps. The HPS luminaries will be arranged to provide the proper threshold and transition levels of lighting. Since the tunnels will operate in a unidirectional traffic mode only, there will be need for only one set of threshold/transition zones per carriageway.

17. The fluorescent lamps will be the only lighting used during night operations when a reduced level of illumination will be maintained. The lighting of the threshold and transition zones are switched in stages as per the external lumans as sensed by a photometre. The various required stages of lighting can be achieved by selectively switching of the HPS and fluorescent lamps.

Surveillance, Communications and Control Systems

18. Systems for the control and monitoring of the tunnel operations and services are included in the tunnel design and have an impact on the IMT design.

19. The surveillance, communications and control systems include the following systems:

Traffic Loop Detectors	Radio Rebroadcast Antenna
CCTV Cameras	Fire Detectors
Lane Use Signals	Motorist Emergency Telephone
Fixed Message Signs	Strobe Location Lights
Carbon Monoxide Monitors	Tunnel Operations Radio
Air Velocity Monitors	

20. All of the above listed equipment is located within the IMT portion of the Sydney Harbour Tunnel. They are either embedded in pavements (loop detectors), mounted in niches (CO monitors) or surface mounted (signals). The surveillance, communications and control system cabling is installed in conduit embedded in the tunnel walkways.

Tunnel Drainage System

21. The tunnel drainage system consists of a closed pipe system installed to collect and remove flow from rain, vehicle drippinngs, tunnel washing and fire fighting operations. The gravity system includes roadway inlets and flame traps spaced at regular intervals in addition to pump stations. The flow collected at a low-point pump station and two intermediate pump stations (external to the IMT) is monitored for hazaradous vapors before being treated and pumped to the sewer system or the harbour.

Fire Protection System

22. The primary element of the tunnel fire protection system is a 150 mm diameter hydrant main enclosed within the tunnel cross section. A public emergency niche is provided in the wall of the carriageways, housing a hose reel containing 36 metres of 19 mm hose, a hydrant valve (65 mm) and two fire extinguishers. The connections to the hose reels and the hydrant valves are embedded in the tunnel duct walls.

IMPACTS

23. There are two levels of influence of the development of tunnel services on the planning for an IMT design: those services which impact the tunnel alignment, cross section configuration and size; and those which can be accommodated without changes to the actual configuration but must be addressed early in the process to permit the proper design to protect the structural integrity of the IMT unit.

IMT Configuration, Size and Alignment

24. The tunnel ventilation system is the one tunnel service which has had the most significant influence on the overall tunnel design and in particular on the size and configuration of the IMT cross section. The Sydney Harbour IMT, which is the center underwater portion of the Sydney Harbour Crossing (see Figure 2), connects the TVS on the north shore and the transition structure on the south shore. The selection of a ventilation system type had a major impact on the development of the IMT cross section (see Figure 6) and is also influenced by the land available for ventilation structures. Early in the planning stage, numerous ventilation system configurations were considered. For the Sydney Harbour Tunnel, the ventilation systems shown on Table 3 were considered.

25. The preferred arrangement for a full transverse or a semi-transverse ventilation system for a tunnel of this length is two ventilation structures, one located at each end of the IMT portion or at quarter points of the overall tunnel. Such an arrangement permits the ventilation ducts to remain as small as possible, thus keeping the IMT cross section as small as practical. The chart on Figure 7 clearly shows the impact of tunnel length on the IMT cross section.

26. During the concept development phase of the project, numerous potential alignments were evaluated. All of the proposed north portal locations and potential ventilation structure sites were on Milson's Point. On the south shore, all but one alignment portaled in the Dawes Point area. The exception was alignment 4a, which

DESIGN

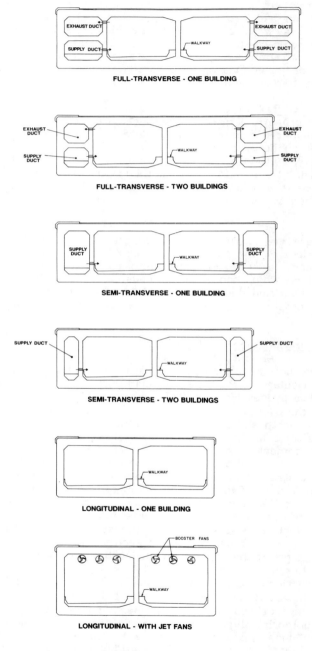

Figure 6 IMT Cross Section Comparison

Table 3. Ventilation System Configurations Considered

	System Type	System Designation	Number of Air Ducts In IMT Cross Section	Number of Ventilation Building Locations
I.	Full Transverse			
	Uniform Supply and	A	2	2
	Uniform Exhaust	B	2	1
II.	Semi-Transverse			
	Uniform Distribution of Supply Air or	A	1	2
	Uniform Collection of Exhaust Air	B	1	1
III.	Longitudinal Supply or Exhaust at Single Point	A	0	1
IV.	Longitudinal w/Jet Fans Uniform Distribution of Jet Fans in Tunnel	A	0*	0

* Requires added space in IMT cross section to accommodate the jet fans.

portaled adjacent to Bennelong Point.

27. Early in the alignment selection process, all tunnel alternatives being considered were configured with two ventilation structures located at approximate quarter points. The only exception to this was alignment 4a which was the easterly most alignment crossing to the south shore adjacent to the Opera House and the Botanical Gardens.

28. While examining the sites for ventilation structures for the 4a alignment, it became evident there were many details regarding the siting of the ventilation structure in North Sydney to be resolved. With the elimination of the possibility of a ventilation structure near the Opera House, an approximate center point ventilation structure site was sought. It was determined that if the tunnel gradient could be lowered in the vicinity of the north shore, an underground ventilation structure was feasible thus preserving the visual environment near Bradfield Park. It was at this point that the use of the bridge pylon was considered. The hollow bridge abutment structure could be modified to house the exhaust ventilation fans.

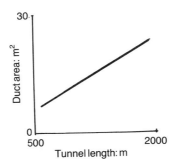

Figure 7 Impact of Length on Immersed Tunnel Cross Section

DESIGN

This permitted protection of the visual environment and the air quality with the discharge of the tunnel exhaust some 88 metres above the shoreline parkland.

29. With the alignment selected and the ventilation structures sited, all efforts focused on establishing a structural dimension for the IMT. Then followed a continuing tradeoff of structural shape adjustments to accommodate tunnel services. This required a continual interplay between the structural, highway, mechanical and electrical tunnel engineers to formulate the basic design.

30. To permit preliminary evaluation of the tunnel ventilation equipment, it was necessary to reach an understanding on basic system criteria for such elements as traffic volume, carbon monoxide (CO) emissions rates, ambient CO levels, allowable CO levels and maximum air velocities (see Table 1). With these criteria in hand, efforts began to evaluate the air flow necessary. Once this analysis was completed and the ventilation system type selected, specific dimensions were developed and transmitted to the tunnel structural engineer.

Accommodation of Tunnel Services in the IMT

31. There are three basic means to accommodate tunnel services within an IMT. The services can be surface mounted, embedded in the structure or contained in a dedicated service duct. Descrete items such as monitors and distribution panels are placed in a previously prepared blockout. The precast nature of an IMT necessitates that all services themselves are installed in their ducts or blockouts after laying and joining of the IMT units.

33. Embedment of tunnel service items provides protection both from physical damage and from high temperature (fire). The use of blockouts for descrete items allows for variation in the size of such items as finally supplied and reduces the possibility of damage/corrosion in the early construction stages. A list of the tunnel services and the method of accommodation in the Sydney Harbour IMT is shown in Table 4.

33. Many of the accommodations are located on a modular basis thus providing the tunnel engineer with a relatively uniform configuration in each IMT unit. Table 5 shows the approximate spacing of each of the accommodated service items in the Sydney Harbour IMT.

34. The CCTV cameras, lane use signals, fixed message signs, fire detection system, air velocity monitors and tunnel lighting are surface mounted within the Sydney Harbour IMT and as such required the following considerations:
    i   Allowance for these items be added to the internal dimensions of the carriageways dictated by the dynamic clearance envelope of vehicles. The allowance for tunnel services at ceiling level of the Sydney Harbour IMT is 200 mm. The profile of the walkway and barrier in each carriageway provides sufficient clearance for tunnel services located on the tunnel walls (walkway width is 975 mm, barrier width is 375 mm). Reducing the depth of tunnel services at ceiling level, whilst reducing civil costs, does increase the cost of tunnel services due to the reduced horizontal spacing required

Table 4. Tunnel Services Accommodated in IMT

Block Out	Imbedded	Surface Mounted
Supply Air Outlets	Gravity Drain Pipes	CCTV Cameras
Emergency Exhaust Dampers	Rising Mains	Lane Use Signals
Electrical Cabinets	Hydrant Mains	Fixed Message Signs
Control Cabinets	Balance Pipes	Fire Detection System
Hydrants/Hose Reels	Loop Detectors	Air Velocity Monitors
CO Monitors	Electrical Conduits	Tunnel Light Fittings
Access Doors	Control Conduits	Strobe Location Lights
Flame Traps		Tunnel Operations Radio
Collection Pits		Radio Rebroadcast Antenna
Visibility Monitors		
Telephones		
Emergency Direction Lights		

for drivers to read smaller height signs and signals.

ii Conduits are being embedded in the structure of the IMT for vertical distribution of cables. The horizontal cabling will be installed in surface mounted conduits. This allows for flexibility in the final design and installation of lighting and traffic control equipment. By this simplification of the construction of the IMT unit, fabrication time is reduced at this critical phase.

35. Electrical and Control Niches. The blockouts for the electrical and control niches are located at walkway level to allow access to equipment housed within each niche. These blockouts are connected via embedded conduits to the main services duct in the walkway and at the top of the blockout terminating at a level below the ceiling for outgoing cables.

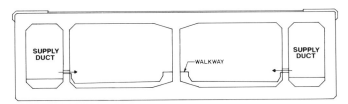

Figure 8 Immersed Tunnel Cross Section

36. Emergency Dampers. The emergency exhaust dampers are to be installed in 2m x 2m blockout openings, grouped in threes in the duct wall. The selection of this opening configuration was arranged after discussions with the tunnel engineer who indicated that the originally proposed 2m x 6m opening was not feasible. This solution allows for the overall total requirement of $12m^2$ at any location and in addition met the needs of the tunnel structural engineer in that specialised treatment of the structure in this area was not required.

37. Supply Air Outlets. The supply air outlets which are used to distribute air from the duct to the carriageway are located at low level in the wall between the duct and carriageway. The nominal size of these blockouts is 950 mm x 650 mm.

38. Hydrant/Hose Reel Niches. The hydrant and hose reels in these niches are connected to a 150 mm diameter wet fire main embedded in the ballast concrete of the air ducts. This protects the service against possible damage in the event of a fire. Blockouts for the niches have been cast, as a hole in the duct wall, with a fire rated steel insulation

DESIGN

sandwich backwall to be installed at the time of tunnel fit out.

39. <u>Midpoint Pump Station</u>. The maximum depth of the tunnel occurs at the joint between the third and fourth IMT unit and therefore dictates that IMT unit #4 accommodate the pipework, traps and sumps that constitute the midpoint pump station. Three submersible pumps are located in the air ducts of both carriageways. Detailed evaluation of the IMT structure established that the thickness of the unit could be reduced to 300 mm locally in the sump areas, to allow hydraulic gradients to be established. The layout of pump station sumps, traps and pipework has been designed after detailed system hydraulic analyses. These elements are incorporated in the structural design of this specific IMT unit. The midpoint pump stations located in the air ducts are connected by a 300 mm diameter balance pipe embedded in the construction of the IMT.

Table 5. Tunnel Services - Location and Spacing

Description	Approx. Spacing (M)	Air Duct NB/SB	Ceiling NB/SB	Centre Wall NB/SB	Duct Wall NB/SB	Roadway Pavement NB/SB
Supply Air Outlets	7.6				x x	
Emergency Exhaust Dampers	60				x x	
Electrical Cabinets	120			x x		
Control Cabinets	120			x x		
Hydrant/Hose Reels	60				x x	
CO Monitors				x x		
Access Doors	120				x x	
Flame Traps	30/120	x x				
Collection Pits	30	x x				
Visibility Monitors				x x		
Telephones	60			x x		
Emergency Direction Lights	30			x x		
Cross Passages	120			x x		
Gravity Draining Pipes	-	x x				x x
Rising Mains	-	x x				
Hydrant Mains	-	x x				
Balance Pipe	-	x x			x x	x x
Loop Detectors	120				-	x x
Mid Point Pump Station		* *				
CCTV Cameras	180		x x		x x	
Lane Use Signals	180		x x			
Fixed Message Signs	70		x x			
Fire Detectors Systm	-		x x			
Velocity Monitors			x x			
Light Fittings	-		x x			
Fire Extinguisher Cabinets	60		x x			
Strobe Location Lights	100			x x		
Radio Rebroadcast Antenna	-		x x			

NB Northbound
SB Southbound
* At tunnel low point

40. <u>IMT Joints</u>. The use of flexible joints at the end of each IMT unit has required that the tunnel service systems which cross this joint be designed to accommodate the working and maximum deflections in both the horizontal and vertical planes. This is particularly critical in the design of the hydrant and rising mains, in that these must not fail under conditions of maximum deflection. This has resulted in the design of a fully articulated service joint to be installed across each of the IMT joints. These have required special detailing of the IMT joint structure to accommodate this provision. In addition, location where IMT joints intersect air ducts, they are sealed with a flexible material to ensure that the ducts remain air tight under all circumstances.

CONCLUSIONS

41. It is extremely important in the development of an IMT design that the tunnel service systems design be developed in sufficient detail to permit the mechanical and electrical engineers to describe dimensionally to the IMT designer the necessary requirements for space within the tunnel units. This would include the need for embedded items such as pipe, conduits, anchors, blockouts for the ultimate addition of equipment such as panels or for sumps as required for the drainage pumps stations within the tunnel. The development of an IMT is such that these decisions must be made at an early stage, one cannot go back in after the tunnel is constructed and begin to cut holes in the structure to provide for tunnel service systems.

42. It may be necessary in the development of an IMT design to prepare a set of tunnel services embedded items drawings. This will depend on the type of materials to be embedded or the number of blockouts required. This must be carefully coordinated with the designer of the IMT so that the dimensional aspects are clearly documented.

ACKNOWLEDGEMENTS

43. The authors kindly acknowledge the strong supportive efforts provided by the Sydney Harbour Tunnel team - the Transfield-Kumagai Joint Venture and Wargon Chapman Partners, the Design Manager. The authors wish to personally thank Alf Neilson, Norm Danziger, Ted West, Dick Naish and Jon Hettinger for their frank and timely reviews of the text of this paper.

44. The tunnel services for the Sydney Harbour Crossing are being designed by GHD-Maunsell-PBI, a joint venture of Gutteridge Haskings & Davey Pty. Ltd., Maunsell Partners, and Parsons Brinckerhoff International.

REFERENCES

1. BICKEL, JOHN O., KUESEL, T. R., Tunnel Engineering Handbook, New York, Van Nostrand Reinhold Company, Inc., 1982.
2. Heating, Ventilating, and Air-Conditioning Systems and Applications, 1987 ASHRAE Handbook, Atlanta, Georgia, American Society of Heating, Refrigerating and Air Conditioning Engineers, Inc., Chapter 29.
3. Permanent International Association of Road Congresses (PIARC), XVII World Road Congress, Technical Committee on Road Tunnels, Sydney, Australia, 8-15 October, 1983.
4. Parsons Brinckerhoff Quade & Douglas, Inc., Subway Environmental Design Handbook, Volume II, Subway Environment Simulation (SES) Computer Program, Version 3, prepared for the Transportation Systems Center of the U.S. Department of Transportation.
5. Permanent International Association of Road Congresses (PIARC), XVIII World Road Congress, Technical Committee on Road Tunnels-Report No. 5, Brussels, Belguim, 13-19 September, 1987.
6. BENDELIUS, ARTHUR G., HETTINGER, JON C., Ventilation of the Sydney Harbour Tunnel, 6th International Symposium on the Aerodynamics and Ventilation of Vehicle Tunnels, BHRA, Durham, 26-29 September, 1988.

DESIGN

7. SAITO, N., NEILSON A.M., The Sydney Harbour Tunnel - The Influence of Local Conditions on the Route and on the Marine and Land Tunnel Types, Conference on Immersed Tunnel Techniques, Manchaster, 11-12 April 1989.
8. MORTON, D.G., MORRIS, M.W., The Sydney Harbour Tunnel Evaluation of Form of Construction for an Immersed Tunnel. Continuous on Immersed Tunnel Techniques, Manchester, 11-13, April 1989.
9. WEST, E.M., Selecting and Designing a Ventilation System for the Sydney Harbour Tunnel, AIRAH Journal, August 1988.

In addition to the specific reference documents highlighted above, there were numerous working papers prepared in the course of design, which were used in the preparation of this paper.

22. The Conwy tunnel — detailed design

P. A. STONE, FICE, and R. C. LUNNISS, BSc(Eng), MSc, Travers Morgan, and S. J. SHAH, BSc(Eng), formerly Christiani and Nielson A/S, Copenhagen

SYNOPSIS. For the Conwy Tunnel detailed designs were developed for two different forms of immersed tunnel construction: a rectangular reinforced concrete element incorporating an external waterproofing membrane and a rectangular steel/concrete composite element where an external steel shell acts compositely with an internal reinforced concrete lining. This paper describes both designs together with the designs for the cut and cover sections and the approaches. The western approach structure is unusual and incorporates an impermeable geomembrane to prevent groundwater ingress. The paper also gives details of the corrosion protection system installed and operational aspects of the tunnel.

INTRODUCTION
1. Conwy is an ancient walled town situated on the estuary of the river Conwy in North Wales. The A55 North Wales Coast Road which passes through it carries traffic never envisaged when the narrow streets and archways were built.
2. Following the feasibility studies into various alternative bridge and tunnel routes to bypass Conwy, described in the paper Scheme Development and Advance Works by Davies, Cramp and Kamp Nielsen, the Secretary of State for Wales announced his decision in July 1980 to promote the construction of a tunnel.
3. This paper is concerned with the detailed design of the reinforced concrete immersed tunnel which is currently under construction. The paper also gives details of an alternative steel/concrete immersed tunnel design which was fully detailed in the contract documents and for which tenders were also invited. The two alternatives are referred to in this paper as the Concrete Immersed Tunnel and the Steel Shell Immersed Tunnel.
4. Travers Morgan and Partners were appointed by the Welsh Office to undertake the detailed design and preparation of contract documents for the Conwy Crossing section of the A55 North Wales Coast Road. This 6km long section includes extensive earthworks, 13 bridges and a number of miscellaneous structures as well as the tunnel under the

Immersed tunnel techniques. Thomas Telford, London, 1989.

DESIGN

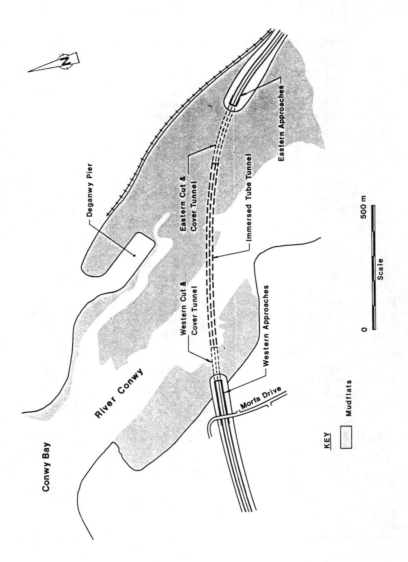

Fig. 1. Plan of tunnel and approaches

estuary and its approaches.

5. Travers Morgan appointed Christiani & Nielsen A/S as specialist sub consultants on the tunnel design. Mott Hay and Anderson were engaged by Travers Morgan as specialist sub-consultants for the detailed design of the electrical and mechanical systems which do not form part of this paper.

SCHEME LAYOUT

6. The general layout of the tunnel and its approaches is shown in Figure 1. The eastern approach to the tunnel is in an open cutting within land reclaimed from the estuary. The tunnel itself which has a total length of 1090m consists of three sections: the East Cut and Cover Tunnel of 260m, the Immersed Tunnel with six elements totalling 710m, and the West Cut and Cover Tunnel of 120m. The western approach is a combination of structural U shaped box and open cutting constructed through the existing saltmarshes.

7. The length of the tunnel and approaches was optimised to achieve a balance between cost and minimising the interference with the hydraulic regime in the estuary. The detailed site investigation revealed a rock pinnacle beneath the Eastern Cut & Cover which together with the thinning of the Irish Sea Boulder Clay however, led to a decision during the detailed design stage to extend the tunnel further eastwards by 60 metres. This made it possible to avoid the need for pressure relief wells in the open earthworks of the Eastern Approach.

Cross Section

8. Traffic flows during August of the design year of 2005 are predicted to reach about 34000 vehicles per 16 hour day (2 way) and dual two lane carriageways are provided throughout the scheme.

9. The tunnel cross section is shown in Figure 2. Each of the two tubes contains two 3.65m traffic lanes with 0.5m wide marginal strips at either side. These marginal strips increase to 1.0m outside the tunnel. The nearside verge, which contains most of the services is a minimum of 1.3m wide and the offside verge a minimum of 0.6m. The wider nearside verge, combined with the marginal strip, enables a stranded vehicle to be passed by two lanes of traffic, albeit at a reduced speed.

10. The vertical headroom is 5.1m over the traffic lanes and marginal strips with an additional 250mm clearance to tunnel equipment mounted on the roof above the carriageway. This additional clearance gives a further safeguard against damage by flapping tarpaulins.

Alignment

11. The tunnel is curved in plan throughout its length. The radius of curvature is 3250m except for a section within the East Cut and Cover Tunnel where it reduces to 875m. This

DESIGN

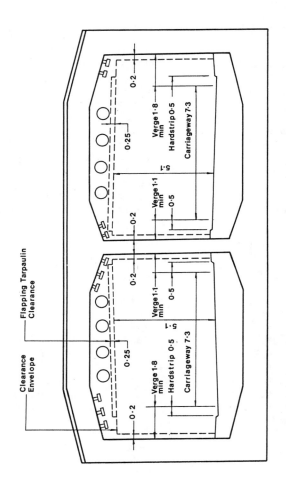

Fig. 2. Tunnel cross-section standards

necessitates superelevating the carriageway in the East Cut and Cover Tunnel.

12. The maximum approach gradients are 4% at the eastern end and 5% at the western end. Within the tunnel these reduce to a general gradient of 0.5% towards the low point which is 502m from the western portal. The original intention had been to have a maximum gradient of 4% on both approaches. However with the gravity type structure of the western approaches the cost saving arising from increasing the gradient to 5% for a short length was so significant that this was adopted. An increase above 5%, although reducing the cost still further would have resulted in an unacceptable loss of service to traffic, particularly HGV's and an increase in background noise levels for local residents. A longitudinal section is shown in Figure 3.

SOILS INVESTIGATION

SI Contracts

13. A preliminary site investigation was carried out in 1979 as part of the Feasibility Study of the three routes under consideration and a comprehensive site investigation was carried out by Soil Mechanics Ltd in 1982. The ground conditions are extremely complex over the whole contract and in all, including several supplementary investigations, over 400 exploratory holes have been bored at a cost of some £2m.

14. The underlying bedrock consists of sandstones and siltstones of the Silurian and Ordovician periods. The area has suffered extensive glaciation from both the North Wales and Irish Sea ice fields. This has led to the deposition of considerable depths of boulder clays and glacial lake clays.

15. On the Eastern Approaches the surface silts are underlain by sands and a considerable thickness of Irish Sea Boulder Clay. Under the Boulder Clay is a thick layer of Glacial Lake Deposits resting on the mudstone bedrock.

16. The bed of the estuary is formed of sand, gravel and cobbles overlying Glacial Lake Deposits. The Irish Sea Boulder Clay is not generally present in the centre of the estuary but the Lake Deposits overlie North Wales Till and bedrock.

17. At the Western end of the tunnel, surface alluvial clays overlie sands and gravels, Irish Sea Boulder Clay, Glacial Lake Deposits, North Wales Till and bedrock. The sands and gravels contain clay lenses and the Irish Sea Boulder clay contains inclusions of sands and gravels.

Pumping Tests

18. The construction of large open excavations in the estuary requires a reduction in the pore water pressures in the Glacial Lake Clays and the underlying bedrock. To confirm the feasibility of depressurising the Lake Clays in the timescale of the contract and provide additional data

DESIGN

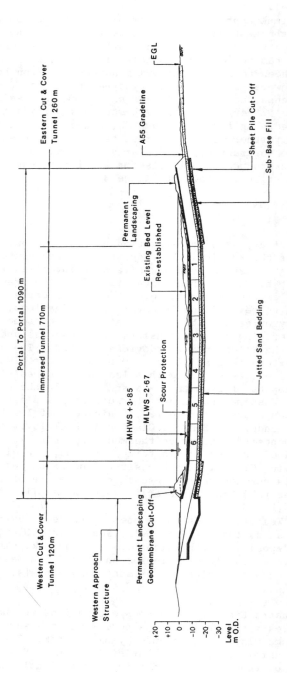

Fig. 3. Long section

282

fully instrumented full scale pumping trials were carried out in advance of the tender.

DESIGN BASIS

19. As stated in the introduction two designs were fully designed and detailed prior to tender. Traditionally steel tunnels are fabricated as twin tubes. This attracts some of the disadvantages of a bored tunnel as the section is deeper and the carriageways wider apart than is possible with a rectangular concrete tunnel. As well as additional dredging in the river the approach structures have to be longer, deeper and wider. In order to avoid these consequential approach costs a steel element was developed which had a rectangular cross section such that the alignment and depth of both alternatives were similar. The approaches and cut and cover tunnels were then common to both the immersed tunnel designs.

20. Preliminary studies indicated that the cost of the steel shell alternative was approximately 14% more than the concrete design. Since this estimated margin was small and the efficiency of the steel industry was improving Welsh Office decided that two alternative designs should be prepared in order to give both the steel and concrete proponents equal opportunity to tender for the Contract.

21. The tunnel structures have been designed for both temporary and permanent loadings in the following stages:

Concrete Immersed Tunnel Tunnel	Steel Shell Immersed Tunnel
1 Casting basin	1 Fabrication
2 Temporary handling & placing	2 Towing to concreting pier
3 Permanent conditions	3 Concreting while afloat
	4 Temporary handling & placing
	5 Permanent conditions

All of these stages except the permanent condition depend on the construction method selected by the Contractor. The loading combinations consisted of self weight, hydrostatic pressures, earth pressures, traffic live load, temperature effects, shrinkage and differential settlements. Drawings were included in the Contract which showed the assumptions which were made with respect to the temporary loading conditions during the design of the permanent work.

22. There are no Codes of Practice which are directly applicable to immersed tunnel structures. The structural design has been carried out on the basis of "permissible stresses" rather than "limit state". CP114 and the Departments Technical Memorandum BE1/73 have been used for reinforced concrete design, CP117 has been used for

DESIGN

composite design and BS 153 for structural steelwork design. The following grades of materials have been used generallly:

Structural Concrete	37.5/20 in tunnel
	30/20 in approaches
Structural Steel	50C
Reinforcement	460/425 Type 2 deformed bars.

23. The tunnel and the approach structures were subject to the procedures laid down in Welsh Office Departmental Standard THG BD2/79 "Technical Approval of Highway Structures on Trunk Roads (including Motorways)". Although a draft Part 3 covering tunnels was available Welsh Office required the submission to conform as closely as possible with Part 1 of the Standard. The tunnel was designated a Category 3 structure and a design check was carried out by Scott Wilson Kirkpatrick & Partners.

Seismic Loading

24. Although the UK is in a quiescent area as far as seismic activity is concerned, the North Western corner of Wales is an area of relatively high activity. On the basis of past activity an intensity of 6MM (Modified Mercalli) was adopted initially. However in July 1984 an earthquake occurred in North Wales which was felt throughout Wales, western England and eastern Ireland. Its epicentre was in the Lleyn Peninsula some 60km south west of Conwy. The magnitude was 5.4 on the Richter scale and the intensity at Conwy was 5MM. As a result of this the design intensity was increased to 6.7MM.

25. At this level of intensity the tunnel structure itself is not vulnerable. The effects of the ground accelerations are adequately covered by the higher stresses allowed for such loadings. Attention was however given to the detailing to ensure the ductility of the structure.

26. The main concern was the possibility of liquefaction in the granular backfill around and underneath the tunnel. While the backfill at the sides of the tunnel could be loaded to increase its relative density and prevent liquefaction this was not possible with the sand fill under the tunnel. As the tunnel is lighter than the soil it replaces its foundation is relatively lightly loaded. To overcome this a percentage of cement clinker has been added to the sand foundation material in order to stabilise it. This method has been used successfully in Japan and Persia, areas of far greater seismic activity.

Thermal Loadings

27. With respect to temperature effects the tunnel has been designed for a gradient between the inside and the outside of the tunnel as well as for overall temperature variations.

Shear and Crack Width

28. In the transverse direction the members forming the tunnel cross section are subject to large axial forces resulting from hydrostatic and earth pressures. The design of the transverse shear reinforcement was based on the principle tensile stress in the concrete in order to take advantage of these axial loads.

29. Although the tunnel has an external waterproofing membrane the flexural crack width has been limited as an additional safeguard to ensure low permeability and high durability.

30. The cover to the reinforcement in the tunnel is 50mm on surfaces protected by the external waterproofing membrane and 40mm in interior surfaces.

Factor of Safety against Floatation

31. The tunnel structures excluding backfill, the carriageway wearing course and other removable items have a factor of safety against uplift of 1.1. The calculations for this considered the range of possible variations in structural dimensions, densities of materials and estuary water in their most critical combination.

Analysis

32. The tunnel structures have been analysed separately in the longitudinal and transverse directions. For the transverse analysis the tunnel cross section has been modelled as a frame. For the longitudinal analysis the tunnel structures have been considered as line beams on a soil foundation. The longitudinal bending moments and shear forces have been calculated by soil-structure interaction modelling, which was developed to take into account the torsion in the tunnel cross section. Torsion is particularly relevant on the East Cut and Cover Tunnel where the rockhead approaches formation level locally and the rock contours are skew to the line of the tunnel.

33. The longitudinal axial forces have been calculated by obtaining compatibility between the longitudinal movements due to temperature variation and shrinkage, and the frictional resistance of the surrounding fill material.

IMMERSED TUNNEL

34. Each of the immersed tunnel designs is made up of six preformed elements approximately 118m long. This element length resulted from a consideration of space available and the forces likely to result on the element during the placing operations. At Spring tides the current velocities can reach 2 m/s and the generally shallow nature of the estuary would make it difficult to handle larger elements.

Concrete Immersed Tunnel

35. The concrete immersed tunnel has a rectangular cross section as shown in Figure 4. The underside and walls are waterproofed with 6mm thick steel plate. As it would cause

DESIGN

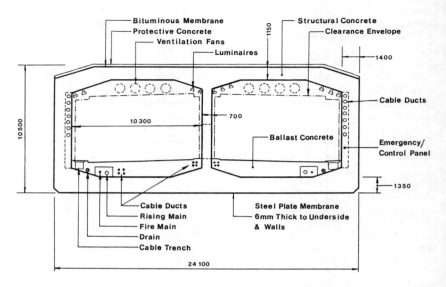

Fig. 4. Cross-section of concrete immersed tunnel

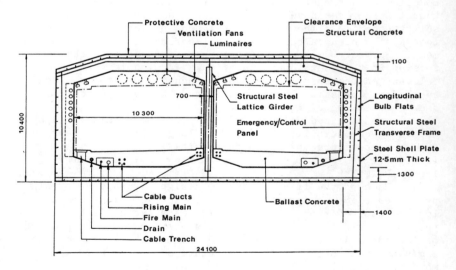

Fig. 5. Cross-section of steel alternative immersed tunnel

unnecessary construction difficulties to continue the steel plate over the roof, the roof slab is waterproofed with two layers of a bitumen impregnated polyester fibre fabric. This in turn is protected by a reinforced concrete layer.

36. For the design it was assumed that these tunnel elements would be precast on site in a casting basin on the west bank adjacent to the West Cut and Cover Tunnel. This presented an ideal location for a casting basin subject to the geology being suitable for the construction of this large dewatered excavation.

37. An analysis was carried out to determine the differential settlements that the elements would suffer during construction in the casting basin as these would result in stresses being built in. This involved a complex soil structure interaction analysis which was further complicated because the stiffness of the structure changes as construction progresses. On the basis of this analysis an optimum casting sequence was derived which, because of the complex nature of the analysis, was made mandatory. Piezocone tests were included in the contract to confirm that the ground characteristics under the elements were within the range assumed in the design.

38. It was decided early in the design that the use of an external steel waterproofing membrane provided the best long term solution in terms of watertightness and durability. All the welds between the plate are tested for watertightness during construction. The membrane is attached to the concrete with shear studs but for analysis purposes no composite action was assumed and the membrane ignored.

39. The contract specification includes requirements for monitoring and controlling the temperature of the concrete. These limit the maximum temperature as well as the differentials between the core and the surface of a section and the mean temperature with respect to a completed adjacent section.

40. When construction work in the casting basin is completed, including the installation of the temporary bulkheads the elements will be checked for leaks. Two tests are required. Firstly a test in air with a vacuum of 0.055 N/sq.mm inside the element. Then the element will be temporarily ballasted, the casting basin flooded and the element inspected to ensure that there is no water penetration.

41. Watertightness of the joints between the tunnel elements and between the immersed tunnel and the East Cut and Cover Tunnel is achieved by Gina rubber gaskets. Once this primary seal has been effected, a second rubber omega shaped gasket is installed inside the Gina from within the tunnel elements after they have been lowered on to their permanent foundation. Thus a double watertight seal is provided at the joints.

42. In order to minimise the effects of temperature and differential settlement, the joints between the elements have

DESIGN

been designed to transfer shear and longitudinal compressive forces but not bending moments or longitudinal tensile forces. Furthermore the joint details are such that they can be used for temporarily supporting the primary end of one element on the secondary end of a previously placed element. The shear transfer mechanism is designed so that it can be released prior to backfilling of the trench and re-established again later. This minimises the stresses caused by differential settlements during backfilling.

43. The tunnel elements are designed on the assumption of a uniform bedding layer. This requires that the elements be placed initially some 800mm above the floor of the trench on temporary foundations. The sand bedding material under the tunnel is then placed hydraulically from outside by the sand jetting method.

Steel Shell Immersed Tunnel

44. The cross section of this alternative design is shown in Figure 5. The basis of the design is that a steel shell is fabricated off site, towed to the site and then a concrete lining placed while the element is afloat. The concrete and steel are assumed to act compositely in the permanent condition. The concept is that the greater material cost of this design is offset by not requiring the temporary works involved in the casting basin.

45. In preparing the design it was assumed that the steel shells would be fabricated off site in a dry dock. The reinforcement for the internal concrete lining was to be fixed, end bulkheads installed and part of the bottom slab concrete placed to act as a keel before the dock was flooded. The floating shell would then be towed to site and moored at a concreting pier where the internal lining would be placed.

46. The design and detailing of the 12.5mm thick shells was of particular importance. As the shells closely resemble quite large ships, shipbuilding techniques were considered appropriate and were applied. Figure 6 shows a typical detail illustrating the bulb flats, scantlings and stiffening frames employed.

47. The details involved a number of intricate weld details and advice was taken from both the British and Danish Welding Institutes. The amount of welding necessitated the specification of a large and detailed program of welding trials and non destructive testing together with the required acceptance levels.

48. The internal concrete lining is placed in a specified sequence in order to limit the stresses in the shell during concreting. The concrete in the roof slab is placed through temporary openings in the top shell plate. These are sealed after use with steel closure plates. Any voids between the steel shell plate and the internal concrete lining are filled by injecting grout.

49. The analysis of the floating shell during construction of the lining is again complex. The structural stiffness had

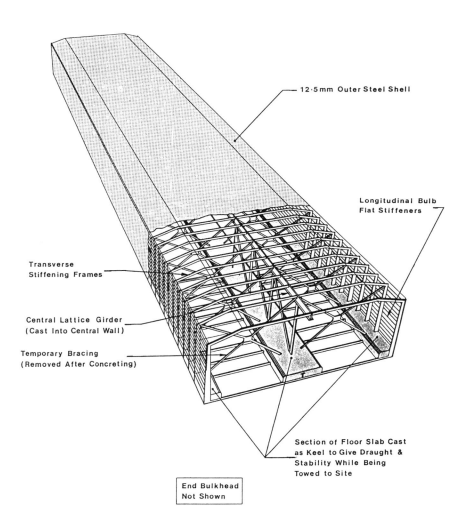

Fig. 6. Detail of steel shell

DESIGN

to be modified at each stage to determine the stress history of the steel shell and composite lining. This casting sequence, which involved 70 separate concrete pours in each element was made mandatory under the Contract.

50. The handling, fitting out, placing, jointing and foundations for this design are all similar to those of the concrete immersed tunnel.

THE CUT AND COVER TUNNELS

51. The cut and cover tunnels are conventional reinforced concrete structures which are being constructed in open dewatered excavations protected from the estuary by temporary bunds. The tunnels are constructed on a compacted layer of gravel with a thickness varying between 1.2m and 1.8m.

52. The cross section is shown in Figure 7. The internal profile above the ballast concrete is identical to that of the immersed tunnel. Externally the section has been squared off as this is more economical. The East Cut and Cover Tunnel has been made overhigh to allow for the superelevation rather than skewing or stepping the section.

53. The main consideration with the design of these sections was the articulation to minimise the effects of differential settlements and temperature. Dilatation Joints were introduced at intervals along the structures. These joints can transfer shear and longitudinal compressive forces but not bending moments or longitudinal tensile forces.

54. Initially a 1.2m wide gap is left in the structure at each dilatation joint. In order to reduce the magnitude of the shear force to be transferred at the joints, the design

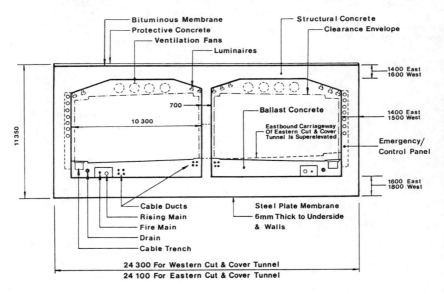

Fig. 7. Cross-section of cut and cover tunnel

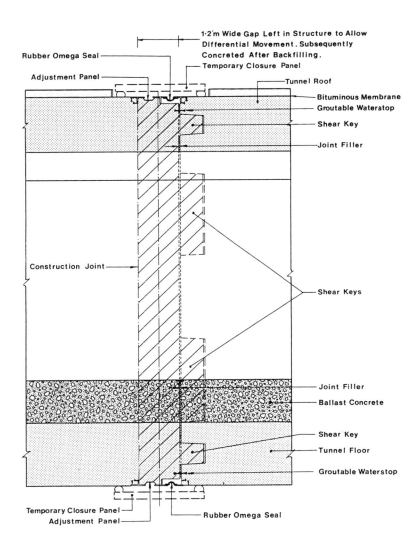

Fig. 8. Section through dilatation joint

DESIGN

requires that a specified minimum amount of backfill is placed around the structure and the ground water levels restored to their natural levels before the joints are made. To achieve this the gap in the structure must be protected from the ingress of backfill and water by sacrificial watertight panels sealed against the outside of the structure. The joint can then be completed from the inside after backfilling and restoring the ground water levels. A section through the joint is shown in Figure 8.

55. A further restriction has been necessary on the West Cut and Cover Tunnel where there is a need to minimise the absolute settlements under the appreciable backfill loading. Here a controlled sequence for backfilling and raising the ground water level has been specified.

56. The only exception to this late completion for shear transfer is the middle joint in the East Cut and Cover Tunnel. This is located over a rock pinnacle which approaches formation level and the shear force to be transferred is not so large. The joint has therefore been designed for completion before backfilling and restoration of ground water levels.

57. Watertightness of the joints has been achieved by either rubber or steel omega shaped sections which can accommodate the required movements. A second barrier consisting of a steel tipped rubber waterstop which can be injection grouted subsequently has also been provided.

58. The general waterproofing system is the same as for the concrete immersed tunnel ie a 6mm thick steel membrane on the external surface of the bottom and side walls and two layers of bitumen impregnated polyester fibre fabric on the roof protected by a reinforced concrete layer.

CORROSION PROTECTION

59. The use of an external steel waterproofing membrane pre supposes that the membrane will last as long as the tunnel. Clearly the corrosion protection of the membrane is of considerable importance. This applies even more to the steel shell alternative because of its structural function in the composite construction. The design life in both cases is 120 years.

60. Various options were available:

i) Provide a small sacrificial corrosion allowance to the steel plates.
ii) Coat the exposed steel surfaces
iii) Coat and provide an impressed current cathodic protection system
iv) Coat and provide a sacrificial anode cathodic protection system with a specified design life.

Opinions varied on the degree of corrosion likely in the long term. Experience in reclaiming bridge caissons and sheet pile quay walls had indicated that very low rates of corrosion occur on buried structures due to the lack of oxygen. However on such a large area of steelwork the possibility of increased local corrosion due to long line effects or fresh water intrusion into the backfill, although unlikely, could not be ruled out. For this reason the sacrificial thickness concept was not acceptable.

61. Coating the tunnel is the single most effective anti-corrosion measure. It was decided, therefore, that the steelwork would be blast cleaned, primed and coated with 400 microns of coal tar epoxy.

62. The coating will however break down in time and there will also be damaged areas from the outset. A cathodic protection system will, therefore, be installed as well based on specialist advice provided by Spencer & Partners. The system is based on sacrificial anodes for the following reasons:

i) The sacrificial anode system protects as soon as it is submerged, ie during construction in the casting basin when the most rapid corrosion will occur. An impressed current system could only become effective when the tunnel is complete.

ii) It is not certain that cathodic protection will be needed in the longer term if the environment proves to be non aggressive.

63. A series of Zinc anodes have been fixed to the steelwork designed to protect it for 20 years. In addition monitoring equipment consisting of reference electrodes, ultrasonic thickness transducers and retrievable coupon holders has been installed to check any deterioration in the steelwork. Should it prove necessary an impressed current system can be installed later. All the connections and continuity cables necessary for this are being installed at the outset.

64. In addition to this the thickness of the steel shell plate included a 3.5mm sacrificial corrosion allowance in view of its importance as a structural member. The 6mm steel waterproofing membrane has an inherent sacrificial allowance as even 1mm would be waterproof.

EARTHWORKS

Tunnel Trench

65. The trench for the tunnel elements crosses the river at an angle and is a relatively large feature in a shallow estuary, as is the tunnel element itself. The Contract, therefore, provided for a 1:50 scale physical model to assess the hydrodynamic forces in the tunnel elements during the float out and sinking operations. The model reproduced the actual configuration of the estuary temporary works proposed by the Contractor.

DESIGN

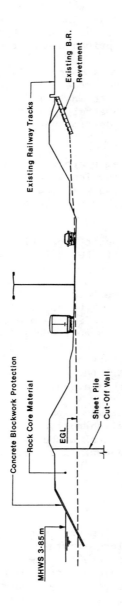

Fig. 9. Section through eastern approaches

66. Although some of the material from the trench is suitable for reuse on the contract after treatment, a large volume is not. A reclamation area was therefore constructed in the upper estuary as an advance contract to hold much of the unsuitable material as well as providing a temporary stockpiling area for suitable materials from the dredging operation. To minimise pollution of the river the return water from area was strictly controlled in terms of dissolved oxygen and suspended solids content as well as overall volume. To minimise pollution of the river and the noise impact on the residents a cutter suction dredger was specified and the hours of operation were limited to those between 0700 and 2200.

Backfilling to Tunnel Structures

67. The backfilling to the tunnel structure comprises ´structural´ backfill which locks the tunnel in position and ´amenity´ backfill which maintains the existing flow conditions in the estuary and restores the visual appearance at low water.

68. The structural backfill is placed at the sides of the tunnel and consists of a lower locking fill of clean sand taken to half the height of the tunnel and an upper loading fill similar to the locking fill but with a higher percentage of fine material. The loading fill loads and compresses the locking fill to counter liquefaction under seismic activity. The structural backfill is retained and protected by a layer of scour protection material consisting of 150mm size stone. This also delineates the permanent tunnel works should there be any future excavation in the estuary.

69. The amenity backfill consists of selected material obtained from the estuary. It is placed above the structural backfill in the intertidal region.

Dewatered Excavations

70. The design of the cut and cover tunnels assumed that they would be constructed in open dewatered excavations and the pumping tests carried out to confirm the feasibility of depressurising the underlying soils have already been referred to (paragraph 18). The size and shape of these temporary excavations was to be determined by the Contractor, however it was important that adequate instrumentation was installed to enable the stability of these excavations and the effectiveness of the depressurisation to be monitored. To this end minimum levels of instrumentation were specified in the Contract for the major excavations and minimum factors of safety for temporary dewatered excavations were specified using both total and effective strength.

APPROACHES

Eastern Approaches

71. The eastern approach ramp is constructed in open

DESIGN

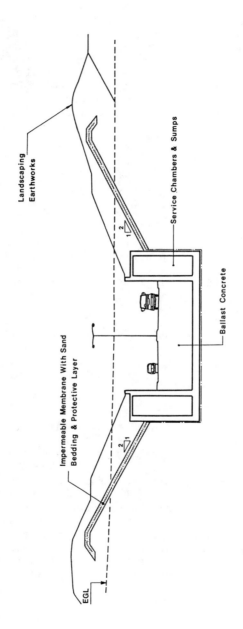

Fig. 10. Section through western approaches

cutting within a reclaimed area in the estuary. Figure 9 shows a typical cross section. The perimeter bund contains a sheet pile wall which is sealed into the underlying Irish Sea Boulder Clay. This is relatively impermeable and forms a permanent cut off to ground water. The sheet piles are welded and driven in pairs. To reduce seepage as far as possible the wall was excavated to a depth of 2.5m and the unsealed clutches welded.

72. The perimeter bund is essentially of rockfill construction. The seaward face is protected by cellular concrete blockwork. This was chosen in preference to rip rap to give a softer landing to any boats driven from their moorings by the prevailing westerly/south westerly winds. The top level of the bund is at 6.5m OD which gives an adequate margin against the predicted 1 in 100 year water level of 5.2m OD. Some minor overtopping due to spray and wave run up is anticipated in severe storms. The upper slopes are grassed but protected with an erosion control matting and the drainage on the inner slope has been designed accordingly.

73. To maintain the integrity of the sheet pile cut off it has to be sealed into the tunnel where it crosses close to the portal. A large collar is constructed round the tunnel to form this connection whilst electrically isolating the sheet piles from the tunnel waterproofing membrane. If they were electrically continuous the sheet piles would act as a considerable drain on the cathodic protection system.

Western Approach

74. The open earthwork solution was not possible on the western approaches. The lack of continuity of the underlying Irish Sea Boulder Clay which would provide a groundwater cut off and the presence of deep buried channels weighed against the use of sheet pile or cement bentonite permanent cut off walls. The solution adopted is shown in Figure 10.

75. The western approach ramp consists of a reinforced concrete U shaped box which is ballasted with mass concrete to prevent uplift. The base slab is generally 1.5m thick with the mass concrete ballast reaching a maximum thickness of 5m adjacent to the portal. The ramp is constructed in a dewatered excavation which is a continuation of the excavation for the West Cut and Cover Tunnel.

76. We were particularly anxious to preserve as far as possible an open aspect similar to the earthworks solution at the Eastern Approach. The high retaining walls of the U box that would have been required to prevent tidal inundation would have formed a deep constrained cutting. The walls are therefore kept low and the earthworks slopes are battered back above them.

77. It is therefore necessary to prevent ground water overtopping the approach walls and this is achieved by incorporating an impermeable membrane within the earthworks side slopes. This membrane extends from the maximum

DESIGN

predicted tide level at the top down to seal against the outside walls of the approach structure. The hydrostatic uplift on the membrane is countered by ensuring there is always a sufficient depth of fill above to provide an adequate factor of safety. This technique has been used on underpasses in Holland but as far as we are aware this is the first such use in the UK.

OPERATIONAL ASPECTS
   78. The Conwy Tunnel is one of three tunnels being constructed along the A55, the other two are the Penmaenbach and Pen-y-Clip tunnels to the west of Conwy. In order to minimise the cost of operating these tunnels it was decided that all operations should be automated employing computer controls and monitoring systems housed in unmanned service buildings as far as possible. This enabled manning to be kept to a minimum and maintenance facilities to be centralised in one location.
   79. Regular meetings have taken place with representatives of the Maintenance Authority and of all the Emergency services in developing the operational proposals for the tunnel. The control of traffic in emergency and for planned maintenance has been given detailed consideration and discussed with the North Wales Police who will be responsible for traffic management.

Ventilation and Lighting
   80. Each tube is ventilated longitudinally by 36 fans in banks of four mounted above the carriageway. The ventilation system is designed to operate automatically using data from carbon monoxide and visibility monitors within the tunnel. Air flow will normally be activated in the direction of traffic. The fans will be fully reversible to provide an efficient ventilation system during two way operation.
   81. Lighting is provided by fluorescent luminaires located on the haunches each side of the tunnel tube. At the threshold zones adjacent to the portals lighting is boosted by high pressure sodium luminaires. The lighting levels are controlled by external photo cells measuring ambient light levels.

Other Aspects
   82. The tunnel equipment is fed from two separate power sources so that in the event of a failure only half the lights and fans should go down. In the event of a major failure the two Conwy Service Buildings have battery back up systems to protect the computer systems and maintain emergency lighting until the diesel generators in each service building can be brought on line.
   83. A pumped drainage system is provided with sumps located at the low point in the tunnel and adjacent to each portal. These are controlled automatically with electrode type sensors. A gas detection system in the mid river sump

initiates automatic foam and halon systems in the event of a hazardous spillage.

84. Emergency telephones and fire extinguishers are located in panels in the near side wall at 118m intervals throughout the tunnel. In addition there are emergency pedestrian exits through the central wall should it be necessary to evacuate one tube.

85. Variable message traffic signs are installed on both approaches. These can direct traffic to share one bore or to close the tunnel completely in the event of an emergency. Traffic is monitored by a system of closed circuit TV cameras which are linked by optical fibre cables to the operations room of the North Wales Police in Colwyn Bay.

86. The inside of the tunnel is painted as this reduces the power required to provide the necessary lighting levels. The approach walls are also painted a dark colour to assist the accommodation of the drivers eyes as they approach the tunnel.

87. The electrical and mechanical systems are being installed under a separate contract the approximate value which is £6m.

CONTRACTUAL ASPECTS

Contract Documents

88. The tunnel forms only part of an overall highway contract and therefore the ICE 5th Edition Conditions of Contract were adopted and the measurement was based on the Departments Method of Measurement for Road and Bridge works. Clearly these were not written with an immersed tunnel in mind and several amendments and additions were required.

89. The Conwy Estuary is a quiet unspoilt estuary and several measures were included in the contract to limit construction nuisance and ensure that the navigation rights of the public were maintained. The contractors access routes to the site, already limited by the physical features of the area were further restricted to reduce the effects on schools and housing estates. The contract also contained severe limitations on the number of contractors vehicles permitted to pass through Conwy. None at all were permitted during summer weekends and morning and evening peak hours. Even during winter only 45 vehicles per day in each direction are permitted.

90. Tenders were invited in February 1986 from a select list of five consortia or joint ventures. These had been selected from the response to the EEC advertisement placed in Autumn 1985. The tender documents amounted to 1331 drawings in some 27 volumes with 19 volumes of Conditions of Contract, Specifications, Bills of Quantities and Bending Schedules.

91. To ensure tenderers had an adequate opportunity to familiarise themselves with the scheme yet avoid an extended main tender period, the five tenderers were given an advance package of data some 6 months before tenders were invited.

DESIGN

This consisted of the factual site investigation data, together with overall scheme layouts and general arrangement drawings which enabled the tenderers to start their overall planning and gain an appreciation of the complex geology and earthworks involved.

92. Tenderers were obliged to price one or other of the two conforming immersed tunnel alternatives, but could also submit one of their own. Although the steel shell alternative was explored by several of them during the tender period none chose to submit a price for it. Several other alternatives were submitted, generally varying the cross section or length of the immersed tunnel but the lowest tenderer elected to construct the concrete immersed tunnel as designed by the Engineer. The Contract was let to the Costain Tarmac Joint Venture for £102m. Construction started in November 1986 with a contract period of 4.25 years.

ACKNOWLEDGEMENTS

94. The Authors are grateful for the permission of the Welsh Office Director of Highways, Mr K J Thomas, to publish this paper.

Discussion

MR E.J.M. HEPPER, Sir Alexander Gibb and Partners
Reference has been made to the difficulty of evaluating
alternative design time savings. The environmental
advantages of alternatives are also difficult to evaluate.
The Conwy estuary is a very sensitive area environmentally;
it is a fishing centre (salmon and mussels); and it is also a
holiday centre, dependent on lack of noise and water
pollution, as well as on lack of visual intrusion.

The Gibb-Tomlinson design alternative would have had the
following advantages

(a) off-site fabrication, so a casting basin would not have
    been necessary
(b) shallow draft, so less dredging would have been required
(c) quick installation of units, which would have meant far
    less visible on-site construction.

These factors would all have had major environmental
advantages for the tourist economy in the estuary. However,
there is no contractual/institutional way of considering them
in the tender evaluation.

MR D.R. CULVERWELL, Consulting Engineer
Paper 17 does not give any information on the thickness of
the section, although scaling Fig.4 suggests that this is
about 0.6 m. Would the Authors indicate typical thicknesses?

Would they also comment on the costs of fabrication of the
steelwork, which looks as though it might be awkward on
account of the double skin, the studs and the general
dimensions?

Although the section is interesting, it is not clear from
the Paper what real advantage it offers over the conventional
single-skin steel shell, which also offers shallow draft for
easy transportation. Would the Authors elaborate on this?

MR G.W. DAVIES, Travers Morgan
With reference to Paper 17, those of us who were involved in

Immersed tunnel techniques. Thomas Telford, London, 1990.

DESIGN

the tender evaluation stage of the Conwy Tunnel were extremely interested in the dual stressed skin concept when we were approached by the Authors on behalf of one of the tenderers. We were also disappointed when the idea was withdrawn before tenders were submitted, as it would have represented a radical advance in immersed tube technology. It seemed to me that the principal disadvantages were as follows.

(a) There would be the risk of damage to the exposed inner skin from traffic accidents and, in particular, from fires; some form of inner skin is probably needed over the whole of the exposed steel shell.
(b) There would be a lack of space within the section to accommodate M & E equipment - particularly the emergency equipment panels and the electrical distribution panels. One answer to this may be to widen the central wall and to accommodate these panels within it; although I appreciate that this would be a less than satisfactory solution, as the panels would then be on the offside.
(c) There would be construction difficulties associated with fabricating and erecting the shells and, in particular, with the concreting operations, on account of the congestion of shear connections between the skins. Perhaps the answer here would be to cast the sections vertically in short lengths, then to rotate them to the horizontal; they are quite light in weight. The inner and outer shells could then be welded at the joints and the joints then grouted up.
(d) There would be a lack of any corrosion protection system.

In spite of these drawbacks, the concept seems practical and may be the way immersed technology develops in the future. Unfortunately for Conwy, the scheme was probably 10 years ahead of its time. I wonder whether Mr Tomlinson could elaborate on why the scheme was withdrawn - was it cost, risk or insufficient time to resolve the problems?

MR M. CHAPMAN, Sir Alexander Gibb and Partners
It would appear that there is a need for codes of practice to be produced for the design of immersed tube tunnels. This would avoid each firm having 'to re-invent the wheel'. If nothing else emerges from this conference, I do hope, at least, that a steering committee will be set up to examine this.

CAPTAIN C.A. JENMAN, Global Maritime
Mr Tomlinson (Paper 17) mentioned going out to competitive tender for the construction of the tunnel units and towage at sea; Mr Ingerslev (Paper 18) lists some criteria for the

DISCUSSION

floating stages of tunnel units. It is to these matters that I should like to refer, as my company, Global Maritime, works on marine operations world-wide in shipping, offshore and civil engineering marine works. One would think that the criteria for each would be the same, as marine risks are the same for all projects with a seagoing phase. At least, marine underwriters, for whom we also work, would think this.

Ever since people have been going to sea in ships, it has been known that this is a risky business. The first point I must make is that, with these immersed tube tunnels, we now have civil engineers going down to the sea ... not in ships but in huge boxes, tubes, binoculars, etc. ... none of which appear to be designed with any reference to the painful years of evolution of ship design: nothing about Samuel Plimsoll who spent most of his parliamentary life fighting and achieving a load line rule to reduce the number of ships lost through lack of freeboard or overloading; nothing about damage stability which all seagoing units in shipping and offshore normally have by rules and/or normal good practice; nothing about impact strength - and here there is a great difference between steel and concrete. A few millimetres of steel cannot resist a mere kiss of a seagoing tug, but a metre of concrete can. A tunnel is not much good with nine out of ten sections, even though that is a 90% success rate.

My interest in this part of immersed tunnel techniques has arisen because Global Maritime are the certification surveyors for the marine works on the Sydney Harbour tunnel, which is to be towed from Port Kembla to Sydney. It seems that most tunnel units are towed short distances in sheltered waters. The one major exception appears to be the Chesapeake Bay tunnel sections which were towed from Orange in Texas to Virginia: an ocean tow of two thousand miles. It appears that these units had about four feet of freeboard and that one unit was lost for a couple of days after a towline parted in heavy weather, perhaps off Bermuda? I am interested in all such operations from both a technical and practical viewpoint, and if, as has been suggested, tunnels for the UK are to be built in Spain, and units for New York in Rio, then designers should consider the criteria for transportation across hostile oceans. Finally, Global Maritime would be happy to contribute to any guidelines produced by the Institution of Civil Engineers concerning the design and construction of immersed tunnels.

DR R. NARAYANAN, The Steel Construction Institute
The double-skin composite construction proposed by Tomlinson et al. (Paper 17) represents an innovative method, which is perhaps 10 years ahead of its time. As I understand it, it was not possible, in the time available, to evaluate the risk factor and to quantify it in money terms; hence, although the concept was developed and validated with the funding support of Messrs Cementation Ltd, the Authors were not able to

DESIGN

tender for the alternative design using the steel solution.

At the Steel Construction Institute, we are convinced that the double-skin composite solution proposed by the Authors is technically feasible and economically viable. We have plans to carry out further studies on the concept to provide design guidance to consultants interested in using this novel method.

With regard to corrosion, we do not really regard it as very relevant for immersed steel structures. Investigations carried out by CONSTRADO on the corrosion of bridges and buildings built at the turn of the century revealed that there was very little evidence of deterioration of steel structures. We would regard the use of steel as technically superior from the points of view of corrosion resistance, toughness, ductility and many other inherent advantages of the material.

MR J.F.L. LOWNDES, Mott MacDonald
The operation of trains in tunnels generates a considerable amount of heat. This heat evaporates such moisture as migrates through the concrete, if special measures are not taken. It is likely then that humidity levels in these tunnels may be very low. This encourages further migration of water, thereby depositing further chlorides. We would therefore need to satisfy this phenomenon by producing a very high humidity in the tunnel to reduce evaporation. A high humidity is at variance with the requirements of train operators and electrical engineers. It therefore becomes desirable either to protect the reinforcing or to provide an impermeable membrane.

MR J.D.C. OSORIO, Posford Duvivier
I should like to hear views on corrosion and design life, as I am now extremely confused by current opinion on these matters.

We used to believe that well-made reinforced concrete immersed in sea water with no leaching or transverse pressure gradient lasted for ever.

I am now told that, even if the very best concrete is used, chloride ion will diffuse inwards at about 1 mm per annum, so that with commonly used covers and no waterproof membrane, corrosion will start after 40 to 50 years. This is very uncomfortable if we are aiming for a design life of 100 years.

Similar variations occur in the estimate of corrosion rates for steel plate.

DR M. GRIFFITHS, Pak-Poy and Kneebone Pty Ltd, Crows Nest, NSW, Australia
Reference has been made to the need for a code of practice

for immersed tunnels. It is worth noting that reinforced concrete immersed tunnel units do not, in general, meet the ductility requirements of the Concrete Codes, at least in the longitudinal direction.

An immersed tunnel in service is a beam on elastic foundation. Longitudinal reinforcement is provided which controls cracking in the serviceability state and resists bending moments in the ultimate state.

Because of the massive size of the immersed tunnel unit, its cracking moment in general would far exceed its moment capacity as a reinforced concrete beam. Hence any overload would not result in a redistribution of forces - a beam on elastic foundation is infinitely redundant - but would snap like a carrot at the critical section, with catastrophic results.

MR D. VAN OLST, Promat BV, Hilversum, The Netherlands
Every day, freight of all kinds is passing through our tunnels; some of this freight is highly inflammable. It is precisely the ever-present danger of fire that illustrates how vulnerable the tunnels can be in a complete road system (Fig.1).

How can these expensive tunnels be made safe in the best and most efficient way? How can they be made suitable for loads that may be a fire hazard, such as loaded tankers? The only right answer is fire-resistant protection for all important parts of the tunnel which are: ventilation; electricity cables; and, most importantly, the load-bearing elements such as tunnel ceilings.

Promat Holland, part of the international Eternit Group and a specialist in the field of structural fire prevention, did some research on the subject of fire. The standards and regulations in Holland are very clearly presented by the Rijkswaterstaat.

We have developed a protective system for tunnel ceilings that is based on the use of Promatect panels of 27 mm thickness (Fig.2). In official tests the system was exposed to a simulated hydrocarbon fire for two hours in a closed space up to 1350°C. During the two hours, the surface temperature of the concrete had to remain below 380°C. The reinforcing steel inside the concrete had to stay below 250°C (Figs 3 and 4).

Has there been any experience with fire in tunnels, and what kind of standards are used in other countries?

MR H.A. KAMP NIELSEN, Comar Engineers A/S, Virum, Denmark
With reference to Paper 18, it is necessary to include the non-uniform reaction distribution from foundation of tunnel, both transversal and longitudinal, in the detailed design.

With reference to Paper 19, how is the tensile strength of concrete related to age?

DESIGN

Fig. 1. Tunnel ceiling after fire

Fig. 2. Promatect-H fire resistant material fixed with stainless steel bolts (in the Zeeburgertunnel, Amsterdam)

DESIGN

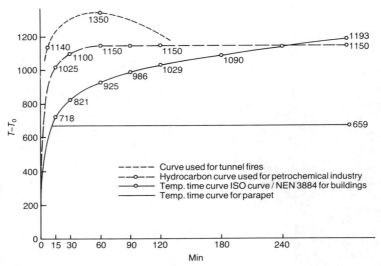

Fig. 3

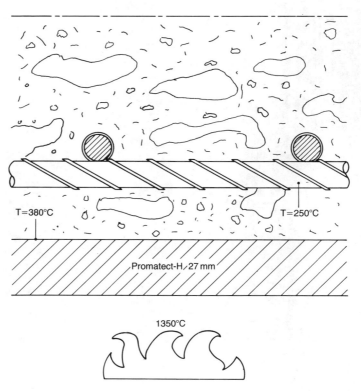

Fig. 4. Maximum accepted temperature after two hours' testing

DISCUSSION

MR R.G. THOMSON, TH Technology Ltd
With reference to Papers 18 and 19, and the remarks on the use of blast furnace cement durability on the Saudi-Bahrain causeway.

I worked for the contractor on this project. Problems were experienced with surface spalling of some mass concrete units caused by salt crystal growth in the larger pores of the surface layer. Mercury porosimetry tests on the surface and deep samples showed that the surface had a higher porosity, which was the result of less than perfect curing of the surface of the units - directly attributable to the slower curing rate of Portland blast furnace slag cement (PBFC) compared with ordinary Portland cement (OPC). The pore size of the core concrete was much finer than, for instance, type 5 SRC concrete.

Other research showed that the use of shutter vibrators may have resulted in less aggregate in the very thin surface layer. There was no evidence that the water-cement ratio was higher than in the core concrete.

MR H.D. OSBORN, HAKA UK Ltd
In paragraph 23 of Paper 19, the Authors refer to the use of Dutch blast furnace cement. The quality of all concretes will depend on the cements used. Has consideration been given to the use of concrete where part of the cement has been replaced by pulverized fly ash?

The problem of steel corrosion has also been mentioned. I recently saw an alternative to steel: namely, a plastic indented bar. I am not sure of its composition. Have any of the Authors come across this material?

MR A. TOMLINSON, Paper 17
In reply to Mr Culverwell's question, the thickness of the sandwich cross-section was 516 mm in in-span areas, increasing to 750 mm at haunches. This includes an outer steel plate 10 mm thick, and an inner steel plate 6 mm thick, the remainder consisting of the concrete core.

The fabrication of the inner and outer steel shells, and the construction costs thereof were studied in some detail by our client, Cementation Construction Ltd. In their opinion, fabrication did not pose any insurmountable problems and could be undertaken successfully, and economically.

The advantages that we perceived our design to possess, as compared with the more conventional steel or reinforced concrete immersed tube tunnel, are discussed in our Paper. Shallow draft of an unballasted unit was only one parameter of many used to arrive at an optimum design. Indeed, it was recognized that in some potential tunnel locations, shallow draft would not be of major benefit; in such a case, different parameters would predominate, leading possibly to a different cross-sectional shape from that proposed at Conwy.

DESIGN

The intention of the development programme was to prepare a design which offered this greater flexibility, combining the best features of conventional concrete and steel tunnels, while realizing savings in project length and cost.

In reply to Mr Davies's question, it was intended to fire protect the exposed inner steel shell plate with a proprietary fire protection system.

The provision of fire protection was not seen as a disadvantage, since (as discussed by Mr van Olst in this session) in the Netherlands, similar fire protection is also now provided to the inner face of concrete tunnels.

We were also concerned with the difficulties associated with the provision of certain M & E services and equipment within our proposed cross-section. These difficulties were an inevitable consequence of preparing an alternative design which differed significantly from a conforming scheme. The limited time available during the tender period necessitated that the existing service layout of the conforming scheme had to be mirrored and adapted. It was therefore necessary to solve rather than to avoid problems, which is obviously much less desirable. The service layout which was produced, although it was both practical and workable, was not ideally suited to our scheme. Later variants of our design incorporate service provisions which successfully avoid or overcome the problems identified in our tender scheme, while still providing facilities in the required locations.

With reference to corrosion protection, it was intended to provide, in addition to a paint system and corrosion allowance to the outer steel plate, an impressed current system, for which a scheme was prepared on our behalf by the University of Manchester Institute of Science and Technology.

This scheme consisted essentially of a series of permanent impressed current anodes on either side of the tunnel. Power consumption was estimated to vary from 2.25 kW initially, rising to 45 kW at the end of the 120 year design life. Sacrificial anodes were rejected because they would not only be very difficult to renew, but would also represent a major pollution hazard in estuarine conditions.

Our proposed alternative scheme was withdrawn from the tender process primarily because there was insufficient time to complete design works, although costs were seen as being competitive.

The Authors would, however, like to take the opportunity to thank both the Welsh Office and Travers Morgan and Partners for their consideration and assistance during the preparation of our alternative design.

With regard to Captain Jenman's contribution, we would support his statement.

The Authors consider that the inherent longitudinal strength of the double composite coupled with its shallow draft were most important features. These would allow sea tows of considerable distances to be safely countenanced.

A further feature which was considered important was

DISCUSSION

compartmentation of the tunnel unit using two internal bulkheads, which would not only be used in unit placement operations but would also provide single compartment damage stability. This would allow the unit to suffer penetration damage without resulting in a total loss, which would not only have the obvious benefit of greatly reducing the hazards inherent in marine operations, but would also presumably reduce insurance premiums.

MR INGERSLEV, Paper 18

I would heartily agree with Mr Chapman's comment on the need for codes of practice. I intended my Paper to be the start of just such a code, but the permissible length of the papers did not permit further elaboration. I would be most interested in the findings of such a steering committee.

In reply to Captain Jenman, I have proposed criteria for the stability of elements in paragraph 16 of my Paper, and have mentioned aspects of towing in paragraph 13. A temporary bow and stern for longer tows would also perform the double duty of a second skin to protect the exposed temporary bulkheads, the only really vulnerable location on concrete immersed tunnel units. Steel immersed tubes, before outfitting, can ride very high in the water and would be eminently suitable for the use of additional bulkheads for damage stability. However, all immersed tunnels after outfitting are always very low in the water, and additional internal bulkheads would not give damage stability because the extra water entering any compartment would be sufficient to sink any unit. Similar conditions can be seen on virtually any waterway where loaded barges ply with only inches between floating and sinking.

If a code steering committee is set up, as suggested by Mr Chapman, I am sure that any contribution that Global Maritime could make would be welcome.

With regard to the questions by Mr Lowndes and Mr Osorio, my reply is as follows.

Mr Lowndes's comment on the need either to protect the reinforcing or to provide an impermeable membrane refers presumably to the inside face of the tunnel. I should like to draw his attention to paragraphs 19-26 of my Paper. It is my understanding that inward diffusion of the chloride ion from the external surfaces, the rate of which - according to research in the UK - varies with pressure, will occur at a rate slow enough that water should not reach the inside face of the tube through the concrete within the design life. This is therefore not a problem. However, precast concrete segmental tunnel liners, which are widely used in subway systems, have exhibited widespread and increasing spalling in salt saturated soils. These liners are generally up to 330 mm thick and have low cover. Less than perfect curing may result in a higher porosity at the surface of precast concrete, as has been observed on various precast units

DESIGN

fabricated for the Saudi-Bahrain causeway.

Moist salty air drawn in through the portals of a tunnel can be a problem, although the amount of salt on the surface of the wall is not likely to be a problem when there is very low humidity, as saline solution which could diffuse into the wall would not be present. A serious problem is the use of salt on roads to prevent the formation of ice, as this can be carried significant distances on the wheels and body of vehicles, before being deposited within the tunnel. It is intended that waterproofing should be used beneath the roadway surfacing for tunnels on the Central Artery I-93 / Third Harbour Tunnel I-90 Project in Boston, Massachusetts, to overcome this issue. In general, high cover is used within immersed tunnels when exposed to moist air.

Suggestions for ways to mitigate this type of corrosion include

(a) extensive geotechnical and chemical analyses of ground and water around the tunnel, so that the potential problems may be determined.
(b) reduction of the permeability of concrete to water vapour by prestressing, or as suggested in paragraph 24 of my Paper.
(c) use of external waterproofing membranes, and use of double seals at points of weakness such as joints (paragraphs 21 and 22).
(d) immediate action to seal off leaks if they do occur.

In reply to Dr Griffiths, I agree with the need for ductility in immersed tunnel units, and have addressed this in paragraph 39 of my Paper. Where ground movements are expected, such as in earthquake regions, it would be possible to introduce additional joints and to design the joints to yield first, thus preventing inundation through use of elastic membranes and seals.

This approach is not very different from that proposed for earthquakes in the USA, where hinges at the tops and bottoms of columns are encouraged; whereas in beams such hinges should not occur. An immersed tunnel can be designed to become a chain than can flex without breaking. Flexible joints have been introduced in some Dutch tunnels at frequent intervals, being temporarily held rigid during construction and installation. Flexible joints were also used for the mass transit tubes adjacent to the ventilation buildings at the landfalls, although these joints were later fixed when adequate rotation had taken place.

This type of approach would prevent catastrophic failures in the longitudinal direction, unlike fresh carrots.

In reply to Mr van Olst, fire testing of a disused tunnel in the USA is in progress. Presumably, results will be published in due course.

I agree with Mr Kamp Nielsen that it is necessary to include the non-uniform reaction distribution from the foundation of the tunnel in both the transverse and longitudinal directions, especially at the detailed design stage. Furthermore, it is necessary to consider loss of support (subsidence) below or to one side of the tunnel over an appropriate distance, depending on the type of foundation used, the underlying soils, and the applied loading.

In reply to Mr Osborn, recent innovations in limited use are: prestressing tendons manufactured using either stainless steel or fibreglass; polypropelene fibres used in fibre reinforced concrete; carbon fibre is also in limited use. I am not aware of the use of a plastic indented bar.

IR HORDEN, Paper 19
In reply to Mr Osorio, it has to be borne in mind that several conditions have to be met simultaneously before corrosion will occur; these are

(a) loss of alkalinity of the concrete around the steel (attributable to carbonation) and/or presence of aggressive ions (such as chloride)
(b) sufficient supply of oxygen; pore water is a very efficient oxygen barrier
(c) sufficient moisture to create an electrolyte.

It is true that, after a period of time, the first condition will be met in an aggressive maritime environment. The length of the period will depend on thickness of cover, type of cement (-matrix), original water/cement ratio. However, corrosion will not occur unless the other two conditions are also met. In outside atmospheric conditions, as well as under water, sufficient pore water will be present to meet the third condition.

Beyond a certain depth from the surface, so much moisture will be retained within the concrete under outside atmospheric conditions that initial corrosion will cease as a result of oxygen starvation, after the oxygen contained in the pore water in the immediate vicinity of the steel has been consumed. Provided that a well-graded mix and a sufficiently low water/cement factor has been used, a cover of 50 mm should lead almost to an unlimited lifetime with respect to steel corrosion.

In permanent underwater immersed conditions, the situation is even more favourable.

With regard to the remarks by Mr Thomson, I should like to comment as follows.

In the case of maritime structures, the use of Portland blast furnace slag cement offers the following distinct advantages over Portland cement.

(a) The total pore volume in the cement matrix is distributed over pores with a much finer size. This leads to a slower diffusion ingress of chloride.

DESIGN

(b) The cement matrix is highly resistant to sulphate attack from sea water.

Flaking and/or scaling of the uppermost surface (order of magnitude of millimetres) is known to be a problem under highly unfavourable curing conditions. Also, under conditions of de-icing salts, flaking has been observed under the combined influence of temperature shock stresses and the growth of expanding salt in the surface skin. Apart from the point of view of aesthetics, flaking does not pose a problem with respect to technical durability. Flaking can be avoided by proper curing.

In reply to Mr Osborn, the replacing of part of the cement by fly ash has three effects on the heat development

(a) reduction in heat development as a consequence of omitting part of the cement
(b) the remaining cement may hydrate better and further, which means an increase in heat development (a possible explanation is that fly ash acts as precipitation nucleus for calcium hydroxide)
(c) delayed heat development due to fly ash reaction (after one or more weeks and thus not relevant).

The replacing of Portland cement by fly ash is effective in reducing the heat development; the first effect is the dominant one. The replacing of blast furnace cement by fly ash appears to be hardly effective in practice; probably the first and second effects are more or less in balance.

In reply to Mr Kamp Nielsen, the development of strength (compressive and tensile) is related to the development of the degree of hydration. This relation is established by way of tests (see paragraph 16 of Paper 19). The tensile strength is usually derived from the compressive strength, using standard formulae which appear to be valid from about one day. Occasionally, these formulae are verified with tensile splitting tests.

MR STONE, Paper 22
In reply to Mr Hepper, we are not convinced that there is a difficulty with evaluating alternative designs that provide a time saving. The real benefits of time saving occur within the contract and would be reflected in the tender price. In addition to these there are traffic benefits of early opening, but it is not difficult to put a value on these and to take account of them in tender appraisal.

With respect to environmental advantages, however, this is more difficult. We must remember that environmental benefits are subjective and that something which benefits boat users may not be a benefit to fishermen and to bird habitat. There is very often a commercial consequence, too. A number of assessment frameworks are available but, at the end of the

DISCUSSION

day, if there is a cost penalty to balance against any environmental advantage, it is a matter of judgement. In the public sector, such judgements can usually be made only at a very senior level.

With specific reference to Conwy, we are not convinced that any of the claimed environmental advantages of the Gibb-Tomlinson alternative really exist. With respect to the advantages claimed, I would comment as follows.

(a) A site for the casting basin was freely available for the concrete elements on inter-tidal land and in a location likely to cause minimal disturbance.
(b) The double skinned alternative would, we believe, have required some dredging of the approach channel through the narrows, whereas the concrete scheme clearly requires none. The main trench would be similar in both cases; therefore, overall there might be more dredging with the alternative.
(c) The greater 'visible on-site construction' of the concrete tunnel units was not a significant environmental disadvantage where the casting basin was situated, and it was, in fact, a tourist attraction.

In the event, the alternative was not put forward, but we believe that environmental consideration would not have been sufficient to offset any additional tender sum in this case.

23. The design and construction of steel immersed tube tunnels: an American technology

N. A. MUNFAH, PE, and Y. M. TARHAN, PE, Parsons Brinckerhoff Quade and Douglas, Inc., New York

SYNOPSIS. Two distinct approaches to immersed tube tunnel construction have developed: steel immersed tube tunnels in the United States and concrete immersed tube tunnels in Europe. Tradition keeps these systems apart, yet each is a valuable option, with steel's greater ease of fabrication and outfitting weighing against concrete's typically lower materials cost. This paper presents the concept, the design, and the construction techniques associated with steel immersed tube tunnels, clarifying the steel tube's major divergent features and advantages.

INTRODUCTION: THE TWO TRADITIONS

1. In the world as a whole, immersed tube tunnels are built almost equally of steel and concrete construction, 48 percent of steel and 52 percent of concrete, according to a 1988 survey by D.R. Culverwell (ref 2, 3) of road and rail immersed tube tunnels throughout the world. This almost equal worldwide distribution conceals the fact that in the United States over 90% of the immersed tube tunnels were built using steel shell, while in Europe 100% of the immersed tube tunnels are concrete. For the rest of the world, the steel immersed tube tunnels represent 55% of the total number of immersed tunnels. See Table 1. This indicates that two distinct approaches have taken place since the construction of the first immersed tube tunnel: an American one which follows the steel shell concept, and a European one using concrete elements.

TABLE 1. TYPES OF IMMERSED TUNNELS

LOCATION	TOTAL	CONCRETE Number	% Total	STEEL Number	% Total
U.S.A.	22	2	10%	20	90%
Europe	24	24	100%	--	0%
Japan	14	3	22%	11	78%
Elsewhere	7	6	86%	1	14%
TOTAL	67	35	52%	32	48%

Immersed tunnel techniques. Thomas Telford, London, 1989.

CONSTRUCTION

2. We will attempt, in this paper, to present some of the design aspects and the construction techniques of steel immersed tube tunnels. The Second Downtown Elizabeth River Tunnel between Norfolk and Portsmouth, Virginia, U.S.A., will be used for illustration.

HISTORICAL DEVELOPMENTS

3. The tradition of immersed tube construction began in the United States, and began with steel. The first immersed tunnel was the 1894 Shirley Gut siphon in Boston, Massachusetts, 79m long and 1.8m in diameter. The first railroad immersed tunnel was a crossing between Michigan and Ontario completed in 1910. The immersed tube was 800m long and had two tracks. The first highway immersed tube was built also in the United States and was the Posey Tunnel between Oakland and Alameda, California. This tunnel, completed in 1928, was a two-lane crossing approximately 740m long, and interestingly, was the world's first concrete immersed tube tunnel. This was followed in 1930 by the steel Detroit-Windsor tunnel between Michigan, U.S.A., and Ontario, Canada. See Fig. 1. This tunnel was the first road tunnel between two countries. The immersed tube length was 670m. The significance of this tunnel was the development of the octagonal double shell cross section, which became a typical cross-sectional shape for most of the steel shell tunnels in the United States.

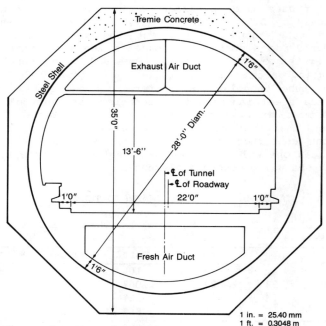

Figure 1. Detroit-Windsor tunnel

4. The first immersed tunnel built in Europe was the 1941 Maas tunnel in Rotterdam, The Netherlands. See Fig. 2. It had two bores, two lanes each. It was a reinforced concrete tunnel and had a length of about 590m. It is interesting to note that the Maas tunnel was originally designed as two separate bores of the steel shell construction type, similar to the American technique developed in the Detroit-Windsor tunnel. The Contractor, however, proposed alternatively that the two bores be combined in one rectangular cross section. This was made feasible by the development of the sand jetted foundation. It became the trade-mark of the concrete immersed tunnels in Europe.

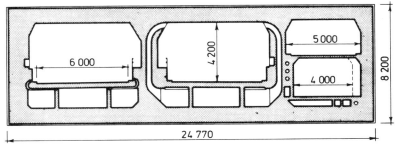

Figure 2. Maas tunnel
(Courtesy H. Van Tongeren—*The Foundation of Immersed Tunnels*)

5. During the 1950's several immersed tunnels were built in the United States. They were all road tunnels and they were all of the steel shell construction. The most notable tunnel in this era was the First Hampton Roads Tunnel in Virginia, which was the world's longest immersed tunnel at the time (1957) with a length of 2,091m.

6. In the 1960's the construction of immersed tunnels flourished in Europe using the typical concrete rectangular box cross section, and was made possible by the development and improvement of the sand jetted foundation technique. It is believed that in this era, with a sudden rise in European immersed tunnel construction - all of it using concrete - the development of immersed tunnels split into two distinct branches: steel immersed tunnels in the USA and concrete immersed tunnels in Europe, while the rest of the world followed either one of these two distinct methods.

7. Many developments occurred in the last 20 years that led to larger cross sections, longer elements and deeper tunnels. At present, the records are shared equally between the two techniques. The concrete construction technique holds the record of having the widest tunnel and the longest element, while the steel shell construction technique holds the record of having the longest and the deepest immersed tunnel. The widest tunnel is the 1977

CONSTRUCTION

Drecht tunnel in the Netherlands with a total width of 48.8m, consisting of four bores, two lanes each. The longest element in a tunnel is in the 1980 Hemspoor tunnel in The Netherlands, with element length of 268m. The longest and the deepest tunnel is still the 1970 Bay Area Rapid Transit (BART) Trans Bay Tube in San Francisco. See Fig. 3. It has a length of 5,825m, and the deepest point is about 40m below water level.

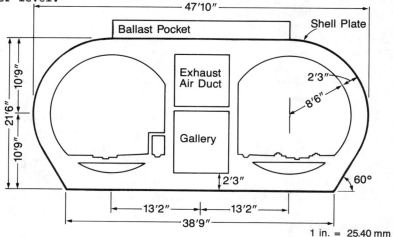

Figure 3. Bay Area Rapid Transit (BART) tunnel

1 in. = 25.40 mm
1 ft. = 0.3048 m

STEEL SHELL IMMERSED TUNNEL CONCEPT

8. Steel immersed tube tunnels can be divided into two categories:

o Single Shell Tube and
o Double Shell Tube.

9. The cross section of a single shell tube consists of an outer watertight steel shell plate continuous over the length of the element, usually 10mm thick, stiffened with internal transverse members spaced at about 1.5m and sometimes with longitudinal stiffeners. Inside the element a cast-in-place reinforced concrete liner is placed to act compositely with the steel shell. The BART tunnel is an example of the single shell tunnel. Traditionally, the single shell tube is protected against corrosion using a cathodic protection system, although a pneumatically applied concrete protection course has also been used.

10. The cross section of a double shell tube includes an interior steel shell plate, usually 8mm thick, stiffened by external diaphragms spaced at about 4.5m and external longitudinal stiffeners. The interior is lined with a minimum thickness of reinforced concrete acting compositely with the steel shell and the stiffeners. An outer form plate, 6mm thick, encloses the exterior diaphragms. The

space between the exterior form plate and the interior shell plate is filled with tremie concrete to provide the ballast needed to control sinking and also protects the interior shell plate against corrosion. Fig. 4 shows a cross section of a double shell tunnel.

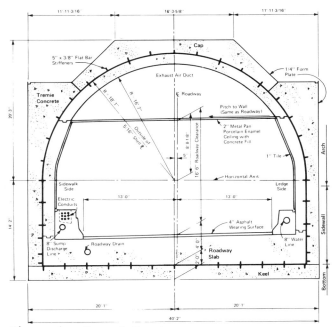

Figure 4. Second Downtown Elizabeth River Tunnel

1 in. = 25.40 mm
1 ft. = 0.3048 m

11. The tunnel element is usually fabricated in a shipyard using the modular technique. In this technique, modular subassemblies are fabricated and then fitted and welded together on the shipways to form an element. The reinforcing bars and all other embedded items are installed prior to launching. The ends are then closed by watertight bulkheads and the keel concrete is placed to provide stability during launching and towing. The element is then launched into the water and towed to an outfitting site where the interior concrete is placed while the element is afloat. Upon completion of the outfitting, the element is towed to the project site, lowered into its final place and joined with the previously placed element.

BUOYANCY AND STABILITY

12. A chief feature of steel tube construction is that the relatively light steel shell is fabricated first - on land - and the massive concrete interior (ballast) is added later, after the tube has been launched and is afloat. Thus, the stability of the tube during outfitting, during placing, and in its final position is of vital importance.

CONSTRUCTION

The uplift stability is evaluated for a range of unit weights for the concrete, water, and backfill. The buoyancy analysis is conducted for upper and lower bounds, considering the extreme cases of a "heavy" element with "light" water during outfitting, and a "light" element with "heavy" water during placement, and in the final position. The traditional compact shape of the element increases its sensitivity to variation in water and concrete densities. Hence a continual monitoring of the material unit weights during construction is usually specified.

13. Handling an element is relatively easy since most element's weight is not added till the outfitting stage. This keeps the element light throughout the fabrication phase, and adds weight only once it is already floating and easier to move. Upon launching, a two-lane road tunnel element usually weighs 2,000 to 3,000 tons and draws about 2.5 to 3m of draft. At the completion of the outfitting, the element weighs about 10,000 tons and has a freeboard of about 0.5 to 0.75m. This is usually sufficient for the safe transfer of the element to its placing site.

14. During outfitting, the interior concrete is placed in a predetermined sequence designed to keep the element on an even trim and list and to control the hydrostatic pressure and the longitudinal (hogging) bending moment.

15. The lateral and longitudinal stability of the element is evaluated at each outfitting stage. Draft-displacement curves are usually established for the element. Fig. 5 shows the draft displacement curves of the Second Downtown Elizabeth River Tunnel.

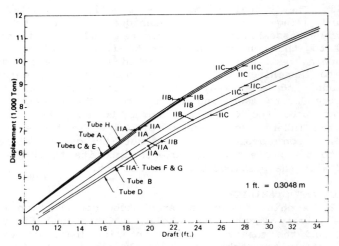

Figure 5. Draft displacement curves for the Second Downtown Tunnel

322

16. The actual safety factor against uplift in the final position is usually about 1.20 to 1.25, not relying on the side friction between the element and the backfill. However, the controlling buoyancy condition occurs during placing and upon dewatering the joints. Usually the minimum safety factors are specified to be 1.05 and 1.02 respectively.

STRUCTURAL ANALYSIS AND DESIGN

17. Unlike concrete elements, which are sensitive to settlement, shrinkage and creep effects, steel-shelled elements have sufficient flexibility and ductility that these factors do not control the design. However, the structural analysis of a steel element must consider separately each stage of construction:

o Fabrication and launching
o Outfitting
o In-Situ.

During fabrication, the element is analyzed for the type of fabrication, using either the modular method of fabrication or the subassembly method. In the modular fabrication, each module is designed to be self-supporting and to span the maximum obtainable plate width, which is usually 4.5 to 5.0m.

18. In the subassembly method, the subassemblies are usually less rigid than the modules, thus requiring special handling methods, and additional temporary stiffeners to maintain the fabrication tolerances. Fabrication tolerances are established to provide adequate design and practical fabrication. The shrinkage and warping effects due to uneven temperature and weld heat are considered.

19. Once fabricated, the element is usually side or end launched, although in The Second Downtown Tunnel, the elements were skidded onto barges. Structurally the element is more suited for side launching than end launching.

20. After launching, the floating element's position in the water varies with each stage of outfitting as the interior concrete is placed in a predetermined sequence. The ends of the elements, which accommodate the joint design, and the end bulkheads are heavier than the rest of the element. Hence, they impose concentrated loads which are balanced by uniformly distributed buoyancy forces. This results in negative (hogging) longitudinal moment on the element. Additionally, circumferential bending moments are applied on the element by the hydrostatic pressure. The longitudinal and the circumferential moments vary with each outfitting stage. Fig. 6 shows the changing moment, and

CONSTRUCTION

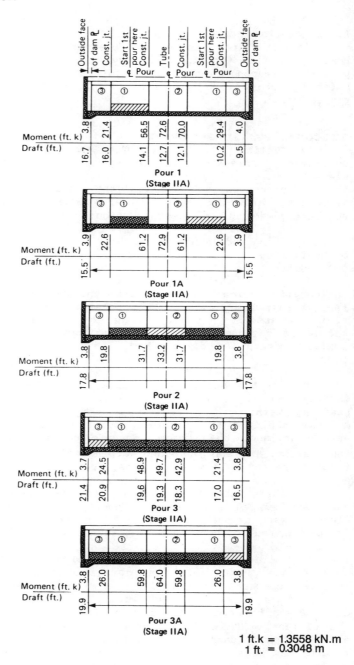

Figure 6. Moment and draft during outfitting – Second Downtown Tunnel

draft of an element of the Second Downtown Elizabeth River Tunnel at five successive operations during outfitting, as concrete is added at different locations along the tube.

21. Designing the tube to resist these stresses uses distinct structural systems. The longitudinal bending stresses are resisted by the shell plate, the longitudinal stiffeners (if present), and the keel concrete. The circumferential stresses are resisted by the shell plates and the transverse stiffeners or the diaphragms. The shell plate and the longitudinal stiffeners are usually subjected to global stresses due to the longitudinal bending and to local stresses due to the circumferential water pressure. Hence, an analysis of the interaction between the global and local stresses is usually performed. In addition, the principal stresses and local buckling of the shell plate are evaluated.

22. In its final in-situ position, the tunnel is essentially a long cylindrical structure, thus a state of plane strain exists. Therefore, a transverse analysis is usually sufficient. An elastic linear frame analysis with soil-structure interaction is usually used. The tunnel is then designed using the composite action among the shell plate, the stiffeners, and the reinforced interior concrete liner.

FABRICATION OF A TUNNEL ELEMENT

23. One of the most advantageous features of the steel shell tunnel technique is its ease of fabrication - at a convenient site that can be as far as several thousand kilometers from the tunnel site. The tunnel element is usually fabricated in a shipyard. However, any other fabrication yard with a reasonable access to the water is adequate to perform the work. Hence, no special site is needed, nor any special site preparation. The fabrication of the tunnel elements is usually performed using standard welding techniques and common welding equipment readily available at all fabrication yards. The proximity of the fabrication yard to the tunnel site has less importance than the proximity of the casting basin to the tunnel site in the case of a concrete tunnel. The only factor influencing the selection of a fabrication yard, besides its technical capability, is the economic factor. This is evaluated by comparing the transportation cost versus the labor rate in the remote fabrication site, the workload and the overall general economy in the fabrication yard area. This issue is extremely important for tunnels in urban areas where the availability of land for casting basins is scarce. This has been proven to be the case in the construction of the 63rd Street Tunnel in New York City, where the fabrication was done in Port Deposit, Maryland; and the construction of the Second Downtown Elizabeth River Tunnel in Virginia, where

CONSTRUCTION

the fabrication was performed in Corpus Christi, Texas. The latter was fabricated at a location over 3000 km away from the project site. This point is important in a competitive bidding situation, since the contractor can locate a shipyard with a low level of workload, or between orders placed for shipbuilding, to fabricate the tunnel elements at a very favorable price.

24. Furthermore, the fabrication of the steel element expedites the overall construction schedule because several construction activities are performed simultaneously. While an element is being fabricated, other elements are being outfitted, the trench is being dredged, the foundation is being placed, and elements are being placed. This is particularly true in a case when an immersed tube section of a project is on the critical path of the overall project schedule. During the BART tunnel construction for example, while some elements were being lowered and connected in the trench, others were still being fabricated. Even the end bulkheads were designed for multiple uses.

25. Finally the fabrication of the steel shell in an established yard, instead of in a specially built casting basin required for concrete element construction, greatly reduces the possibility of adverse environmental impacts and consequently reduces the process of assessing impacts and obtaining permits. Such simpler assessment and permit process can significantly lower the cost and advance the schedule of the project.

Fabrication Techniques

26. In the modular fabrication technique, the shell plate is fabricated in modules of about 4.5m long. The plate is laid flat and the longitudinal stiffeners are welded onto it, then the plate is rolled to the required diameter to form a circular cross section. While the module is held in a jig, the transverse stiffeners are welded onto it. The modules are then fitted and welded together to form the element. This method is particularly useful in the case of a circular cross section. In the subassembly technique, the cross section is divided into several subassemblies of a length approximately 20 to 25m. Each subassembly is fitted with its stiffeners, then they are all joined and welded together to form the cross section. Several assemblies are then joined together to form an element. The watertightness of a steel element is of vital importance. Therefore all the welds are tested for watertightness.

27. Several launching and lowering techniques have been used: Side launching, end launching, or skidding onto barges. The application of each technique depends on the fabrication yard set-up, the equipment, and the available water depth. The element is then transported to the outfitting site where the interior concrete is placed while the element is afloat.

28. The outfitting site is usually adjacent to the project location, although, in the case of the 63rd Street tunnel in New York for example, the elements were outfitted in Maryland, several hundred kilometers away from the project site. The critical factors of an acceptable outfitting site are: available berthing space, sufficient draft, and good land access. These requirements are usually easily attainable within the vicinity of a project site by utilizing or modifying existing facilities. For example, the Second Downtown Tunnel elements were outfitted using an abandoned U.S. Coast Guard pier adjacent to the project. If no outfitting facility is available, a temporary inexpensive facility can be easily constructed for this purpose, and dismantled upon the completion of the project.

PLACING AND JOINING THE ELEMENTS

29. Upon completion of the outfitting, the element is towed to the project site and lowered to its final place in the prepared trench using a lay barge. Tremie concrete or gravel ballast is placed in the ballast pockets to provide the needed weight to overcome the element's buoyancy. Each element is equipped with a set of rubber gaskets, alignment wedges, couplers, jacks, and a dewatering system. See Fig. 7. The gasket set used in most of the tunnels in the United States is composed of two trapezoidal continuous

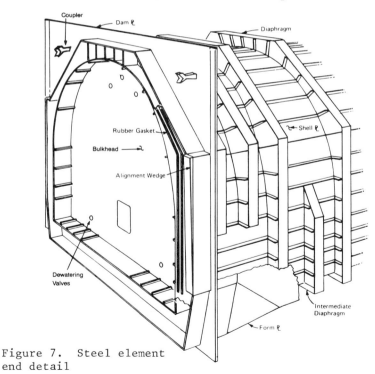

Figure 7. Steel element end detail

CONSTRUCTION

gaskets. Fig. 8 shows a typical gasket detail. It is believed that the double gasket provides added protection, less vulnerability to damage, and greater tolerances in the fabrication of the element ends. The cantilever lip of the outer gasket acts similar to the tip of the Gina gasket in providing the initial seal.

30. After the element is lowered, and while it is still being held by the lay barge, it is moved to the outboard end of the previously placed element so that it can be received by guide plates on the element in place. Couplers extending from one end of the element can be engaged with their mating parts on the other element. Hydraulic jacks pull the element being placed into contact with the previously placed element and give the initial seal of the gasket. The joint is then dewatered, and the hydrostatic pressure on the far end of the element compresses the gasket and seals the joint. The element is then surveyed for horizontal and vertical alignments. If the alignment is found to exceed the placement tolerances, the joint is reflooded, the alignment wedges are reset, and the joint is redewatered. Fig. 9 shows a simplified flow chart of the joining procedure. After the joint is dewatered, the bulkheads are removed, steel closure plates are welded to join the steel shells of the adjacent elements, and the interior concrete is completed.

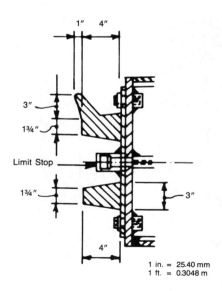

Figure 8. Gasket detail

(Courtesy Bickel and Kuesel—*Tunnel Engineering Handbook*)

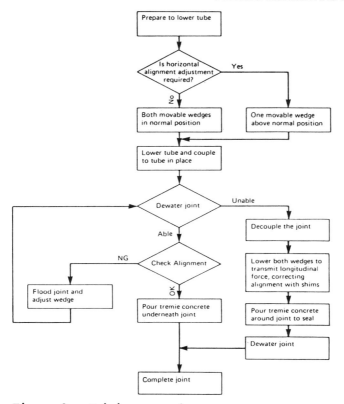

Figure 9. Joining procedure

CONCLUSION: THE ADVANTAGES OF STEEL

31. The advantages of steel immersed tube tunnels are the convenience of:

o The fabrication techniques
o The outfitting procedure.

32. The fabrication of the steel tunnel elements in shipyards instead of special casting basins provides an expeditious construction schedule, lesser environmental impact, and therefore the potential for a considerable cost saving. Actually realizing these savings depends greatly on the local circumstances and regulations. As the European Community unifies economically and removes economic barriers, shipyards in neighboring countries can enter competitive bidding for the task of tube fabrication. This will further enhance the competitiveness of steel with concrete and give European tunnel designers a new opportunity to use steel as a valuable option alongside

CONSTRUCTION

concrete, choosing one or the other for the best choice on a particular project. The outfitting of the steel tube elements while they are afloat allows ease of handling since the elements are comparatively light on land, and reduces the temporary foundation cost greatly since the majority of the weight is added while the element is buoyant.

33. Although the selection of the type of tunnel depends on sound engineering and economic reasons, tradition has played significant roles in the selection process, leading to a total divergence between the two types of tunnel construction. It is hoped that such technical exchange conferences as this will lead to greater opportunity in the future to break the barrier of tradition in selecting a tunnel type.

REFERENCES

1. J.O. Bickel and T.R. Kuesel "Tunnel Engineering Handbook" Van Nostrand Reinhold Co. 1982.
2. D.R. Culverwell "World List of Immersed Tubes" Part I, Tunnel and Tunneling, March 1988.
3. D.R. Culverwell "World List of Immersed Tubes From Hong Kong to Hampton Roads" Part II, Tunnel and Tunneling April 1988.
4. D.R. Culverwell "Immersed Tube Tunnels" Tunnel and Tunnelling March 1976.
5. Proceedings of the "Delta Tunnelling Symposium" Amesterdam November 1978.
6. Proceedings of the "Delta Tunnelling Symposium II" Amesterdam November 1987.
7. N.A. Munfah "The First Horseshoe Immersed Tube Tunnel: Concept and Design Considerations" Proceedings 25th Structures, Structural Dynamics and Material Conference, Palm Springs California, 1984.
8. N.A. Munfah "The Fabrication of An Immersed Tube Tunnel: A Case History" James F. Lincoln Foundation Award, Cleveland Ohio, 1984.
9. D.N. Tanner "Immersed Steel Tube For Elizabeth River Tunnel", Tunnel and Tunnelling March 1983.

24. New approaches to immersed tunnel construction

F. J. HANSEN, MSc, FICE, and J. F. HANSEN, MSc, Hansen and Hansen Ltd, UK and Hong Kong

SYNOPSIS. The Paper reviews the design philosophies and construction methods that have been used for immersed tunnels. The ways in which the steel or concrete elements are prefabricated are compared and some new techniques evaluated. A new method of sinking the units using positive buoyancy is presented which eliminates the need for jacking pads on the seabed. Joining of the elements and the joints themselves are reappraised and finally the foundation method is discussed and a scheme developed which combines the principles of sandplacing and sandflow.

INTRODUCTION

1. Immersed tunnels have in the past been surrounded by a certain mystique and understandably so as they generally were contractors alternatives to major bridge schemes or difficult deep tunnels. Furthermore contractors with experience have felt no strong urge to unveil the mysteries or trying new methods when they already had one that worked. But within the last few years a number of immersed tunnels have come out for tender with several contractors bidding for the work, so there is now a definite incentive to find new and more competitive construction methods. And with new construction methods no doubt new design details will be developed since it is the construction which to a great extent dictates the design.

2. It is often the in-situ construction of the approach ramps and cut-and-cover sections of the project which present the greatest technical problems and take up most space in papers on immersed tunnels. This paper will only deal with the aspects exclusively associated with the immersed section, i.e. the fabrication of the tunnel elements, the placing, joining and foundation method.

DESIGN OF THE TUNNEL ELEMENTS

3. It is quite normal to talk about a "steel" tunnel or a "concrete" tunnel, which is completely misleading when one can find - as in a very recent case in Denmark - that the "concrete" tunnel contained more

steel than the "steel" tunnel. The difference in design philosophy of the two types of tunnel is first of all governed by the different approaches to the fabrication method of the tunnel elements, and since they are the greatest single cost item it is of course extremely important to find the most effective fabrication method.

4. The design philosophy behind the steel tunnel is relatively simple in that the steel shell provides the watertightness as well as the structural strength. Concrete is added to give the necessary ballast and contributes, in varying degrees, to the strength of the element.

5. The design of a concrete tunnel is not so straightforward. While it is generally accepted that reinforced concrete can be made watertight, i.e. service reservoirs and watertowers, there is not the same confidence with immersed tunnels. This is understandable as a reservoir can be emptied and repaired if it is causing problems but a similar approach does not work with a tunnel. Also the thick wall and roof sections and the casting sequence make crack control more difficult. However, leaking itself is not a serious problem and is unlikely to affect the operation of the tunnel. What is of real concern is that the reinforcement may start to corrode and that the tunnel may eventually become unservicable. Over the last decade a great deal of research, funded by the oil companies involved in projects in the North Sea, has been carried out concerning the corrosion of reinforcement in offshore structures. The findings have shown that surface cracks up to 0.6mm do not appear to affect the corrosion of reinforcement in permanantly submerged concrete. Hence flexural cracking should not need any special treatment for an immersed tunnel. However the "hollow leg" situation has been identified as a potential problem due to the presence of both corrosive salts and oxygen on the inside surface. This is the situation on the inside of an immersed tunnel so that if corrosion is going to be a problem it will be on the inside where it is relatively easy to repair. Taking this reasoning a step further a case could be made for putting the steel or waterproofing membrane on the inside face.

6. The major problem in a concrete tunnel is the control of the thermal cracking, due to construction restraints, which penetrate the full thickness of the concrete section. Various design philosophies have been adopted. One approach is to design only for strength requirements and then to provide an outer waterproofing membrane. At the other extreme the membrane is dispensed with by elimination of cracking using cooling pipes to control the temperature. Various intermediate solutions have also been adopted wherby cracking has been minimized by the introduction of contraction joints in the walls or additional reinforcement has been provided to limit the size of the thermal cracks. However, even in these cases a waterpoofing membrane has been added as an insurance. There is no clear winner as far as the design is concerned so that each

case must be judged on its merits. The size of the elements, fabrication facilities and programming constraints are just as likely to determine the final choice.

PREFABRICATION OF THE TUNNEL ELEMENTS

7. For "concrete" tunnels it has been the practice to build a dry dock for the purpose and to construct all the elements in one batch and then to place them in rather quick succession one after the other. But lately the tunnel projects have become longer requiring more elements which make this approach impractical. For the MTR tunnel in Hong Kong and the Eastern Harbour Crossing also in Hong Kong it was decided to have a smaller dry dock and use it more than once. The MTR dry dock was used 4 times and the Eastern Harbour Crossing was used 3 times.

8. The "steel" tunnel concept is based upon the fabrication of the elements one at a time on a slipway and completion of the required concreting a float. The advantage of this method is the saving of a dry dock and an earlier start of the placing of the elements. There are also distinct disadvantage of this method, difficult to assess in clear terms of money so proponents of the two different methods have just pressed on with their schemes without being influenced too much by ideas from the other side.

9. Recently, however, some new thinking has appeared, which could influence the existing methods to such a extent that the difference between them would disappear with a design concept that would indisputably meet the long term structural and durability requirements in the best possible way.

10. On the recently completed cable tunnel in Singapore the tunnel elements were made up of short 3 m segments cast vertically and then assembled into 100 m elements by means of longitudinal prestressing. The joints between the segments were made big enough to be concreted up and the external formwork left in place as a waterproof cover.The elements were assembled above water on a platform, which could lower the elements into the water and the production rate achieved was one element per week.

11. There are two aspects of this approach which make it superior to the "steel" and the "concrete" method. It makes the concreting easier and eliminates the temperature shrinkage problem, so that full advantage can be taken of uncracked concrete watertightness, and it delivers the elements on a production line as fast as the "steel" method without the need for an external steel shell and concreting afloat.

12. On a recent tender for a very large immersed tunnel project at least one of the bidding groups found it competitive to apply vertical segmental construction and to assemble the segments into elements in a small drydock,as shown on Fig.1. The method could be applied to

CONSTRUCTION

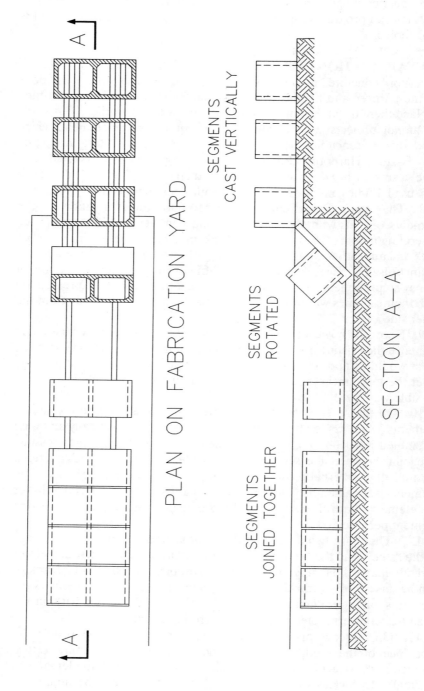

Fig. 1 Unit fabrication using vertically cast segmental construction.

both types of tunnel and the rate of production could be the same, so the choice would not be a question of construction method but rather an assessment of the quality of the two designs.

13. If the vertical construction of the entire cross section in one pour eliminates the the risk of getting cracks penetrating the full thickness of the concrete section it would appear unnecessary to penalise the concrete with a waterproofing membrane.

14. To obtain the required rate of production it is possible that the vertical construction will need more working space than the conventional horizontal method and it is highly unlikely that it could have been applied to the Hong Kong's second harbour crossing. But the method can be developed in many respects. How to organise the concreting and handle the segments and perhaps even divide the cross section into two halves for easier handling even of segments of 1-2000 tonnes can be moved, rotated and lowered onto the dry dock without the need for excessive lifting equipment. For example it would be quite feasible to give the immersed tunnel the same shape as a bored tunnel, i.e two circular tubes with cross connections at suitable intervals, as indicated on Fig 2. No doubt the example of the Singapore tunnel will cause the technique of immersed tunneling and segmental construction to be extended in both directions to larger as well as smaller structures.

HANDLING OF THE UNITS

15. The most common way of placing immersed tunnel elements is to float them to the sinking site on their own natural buoyancy and there to ballast them to negative buoyancy so they can be lowered to the trench bottom suspended from the surface. The surface suspension does not provide adequate position control for the final joining operation so another system of horizontal control has to be introduced before the elements can be safely joined to the previous element and supported on prepared temporary supports.

16. For railway tunnels in particular the internal water ballast complicates the finishing work by making access difficult as it can only be removed piecemeal when corresponding ballast has been placed. It has therefore been considered to make the elements negatively buoyant at the fabrication yard and use external temporary buoyancy during the transport to the sinking site. This method was applied on the Singapore tunnel and greatly facilitated the internal finishing.

17. Another approach could also offer advantages. One could maintain a positive buoyancy and pull the element down towards anchor points in the trench. The possible advantage of this approach could be threefold. There would be less temporary water ballast and therefore better access for the placing of ballast concrete. It would be possible to use the same system for the pulling down operation and the final

CONSTRUCTION

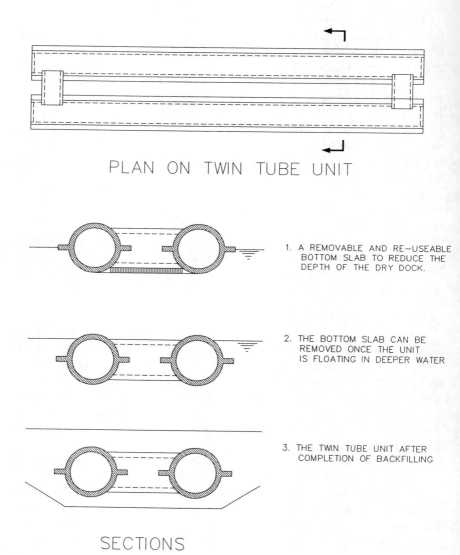

Fig. 2 A development of the segmental construction to allow fabrication of a twin tube unit.

position control, as the control lines will be shorter and tighter the closer the element gets to the bottom and be particularly useful in deep water. Thirdly it eliminates the necessity for temporary supports at the free end of the element. Also a great advantage in deep water.

18. The drawback of this positive buoyancy approach could also be said to be threefold. It is necessary to establish anchor points in the trench and it is necessary to apply the required forces from below and worst of all it could be called unproved technology. It would however appear that the two first mentioned disadvantages are quite easy to deal with and assess. For anchor points in the trench it would be natural to use gravity in the form of the previously placed elements at one end and a moveable ballast box at the other end and to make sure, that the elements at that stage would be heavy enough to withstand the uplift from the element being pulled down. "Heavy enough" is only a question of ballasting and in keeping with the idea of relying on minimum of temporary internal water ballast the answer, of course, must be external solid ballast. Arrangements should therefore be made to place material on top of the elements before the next elements is to be pulled down. That may not be too much of a hardship since backfilling on top of the elements have to be placed in any case and to retain the required amount of fill temporarily on top of the element would only require some quite low removable walls at the end of the element.

19. Following the same line of thinking it would be natural to use a movable ballast box at the other end for the anchor. Here again "heavy enough" would mean to make the box big enough to contain the required amount of solid ballast material. And after the placing of the element has been completed the ballast box would be emptied so that it could be moved forward to the next position, and the ballast material would of course be dumped on top and alongside the element just placed. So the task of providing anchor points would be to provide a box of adequate size and to double handle a small proportion of the backfilling material. A possible arrangement is shown on Fig.3.

20. The forces required to pull the element down would be of the same order as those used for the conventional winching down. It could even be considered safe to use less force. In the conventional method it is necessary to allow for an increase in salinity with depth and hence a loss of negative buoyancy; whereas with positive buoyancy it is the complete opposite. The position control gets better with depth, so even with no increase in salinity it would be safe to work with a reduced positive buoyancy at the start of the operation.

21. The forces required could therefore be applied in the conventional way by winches, block and tackle with only one difference. In the positive buoyancy method the multi-sheave blocks would be attached to the element and the anchor points below with only a single line pull from the winch pontoon to the element. Furthermore, since the pulling down forces

CONSTRUCTION

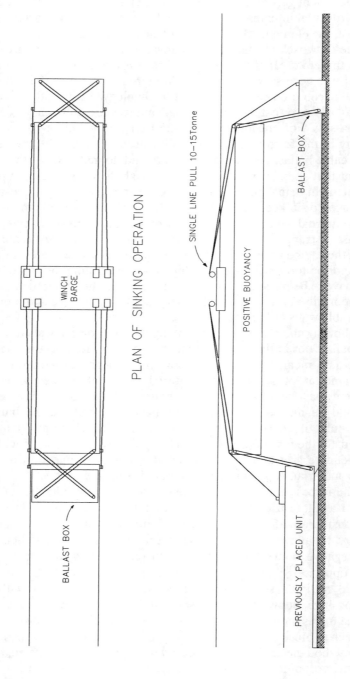

Fig. 3 An arrangement for sinking an element using positive buoyancy.

naturally would be applied at the four corners of the element with the winch pontoon at the centre, the line pull would be at rather shallow angles. The vertical component of the forces acting on the pontoon would be small compared to pulling down forces and conversely any surface movements of the vessel will only have a small effect on the pulling down.

22. To have complete control over the movements of the element during the pulling down two sets of block and tackle will be attached at each corner, one to provide a vertical pull and one to provide a diagonal pull. At the early stages of the operation the varous pulls are interdependent and governed by the horizontal forces; but at the closing stages where the position control is most important they will be independent.

23. A natural development from blocks and tackle is, of course, to apply the forces directly at the corners by means of hydraulic jacks powered through umbilicals from the surface. Special equipment could be developed, as shown on Fig.4, but existing equipment and established prestressing techniques could be used. This approach has the advantage that the pulling down is completely independent of surface conditions and the element could be disconnected from the umbilicals and locked in position if it became necessary for the surface vessel to take shelter.

24. The third disadvantage, of unproved technology, can only be tackled by perseverance and by pointing out that it may be unproven to immersed tunnelers but certainly proven and used in other lines of civil engineering.

JOINING THE ELEMENTS

25. One detail of "european" immersed tunnels attracts particular attention and that is the joint between two elements with the familiar "gina" gasket and the omega seal and at one time even a third line of defense. The great attraction of that detail is that it works and makes it possible to place the elements quickly on temporary supports with a watertight connection to the previous element. The supporting surface vessel can therefore be released quickly and the joint completed in the dry from inside.

26. It is of course important that the element can be made independent of weather conditions quickly and a great advantage that the joint can be completed from the inside, but is it really necessary to accommodate rotation, shear and longitudinal movements and have 100% instant watertightness all in that joint?

27. If a longitudinal analysis of the tunnel structure as a whole shows that some movement joints are necessary or would be beneficial, they

CONSTRUCTION

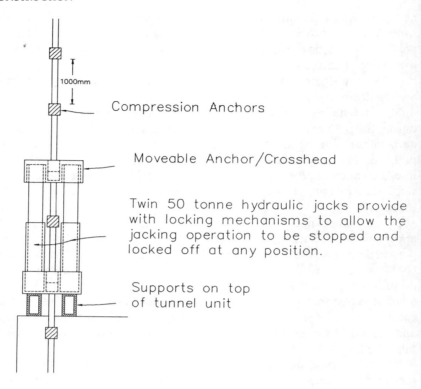

DETAILS OF THE LOCKING MECHANISM

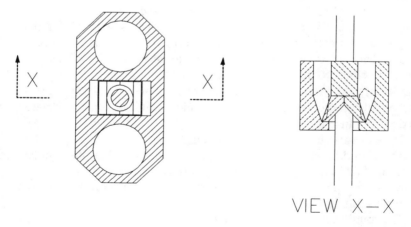

VIEW X–X

Fig. 4   Specialist jacking equipment for postive sinking which can be operated from the surface using an umbilical connection.

should of course be provided, but not necessarily at the the underwater joints. It is already established practice to build movement joints into tunnel elements in the fabrication yard and to lock them rigid by temporary prestressing across the joint, so why not exploit that practice?

28. It is obviously easier to make a rigid joint 100% watertight than a movement joint and obviously also easier to make a match cast movement joint watertight than it is to make it between two precast tunnel elements underwater.

29. Using positive buoyancy for the placing of the elements there is no urgency regarding instant watertightness or release of the surface vessel. Strictly speaking the element is on its temporary supports as soon as it is below water, so the task of joining the two elements together is not a question of instant watertightness. The elements can be pulled together or kept apart for as long as required to put things right. And since it makes sense to aim for a rigid joint the initial seal between the elements serves only the temporary purpose of keeping the leakage at a manageable level so the permanent works can be completed "in the dry".(Which is not 100%)

30. When the hydrostatic pressure in the space between the elements disappears the thrust at the free end has to be carried across the joint. The gina gasket does that by combining a body of rather hard rubber and a soft nose. The soft nose ensures the initial seal and the hard body is strong enough to take the thrust. However, a timber section with a soft rubber nose could do the same thing only better and cheaper, as shown on Fig. 5. The rubber profile cannot be modified on site, so the detailing of the two matching element ends is rather elaborate to achieve the required accuracy of the mounting. A timber section can be machined on site and bolted directly to the concrete outer face.

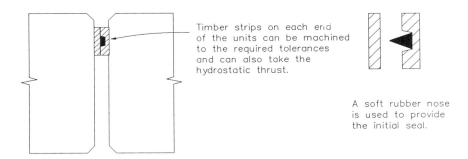

Fig. 5   A simple tunnnel joint which provides the initial seal but is not intended to allow permenant movement.

CONSTRUCTION

31. Since the positive buoyancy method combines the sinking and the temporary supports on the trench bottom into one single operation, it is possible to keep the element in any chosen position for any length of time. That may conceivably influence the method of joining the elements. Is is not necessary to dewater the joint space immediately. It would be possible to have two soft rubber seals and to hold the elements together, while the space between the seals could be grouted up. And the space between the bulkheads would be dewatered, when the grout has attained the required strength to carry the hydraulic thrust across the joint. Several other types of rigid joints could be developed.

FOUNDATION METHOD

32. The foundation methods for the "steel" tunnels and the "concrete" tunnels have differed in the past. The "steel" tunnel would be placed on a prepared screeded gravel mattress whereas the "concrete" tunnels would be placed on temporary supports and have sand pumped underneath afterwards.

33. If the fabrication of tunnel elements develops towards vertical segmental construction and advantage is taken of the improved concreting technique by eliminating the external waterproofing membrane the "concrete" tunnel ought to be competitive in most cases. It would then appear that the "steel" tunnel and its screeded gravel bed would lose out and only be applicable in special cases.

34. The technique of pumping sand underneath a tunnel element has developed into a kind of competition between the so called sandjetting and sandflow. They both work and they both have their proponents and detractors depending on which method they happen to have used. Very little practical information is available to determine whether one method provides a better foundation than the other. But having had opportunity to observe various variants of both methods in action it is possible to draw some conclusions.

35. It is certainly easier to pump sand than it is to lay a screeded mattress and even more so when the water gets deeper. It is also easier and cheaper to pump sand through a fixed hole in the bottom slab, than it is to travel a gantry on top of the element to manoeuvre a jetting pipe in underneath it. It is obviously better to be able to move the point of delivery than to have fixed predetermined points, especially when there is no reliable way of determining the best spacing of the delivery points. It is in any case the characteristics of the sand that govern the results and neither method appear to lend itself to continuous monitoring. It is generally a question of shifting to the next hole or moving the pipe, when it looks like no more progress can be achieved from where they are.

36. There is therefore still scope for improvement and the obvious thing to look for is a method which lends itself to monitoring, has no

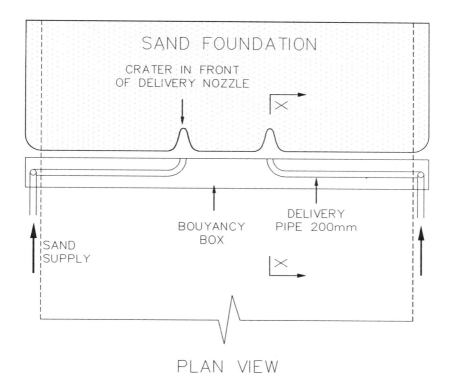
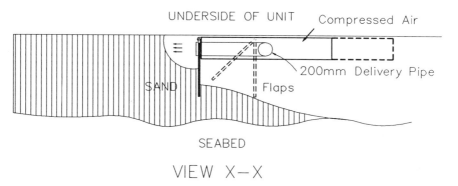

Fig. 6   An arrangement for placing the sand foundation which combines the principles of sandflow and sandplacing.

moving gantry travelling on top but which has moveable delivery points.

37. The first gantries on top projected above the water surface, but later on in deeper water it has been found possible to operate underwater, and for the Sydney Harbour Tunnel presently under construction the gantry should have been just about flush with the harbour bottom to be allowed in the shipping lanes. But with the positive buoyancy method in mind there can be no doubt that anything traveling on the element should be traveling on the underside of the element and be positively buoyant.

38. The principle of travelling underneath the element and being able to shift the delivery pipes transversely has been applied on the Hong Kong,s second harbour tunnel. However, the ability to position the delivery head at any position is something of a luxury and must be paid for by a complex mechanical arrangement which is difficult to maintain under water.

39. From the experience gained both in Hong Kong and in Kaohsiung, where the sandflow method was used, it is clear that the transverse movement is not necessary and a few delivery points moving along from one end of the element to the other can fill the space under the tunnel in a perfectly satisfactory manner.

40. To support a rigid delivery pipe underneath the tunnel is a simple matter of placing it inside a shallow buoyancy box as shown in Fig 6. The dimensions of this box can be chosen to provide as much positive buoyancy as necessary, and it can be reduced to nothing by letting in water. The sand can be delivered through a system of rigid pipes inside the elements out through the last bulkhead and a flexible connection to the delivery pipe or pipes in the buoyancy box. .

41. Being able to deliver the sand underneath the tunnel element is of course the main priority, but to have a reliable monitoring system is also highly desireable. In the sand flow method there is an increase in pressure as the sand "pancake" builds up.

42. To establish a relationship between pressure increases and sandplacing progress the buoyancy box is provided with flaps as shown. It is obvious that the water pressure will increase as the sand gets closer to the flaps. Hence the movement of the equipment can be triggered by a predetermined pressure increase.

43. It is with this approach feasible to turn a continuous monitoring process into an automatic control system of the entire sandplacing operation. The only surface activity where the sandplacing is taking place would be to probe alongside the sides of the elements to check that the sand has reached out to the full width and that progress on the two sides are the same.

26. Practical accuracies of construction of immersed tunnels — a case study: The Fort McHenry Tunnel, Baltimore, USA

W. C. GRANTZ, PE, and A. R. LANCELLOTTI, PE, Parsons Brinckerhoff Quade and Douglas, Inc., Boston

SYNOPSIS. Tolerances which are set excessively tight in contract specifications can severely impact bidding prices, delay construction progress, and may even result in claims for extra costs if a Contractor can prove that the requirements are unreasonable or even unattainable. The purpose of this paper is to provide some guidance regarding practical accuracies for the various stages of fabrication and placement of immersed tube tunnels.

INTRODUCTION
1. Immersed tube tunnels are being used all over the world to construct water crossings. The total number of these projects however remains relatively small as compared to other works, and the opportunity for an engineer to participate in the design, construction or construction management of such a project is a rare occurrence. Consequently it can sometimes be the case that an engineer in charge of an immersed tunnel design or construction project is doing it for the first time.
2. While most of the general structural aspects of these tunnels are relatively well known, some of the practical aspects of fabricating elements and placing them involve decisions and judgements that are often hard to make. The design engineer may have rather optimistic ideas of what can be done and will specify accordingly. Following this, the resident engineer may not understand what the really critical needs are and slavishly try and make a contractor adhere to an unnecessary nicety, at great cost, and which has no significant benefit to the end result.
3. This paper will consider some of the more important types of tolerances that may be encountered and their relative importance, and will examine a case study: the Fort McHenry Tunnel in Baltimore, Maryland, U.S.A. to compare some of these tolerances as specified with actual field-measured and accepted results.

THE FORT McHENRY TUNNEL
4. The Fort McHenry Tunnel was a very large vehicular tunnel project. It carried eight lanes in four bores. There were two two-bore immersed tunnels constructed side-by-side

CONSTRUCTION

with less than 3.0 metres of clearance between them as shown in Fig. 1. These tunnels were each 16 elements long and were in a long horizontal curve for most of the length of the crossing. The total length of the immersed elements for each tunnel was approximately 1,650 metres. Steel shell elements were used with a fully circular "binocular" configuration.

5. For the Fort McHenry Tunnel, the elements were fabricated in a small shipyard as watertight steel shells and were side-launched into the Susquehanna River at Port Deposit, Maryland. They were then towed some sixty kilometres in a floating condition (two metre draught) to an outfitting pier in Baltimore near the project. Here the interior and exterior concrete work was done. Finally they were towed into position in a placement barge and installed. The two tunnels were constructed side-by-side progressively alternating element placement between one tunnel and the other.

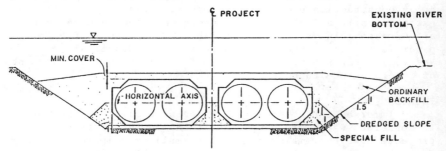

Fig. 1. Fort McHenry Tunnel section

FIELD SURVEYS

6. Field surveys were carried out at various stages of element fabrication, outfitting with concrete, placement and post-placement.

During Fabrication

7. Surveys during fabrication were used to control the flatness of the bearing surfaces, the geometry of the element and establish an initial shape of the element just prior to launching.

8. The bearing surfaces at the ends of each element were carefully validated by survey to establish any deviations from a plane surface spanning both bores. This was done to determine deviations from tolerance and to relate these surfaces to a similar validation survey carried out earlier on the mating element, now already underwater. The accuracy of the angle of the bulkhead on the end of each element was also checked by survey. An example of this is given in Fig. 2 showing the results of the validation surveys made on the mating surfaces of the two bores of Element No.1 (outboard) and Element No.3 (inboard). Note how some of the irregularities compensate between the two opposite faces in this case. This was the result of adjustments made by shimming

and grinding of the bearing surface of Element No. 3 to
improve the match with Element No. 1 already in place.

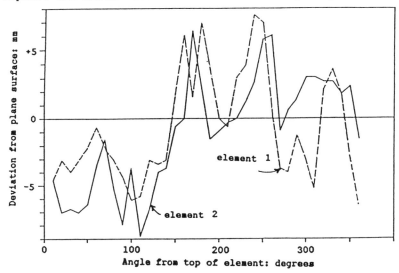

Fig. 2. Validation of Bore No.1 Joint 1 - 3

9. Surveys were made to verify that the roundness of the shell plate was maintained during the various phases of assembly. Surveys were also made to control the straightness of elements or the accuracy of their geometry to maintain interior clearance for the later placement of concrete.

10. Surveys were made at the end of the fabrication process while the element was supported on the shipways to establish a control reference of its initial shape. These surveys consisted in leveling data taken at points on each end of every other transverse diaphragm. This data could be related to an average plane which could later be reestablished even when the element was floating. Subsequent surveys were taken while the element was afloat and being outfitted with concrete. Comparison of this data permitted the changing shape of the element to be monitored throughout the concreting operation, prior to sinking. It provided a field check on the efficacy of the specified sequence of placement in preventing both undue distortions of the elements and potential overstressing. Excessive distortions could potentially affect the orientation of the mating surfaces of an element.

11. Since the element was to have the internal concrete placed while afloat, it was necessary to set out survey references within the element before launching. This was due to the difficulty of doing meaningful survey work inside an element after it was afloat and subject to motion and change in shape. This was done by welding plates to the shell and marking reference centerlines and elevations on these plates. During the outfitting of the element, string lines pulled

transversely between these plates were used to set shuttering (formwork) positions.

During Placement of Each Element

12. During placement of each successive element in construction the tunnel, survey operations for horizontal and vertical position control had to be very carefully planned and executed.

13. The element being placed was engaged on steel "guide beams" cantilevered from the tunnel portion already in place. Vertical adjustment at the joint between elements was accomplished by placing steel shims on the top of the guide beams. Horizontal adjustment at the joint resulted from the sides of these guide beams centering the tube being placed. These guide beams were also shimmed to permit slight lateral adjustments to make up fabrication errors and provide exact registration between the ends of the elements.

14. Underwater closed circuit television (CCTV) was used for visual monitoring from the placement barge control room of the joining of the elements. By using a calibrated staff extending from the element being placed the CCTV clearly showed alignment and position of the element being placed with respect to the outboard end of the tunnel already in place. Positive checks of alignment at the joint were also made by a diver using a steel straightedge to measure the coincidence of from one element to the other. After the joint could be accessed from inside of the tunnel measurements were made at intervals around the periphery of each bore of the space left between the elements to determine actual gasket compression. Fig. 3 shows an example of this verification. In a few of the joints surveyed, it was found that the inner annular gasket had not been compressed at some locations. The water was being kept out of the joint between elements by a flap on the exterior gasket. The reason that the inner gasket was not compressed was due to an unfavorable accumulation of fabrication errors and overall element distortion or misalignment. Despite the apparent lack of gasket compression in some joints, no joint ever leaked significantly.

15. Alignment of the outboard end of the element being placed was accomplished by survey. This was done by projecting the position of the element to the surface by using an optical plummet to sight down through a dry pipe (supported in a survey tower) to an illuminated target located on the top of the element itself.

16. Foundation material for the elements was installed using the "screeding" method commonly used in the United States. Vertical grades were set by accurate screeding of the foundation material using a special piece of equipment designed for this purpose. Soundings taken on a two metre spacing grid verified the elevations on the screeded foundation to an accuracy of about 25 mm. Vertical and rotational alignment was checked from level staffs mounted on the outboard survey tower.

17. Surveys for settlement and rotation were continued during placement until all initial ballast had been placed on the element. See Table 1 for typical alignment results before and after initial ballasting with sand.

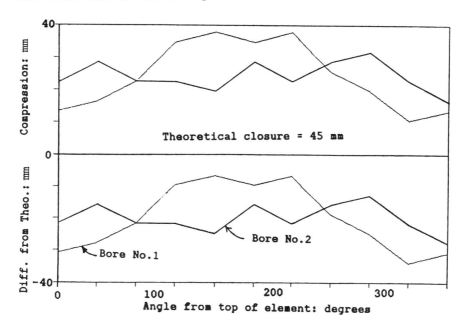

Fig. 3. Joint 1 - 3 Closure verification

Interior Surveys:

18. After the elements had been joined structurally and completely backfilled, a survey was made to establish the basis for a best-fit alignment of the walkways on both sides of each bore. This included not only the horizontal alignment but also the vertical alignment of the top of curb with respect to the concrete structural slab. This survey allowed smoothing of the visual lines of the tunnel both vertically and horizontally and also established a control for both the asphalt paving surface (and maximum and minimum thickness) as well as the control lines of elevation for setting the height of the tunnel ceiling panels. This automatically provided the nominal overhead clearance between the pavement surface and the ceiling in the tunnel. Thus, variations at joints, pavement slab elevation, etc. ultimately were not visible to the motorist.

TYPES OF TOLERANCES TO BE CONSIDERED

19. Tolerances must be developed for many aspects of element fabrication, element outfitting, and element placement. Consideration should include:

CONSTRUCTION

Table 1. Element alignment after placement and backfill
(measurements in mm)

ALIGNMENT OF OUTBOARD ENDS OF ELEMENTS

```
At Time of Placement      ||After Sand Ballast   ||Change                  |
-----|-----|-----|-----|-----||-----|-----|-----|-----||-----|-----|-----|-----|
Elem|Line |Grade|Roll |Sta. ||Line |Grade|Roll |Sta. ||Line |Grade|Roll |Sta. |
-----|-----|-----|-----|-----||-----|-----|-----|-----||-----|-----|-----|-----|
  1  |  9  | 49  | -34 | -15 ||  15 |  12 |  -3 |   3 ||   6 | -37 |  30 |  18 |
  2  |  9  |  0  | -34 |  73 ||  37 | -15 |   6 |  67 ||  27 | -15 |  40 |  -6 |
  3  |  3  |-34  |  0  |  -9 ||   0 | -46 | N/A |  12 ||  -3 | -12 | N/A |  21 |
  4  | 58  | 18  |  55 | 122 ||  18 | -18 | N/A | 122 || -40 | -37 | N/A |   0 |
  6  |-27  | 64  | -70 | -15 || -24 |  46 | N/A | -15 ||   3 | -18 | N/A |   0 |
  7  |  6  | 43  | -30 |   3 || -18 |  21 | N/A |  -3 || -24 | -21 | N/A |  -6 |
  8  |-58  | 37  |  -9 |  18 || -40 |  21 |   3 |  -6 ||  18 | -15 |  12 | -24 |
 32  |  0  | 61  | -18 |  67 ||   0 |  24 |  -3 |  40 ||   0 | -37 |  15 | -27 |
 31  | -9  | 49  |  -9 |  58 || -12 |  24 |  15 |  58 ||  -3 | -24 |  24 |   0 |
 30  | 15  | -6  |  -6 |   0 ||  15 | -18 | -24 |  -3 ||   0 | -12 | -18 |  -3 |
 29  |  0  | 58  |   3 |   0 ||  15 | -12 |   6 |  58 ||  15 | -70 |   3 |  58 |
 28  | -9  |-12  | -21 | -40 || -21 | -52 | N/A | -37 || -12 | -40 | N/A |   3 |
 27  |  0  | -3  | -21 | -12 ||   0 | -27 | N/A |  -6 ||   0 | -24 | N/A |   6 |
```

ALIGNMENT OF FREE ENDS OF TERMINAL ELEMENTS

```
At Time of Placement      ||After Sand Ballast   ||Change                  |
-----|-----|-----|-----|-----||-----|-----|-----|-----||-----|-----|-----|-----|
Elem|Line |Grade|Roll |Sta. ||Line |Grade|Roll |Sta. ||Line |Grade|Roll |Sta. |
-----|-----|-----|-----|-----||-----|-----|-----|-----||-----|-----|-----|-----|
  1  |  9  |  9  | -43 | -15 ||  15 |  -9 | -18 |   3 ||   6 | -18 |  24 |  18 *
  2  |  6  | 30  | -34 |  73 ||  94 |  -3 |  -6 |  67 ||  88 | -34 |  27 |  -6 |
 31  |-18  | 15  |   9 |  58 || -15 |   9 |  15 |  58 ||   3 |  -6 |   6 |   0 |
 32  |-40  |-37  |  -9 |  40 || -30 | -40 |  -3 |  30 ||   9 |  -3 |   6 |  -9 |
```

Sign Convention: Line: North = +   Tolerance: Line = +/- 1 1/2" (38mm)
                   Grade: High = +   Grade = +/- 1 1/2" (38mm)
                   Roll: Toward North = +
                   Station: East = +
                   Change = (After Ballast - At Placement)

(a) Fabrication: Circularity of shells; planeness and smoothness of bearing surfaces; straightness and transverse spacing of bores at the ends of the elements; angles between ends of elements (both vertically and horizontally); shape, spacing, location of supporting guide beams, and others.

(b) Outfitting: Interior concrete surfaces and clearances, roadway pavement position at the ends of the elements, and hog or sag of the element as concrete placement progresses.

(c) Placement: Vertical alignment, horizontal alignment, rotation, settlements, alignment wedges pre-placement adjustments, inboard joint positioning requirements, and outboard joint positioning requirements.

(d) After Placement: Adequate space for walkways and built-in provisions for air flues, conduits, etc., proper ceiling panel horizontal space and vertical clearance, and a suitable means of adjustment (within tolerance) of lighting fixtures to provide smooth, pleasing alignment.

RELATIVE IMPORTANCE OF VARIOUS TOLERANCES

20. It should be noted here that the tolerances which are not critical in a vehicular tunnel may be very significant in a rail tunnel where the dynamic envelope of a train is used to closely set the internal shape of the tunnel bore (here the length of the train units comes into play as well). Even in this case, tolerances must not be specified to be unnecessarily stringent. Some reasonable interior allowances must be made for mismatch in relative line and grade of the element, and variations in concrete placement.

Fabrication:
21. For the Fort McHenry Tunnel elements, the most important area requiring accurate fabrication was the bearing surfaces at the joint area. These joints consisted of an extension of the circular shell and a parallel circular plate both supporting a flat bearing plate. On one side was attached a gasket assembly made up of two gaskets and a bearing bar (used to prevent crushing failure of the gasket) which was fabricated separately and assembled in quarter sectors. On the mating element face, the corresponding area was simply a flat bearing plate to which the gaskets would make their seal. To provide a controlled transfer of load at the horizontal axis of the element during depressurization of the joint, two massive adjustable wedge bearings were provided. These wedges were preset to guide the elements in making corrections to horizontal alignment and could be reset if the first joint depressurization gave unacceptable alignment. Provision was made in the joints for forming and placing tremie concrete in the event the a seal could not be established in the joint. This contingency method of sealing the gasketted joints never had to be used.

22. The need for accuracy in fabricating the joints is due to the error-magnifying effect of the length-to-width ratio of the elements. This is particularly a problem for long, narrow, single-bore elements which may have an aspect ratio of nearly 10 to 1 (Fort McHenry Tunnel elements had aspect ratios of around 4 to 1). An error of 10 mm across a joint on an element with a 10:1 aspect ratio will be magnified to a horizontal positioning error (i.e. a swing) at the outboard end of the element of 100 mm. The adjustable wedges permitted correction of such errors but only to the extent to which the gaskets could tolerate deviations and still accomplish a seal. The gaskets were not only affected by peaks and valleys in the bearing surfaces, but also by the relative vertical alignment, or relative horizontal alignment of the elements meeting at the joint. Hog or sag of the element at placement could also

CONSTRUCTION

skew the bearing faces and affect the efficiency of the seal.

23. The bulkheads were fabricated in a horizontal position on a supporting structure. The joint assembly was then installed on the bulkhead and very carefully surveyed for planeness. These combined bulkhead and joint assemblies were then raised into position on the ends of the element. they were very large and heavy and difficult to keep precisely straight. They were approximately 26.5 metres wide and 13.1 metres high. The specified tolerance was +/- 6 mm from an average plane and +/- 6 mm from the theoretical plane. This was very difficult to achieve. Approximately twice this tolerance was eventually used. Detailed analyses were made of how the two surfaces between tubes would actually match to each other. Even at that, a considerable number of the bearing plates were adjusted by welding on shim plates and grinding them to adjust to acceptable surfaces.

24. It should be noted that this specification was reviewed in advance by a potential fabricator who assured the Engineer that there would be no problem in attaining this tolerance. As it turned out, it was not so easy.

25. The cylindrical steel shell plate was 11.9 metres in diameter and only 8 mm thick. The specification for circularity was +/- 13 mm. This was found to be very difficult to achieve in practice. The actual variation was held to close to that value for the greater part of the length but at the ends of the elements, prior to setting the end bulkheads in place, there was a tendency for the cylinder to sag out of round as much as 50 mm. In the end this was corrected so that the tolerance was finally set at +/- 25 mm. This dimension was critical only in the sense that, with fixed steel shuttering, the thickness of the structural concrete ring would be reduced in areas of negative tolerance. For such thin steel shell material on such a large diameter the +/- 13 mm tolerance was barely attainable at great effort and not worth the trouble. A tolerance of +/- 25 mm would have been more realistic and would not have significantly affected the structure.

26. The exact straightness of each element, was not too critical. On the other hand, the setting of the bearing surfaces to the proper angles with respect to the longitudinal axis of the element was very critical because of its effect on maintaining horizontal alignment of the tunnel within proper placement tolerances.

27. Other fabrication tolerances of importance concerned the location of the guide beam bearing surfaces, the location of the couplers for the jacks used to make the initial gasket seal, the orientation and shape of the wedges, the surface of the lower shuttering plate which supports the elements on the screeded bed, and the center-to-center distance between bores at each end of the element. Normal fabrication tolerances were used. The location and smoothness of the lower plate was important since it supported the element and thereby established its elevation. Screeding was sometimes adjusted to

accommodate this shuttering plate if it was determined to be
too low.

Outfitting
28. Outfitting of a steel shell element involves installing
all the interior concrete structure. This is done by placing
concrete sequentially to avoid distortions and overstresses of
the tunnel elements. If the concrete was simply placed
starting at the ends of the element, a "hog" would result; or
if work was started at the middle, a "sag" would result.
Consequently, concrete is staged in a balanced sequence
longitudinally. Concrete is also placed in stages in a
vertical sense: First the invert, then the haunches, then the
roadway slab, then the walls, and finally the arch at the top.
The changes in shape of a typical element during the
outfitting operation is shown on the graphs of Fig. 5 These
graphs compare element shape at some of these stages for the
two bores of Element No. 7. Four examples are given: between
pre-launch and casting invert; between completion of the
invert and completion of the haunches; same for walls and
arch; The final comparison is made between before launching
and completion prior to placement. The distortions were small
as planned and this element ended with an accumulated sag of
only 21mm. During the outfitting of the Fort McHenry
elements, the contractor made numerous requests to change
casting sequences to permit advancement of the work. These
proposals were checked by computer analyses and some were
accepted; others were not. The survey monitoring mentioned
previously was useful in checking that unacceptable
distortions did not actually occur.
29. After the elements were on the sea bed and the final
surveys had been made for the fabrication of the ceiling
panels, it was found that considerable encroachment of the
upper part of each wall into ceiling space had occurred. The
encroachment amounted to anywhere from 25 to 75mm for each
wall making the space for the ceiling panels as much as 130 mm
too narrow. It was very difficult to explain so much variation
in the walls but it was believed to be due to a combination of
factors including changes in shape during outfitting, and
perhaps deformation due to in situ loading. In the case of
the Fort McHenry Tunnel, provision was made in the
specifications for the tunnel finish contractor to make a
detailed, as-built survey of the tunnel prior to fabricating
the ceiling panels. The result was that this rather severe
departure from tolerance had no significant effect in the
final appearance and utility of the tunnel. No effort was
made to correct the encroachment by removing concrete, as this
would have done nothing but reduce protective cover on the
reinforcing steel and even weaken the structural ring
unnecessarily. Since there was ample space for lighting and
signs, and the alignment at the coves (intersection of wall
with ceiling) was smooth, there is presently no way of
telling, in driving through the tunnel, that the dimension

CONSTRUCTION

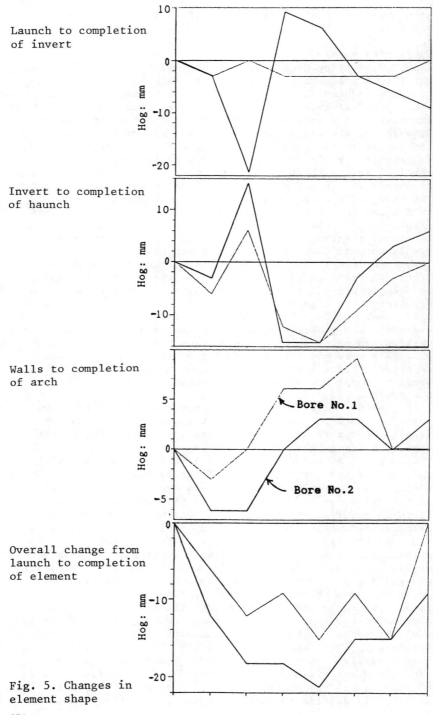

Fig. 5. Changes in element shape

across the tunnel at the ceiling line is less than was planned.

30. This is a good example of where a generous tolerance may be specified; if room is available. Extra allowance should be made outside the traffic clearance envelope to permit deviations which may occur in locations where attempted corrections could actually do structural harm.

Placement of elements

31. The planning for placement of each element was very critical. From the point of view of achieving desired placement tolerances, it involved survey planning, decisions concerning screeding elevations, shimming for alignment at the joint between elements, presetting of the wedges, and careful evaluation of how to achieve the best possible coincidence of the internal surfaces of the mating ends. Of particular concern was the matching of the roadway slabs in each element. Appropriate data were reviewed at preplacement meetings for each element where screeding elevations were set, shim thicknesses were decided on and desired alignment targets were established.

32. In general it can be said that the tunnel was always off theoretical line by some amount: in vertical elevation; in horizontal position; in rotation; or in axial alignment. The effort was always one of correcting back to the project theoretical alignment in a smooth manner both vertically and horizontally so that the tunnel would never get out of control in either direction. This became more critical as each tunnel approached the mid-river closure where the final element was to be placed. Major changes in direction could not be tolerated by the gaskets. This meant that the planning had to solve how to place several elements ahead to come back tangent to line and gradient. This was done horizontally using the wedges and vertically with shimming at the joint and by adjusting the screeding for each element. The preset wedges worked very well and only three joints required readjustment of the wedges during placement (the readjustment bringing the element to within 10 mm of the desired position). It is important to remember two things: First that misalignment across a joint between elements is more visible and difficult to compensate for than being somewhat off-line at the end of the element being placed; and second that vertical alignment is as critical -- if not more so -- than the absolute horizontal alignment (especially near the low point in the tunnel).

33. The screeding had to assure proper seating of the joint and recovery back to line of a diving or a rising tunnel end. This also worked well except for the occasional instance of soft bottom mud flowing into the trench after screeding, causing difficulty with some of the elements in initially reaching final grade. In some cases, this actually resulted in the removal of an element and rescreeding. Table 1 provides a summary of the deviations from line and grade of a sampling of

CONSTRUCTION

the Fort McHenry Tunnel elements, and shows the effect of placing sand ballast on the elements.

34. The interior reference marks installed prior to launching were tied by survey to the location of the exterior guide beams and were later used to relate the interior roadway slab to these exterior beams, enabling the next element to be adjusted to give proper registration of the roadway slabs from tube to tube. This registration was considered more critical than variations in the upper arch surface or in the invert surface of the structural ring. This was because the pavement wearing layer could not tolerate a large variation in slab elevation over the short distance across a joint. Too large a change would lead to an excessively thin wearing layer on one side of the joint and very thick one on the other. Otherwise if not corrected there would have to be a bump in the roadway to get over this change. With the wide, two-bore element, the construction of the roadways within the elements to proper elevations was very important. Proper seating of the element being placed was also very important in this regard, ensuring that the roadway surfaces would not be affected by too high an elevation or perhaps a rotation in the joint. The total deviation from a longitudinal straight line across the joint at the curb lines of the pavement should not be allowed to exceed +/- 25 mm if a 100 mm wearing course is to be used.

CONCLUSION

35. In deciding on tolerances, one must remember what is important and what is not. What is being done physically in prefabricating sections of tunnel and joining them underwater on a foundation prepared in-the-blind, must also be taken into account. As they say, one is "not building a watch". An immersed tunnel takes a while to truly stabilize. This usually will not occur until after placing the cover of protective backfill. An example of this occured with the Second Hampton Roads Tunnel where placement of backfill resulted in an incremental settlement of nearly 0.2 m for one of the elements. Note that subsequent to this large displacement, further settlement of the same element was close to zero.

36. The final result must always be kept in mind: will everything fit together as designed and will the tunnel be pleasing to drive through? An architect once told one of the authors of this paper that while the details of a quality job may not be noticed by a motorist entering a tunnel the overall effect is to give a sense of ease and confidence. Conversely, a structure that has uneven lines, lighting that wanders up and down to the eye, imbues the same person with a vague feeling that all is not well.

27. The Conwy Estuary Tunnel — construction

J. F. McFADZEAN, BSc, and D. PHILLIPS, BSc, Costain Tarmac Joint Venture

SYNOPSIS. The Conwy Crossing Contract was awarded to Costain Tarmac Joint Venture in September 1986 and includes the construction of the U.K.'s first immersed tube road tunnel. The road tunnel is 1090m long and comprises a 260m East Cut and Cover section, a 120m West Cut and Cover section and six 118m long tunnel elements which are placed by the immersed tube method in between these sections. The Conwy Estuary is hydraulically and geotechnically complex and the paper details the design and construction aspects of the various temporary works required to overcome this harsh environment, together with the planning works required to immerse the elements. Details of the construction of tunnel sections are described.

INTRODUCTION
1. The Contract for the construction of the A55 Conwy Crossing was awarded by the Secretary of State for Wales to the Costain Tarmac Joint Venture in September 1986 and work commenced on site in November of that year. At a tender value of approximately £102M the Conwy Crossing is the largest roadworks contract to be let in the U.K. and includes Britain's first Immersed Tube Road Tunnel. Engineer for the Works is Travers Morgan and Partners in association with Christiani & Nielsen A/S.
2. The contract requires the construction of approximately 6km of two lane dual carriageway Expressway from the Glan Conwy Interchange to Penmaenbach Head, bypassing the settlements of Llandudno Junction, Tywyn and Conwy, by initially following the course of the river and crossing it diagonally by tunnel northwest of Conwy Town (Fig.1) before passing westward along the coastal flats to the north of Conwy Mountain.
3. In addition to the associated earthworks, drainage, services, side roads and carriageway works, 14 bridges are to be constructed, the largest being a 10 span viaduct at the eastern tie-in to the existing A55 over the existing Glan Conwy Roundabout.

CONSTRUCTION

Other structures include a reinforced concrete U shaped articulated raft structure at the Conwy Morfa Interchange, a 154m long underpass formed by diaphragm walls, where the new A55 passes beneath the existing Causeway to Conwy and of course the tunnel and approach structures.

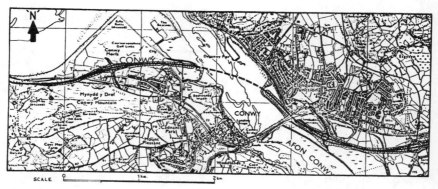

Fig.1 - A55 Conwy Crossing - Route Layout

4. The A55 will pass beneath the Conwy Estuary through a 1090m long tunnel comprising a 260m Eastern Cut and Cover section, a 120m West Cut and Cover section and six Immersed Tube tunnel elements each of which is some 118m long, 24.1m wide and 10.5m high with a mass of 30,000 tonnes.

THE TUNNEL SITE

5. The site for the construction of the tunnel crossing is shown on Fig.2. Access to the East Cut and Cover site is obtained via a purpose built level crossing from Glan y Mor Road over the Llandudno Branch Rail Line; however this restricted access is unsuitable for earthmoving plant and therefore an alternative access beneath the Causeway was constructed early in the contract in order to allow major excavation works to proceed. On the west side of the river access to the tunnel site is gained by crossing the main Holyhead rail line via a purpose built temporary bridge at the western extremity of the site and from there by haul road through the Works.

6. Severe restrictions are placed on the use of the existing A55 through the town of Conwy particularly during the Summer months and consequently railway sidings were constructed at the west end of the site and all reinforcement, cement, sand, steel plate and the like are delivered to site by this means.

7. The tidal range in the estuary varies from 4m at Neap tides to 6.5m at Spring tides, and the associated tidal

currents can exceed 2m/s but these factors do not appear to detract from the success of the estuary as a commercial and recreational boating centre. Consequently disruption to this amenity is to be minimised during all phases of construction.

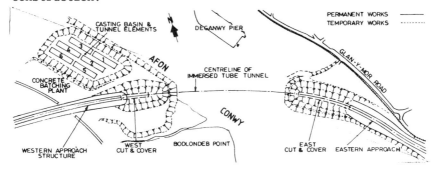

Fig 2 - Tunnel Site

8. The geology of the Conwy Estuary is complex and variable however a simplified structure along the centreline of the tunnel is shown in Fig.3. The underlying bedrock comprises a sequence of mudstones, siltstones and sandstones of Silurian and Ordovician age, which is overlain by the grey stony North Wales Till and Glacial Lake Deposits. This latter deposit, locally in excess of 20m thickness, comprises a laminated sequence of silts and clays with partings and thicker horizons of fine to medium sand. The Glacial Lake Deposits are in turn overlain by stiff brown Irish Sea Boulder Clay. These Glacial Lake Deposits however are frequently overlain by a considerable thickness of poorly sorted granular deposits, the origin of which is uncertain. A variable sequence of recent alluvial deposits is also encountered at the surface including soft organic clay, peat bands, sands and gravels.

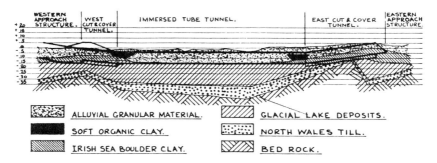

Fig.3 - Tunnel Centreline - Geological Profile

CONSTRUCTION

TUNNEL COFFERDAMS

9. The East and West Cut and Cover tunnels, their approaches and the six immersed tunnel elements are all constructed within bunded dewatered cofferdams.

Eastern Approach and Cut and Cover Tunnel - Cofferdam

10. The open excavation for the approach section to the Eastern portal is protected from water inundation by a permanent works sheet pile cut-off wall driven into the impermeable Irish Sea Boulder Clay which underlies the sands and gravels of the Estuary. This cut-off wall continues into the upper estuary east of the Causeway but the cut-off at the Causeway itself takes the form of two permanent works diaphragm walls constructed to a depth of up to 40m which forms the A55 underpass and the abutments to the four bridge decks which cross the A55 at that location (Fig.4).

11. The Eastern Cut and Cover tunnel and its portal structure are constructed within a cofferdam that is an extension of the above and is protected from tidal inundation by a bund of granular fill, which has a cut-off wall and a comprehensive dewatering system of well points and ejectors together with a surface water drainage system and the attendant pumped discharge system. (Fig.5) The nature of the alluvial clays indicated on the cross section are such that they had to be excavated and replaced with granular material prior to construction of the bund itself in order to ensure the stability of the cofferdam.

Fig.4 - Eastern Approach and East Cut & Cover Cofferdam

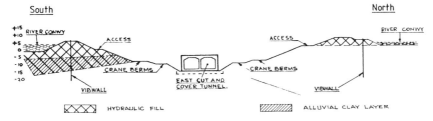

Fig.5 - East Cut and Cover Cofferdam - Cross Section

12. The cofferdam therefore was formed in the following sequence.
   a) The sands and gravels overlying the alluvial clays were dredged by cutter suction dredger (Orion) and stockpiled for subsequent use.
   b) The alluvial clays were then excavated by a pontoon mounted backhoe dredger (Styx) and loaded into split hopper barges for transportation and disposal in licensed dumping grounds at sea.
   c) The bunds to the cofferdam were then formed with hydraulically deposited granular materials dredged from the estuary.
   d) The estuary side slopes of the bund were trimmed by backactor and then protected against current and wave attack with rock armour.
   e) The bund cut-off wall comprises a thin grout diaphragm which is installed using the ETF H pile method. The procedure for formation of a "Vibwall", as it is known, involves sequential insertion and withdrawal of a lance and simultaneous injection of bentonite/cement grout to produce a continuous but flexible cut off diaphragm. The lance comprises a universal beam section 600mm x 300mm with flange and web thicknesses of 30mm and sufficient length to allow a 24m penetration depth. The lance is inserted to the required depth by the use of a vibrator. During the insertion stage a limited amount of grout is pumped from the lance toe to provide lubrication to the sides of the lance. The lance is then withdrawn at a controlled rate whilst grout is simultaneously pumped under pressure to fill the void formed. Throughout lance insertion and withdrawal/grouting cycle a number of parameters are automatically monitored and recorded to ensure the integrity of the diaphragm.
   f) A dewatering system was installed to minimise water ingress through the granular material by means of a ring of well points, and to depressurise the lake deposits by the use of an ejector system. The principle of an ejector system is that ground water is drawn into the ejector by the pressure reduction generated at depth in a well by the ejector nozzle venturi system.

CONSTRUCTION

A high pressure water supply is passed through the venturi in an enclosed pipework system, while the ground water arising is pumped away through a separate return header main. The ejector system is powered by a single high pressure centrifugal pump, with the requisite standby equipment, which delivers high pressure water to the supply main. The discharge water is recirculated into a tank adjacent to the high pressure pump via the return main.
g) Finally the excavation was completed by traditional earthmoving plant to formation level with the provision of crane, access and slope maintenance berms around the structure.

13. The reinforced concrete East Cut and Cover structure is then constructed, backfilled and a steel bulkhead installed at the western end prior to flooding and removal of the bund in readiness for the first immersed tube element.

The Western Approach, West Cut and Cover Tunnel and Casting Basin - Cofferdams

14. The two large cofferdams (Fig.6) on the west bank of the Conwy Estuary are again formed within bunds with associated "Vibwall" cut-off and dewatering systems. The Casting Basin working level is -10.0m OD whilst the Western approach/cut and cover excavation varies from a level of -12.0m to -17.0m OD at the location of the cut and cover bulkhead. The engineering solution and sequence of construction is identical to that described for the East Cut and Cover Tunnel.

15. The Casting Basin is some 300m by 200m in plan and is large enough to allow the construction of all six tunnel elements simultaneously with all the attendant storage areas, crane and access ways. The bed of the Casting Basin comprises a 600mm layer of 40mm single sized crushed rock which is topped with a 150mm layer of 20-5mm crushed rock. This layer is carefully profiled to the required vertical alignment of each tunnel element and a grid of concrete beams are placed within this layer on the line of all site welds again to suit both horizontal and vertical alignment of the tunnel elements. The underlayer will also allow the hydrostatic water pressure to be generated beneath the elements and therefore allow subsequent flotation.

Soils Instrumentation

16. Excavation of the temporary works in complex ground conditions has been necessarily controlled by extensive soil instrumentation monitoring pore water pressures and soil movements. To date 148 pneumatic piezometers, 79 standpipes, 14 extensometers and 24 inclinometers have been installed generating approximately 4000 individual readings per day.

The data was used to control the rate of excavation and of course to ensure the safety of the works consistent with stability requirements and to monitor the efficiency of the ground water control systems and "Vibwall" cut-off. Details of the soil instrumentation is given in Ref.1.

Fig.6 - West Bank Cofferdams

TUNNEL CONSTRUCTION
17. The tunnel elements and the cut and cover tunnels are of similar geometry and dimensions, the tunnel element cross-sectional details are given in Fig.7. The cut and cover tunnels are constructed in-situ in their respective cofferdams while the tunnel elements are constructed simultaneously in the Casting Basin, prior to the subsequent flooding of the Casting Basin and sequential float out of the elements and the immersion operations. The construction programme also dictates that the elements and cut and cover tunnels have to be constructed simultaneously.

CONSTRUCTION

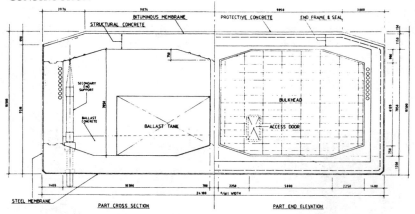

Fig.7 - Tunnel Element - Cross Section

18. The tunnel sections are formed in reinforced concrete with some 64,000m^3 of concrete required for the six tunnel elements and some 12,500m^3 and 26,200m^3 required for the West and East Cut and Cover tunnels, respectively. The external surfaces of the tunnel bases and walls are lined with a 6mm steel waterproof membrane with structural connections provided by shear studs. The membrane in turn is protected by a solvent free urethane tar two part paint compound, Protegol 32-10, and cathodic protection in the form of zinc anodes. Provision is made in the structure for an impressed current system of protection at a future date,if required. Externally the roof of the tunnel is sealed by a two layer polyester fibre based bituminous membrane which is topped with a 150mm layer of mesh reinforced protective concrete. The steel plates forming the steel membrane are fabricated into 12m x 12m base and wall panels in the on-site workshop, grit blasted and painted prior to transportation to the casting positions were they are set out on the casting beds and welded together. Corner panels are formed off site then transported to site, painted, then placed and welded into position, using the base forms for temporary support. Temporary support for the wall panels is provided by the wall formwork shutters see Fig.8. The panels for the East Cut and Cover Tunnel are transported across the river on trailers which in turn are transported by a barge, appropriate ramps are constructed on each bank of the river. In all some 1,400T of plate is required for the tunnel elements and some 800T for the cut and cover tunnels. Each element typically has a cast-in steel end frame at each end which has provision for a steel plate to be accurately fitted that will incorporate the special rubber Gina Gasket manufactured by Vredestein. This will form the initial seal between each element when they are placed in position beneath the River Conwy.

Fig.8 - Tunnel Element - Under Construction

Formwork

19. While the various tunnel sections are geometrically similar the design of the formwork systems for the tunnels were determined taking cognisance of a number of other factors which included cranage restraints, number of potential form re-uses and the necessity to provide temporary support to the steel membrane panels and allow subsequent insitu welding of the panels.

20. For the wall forms various options of form restraint were considered but finally the use of through going form ties at a regular pattern with a waterproof coupler, welded to the waterproof steel membrane in the workshop, was adopted.

21. Base side shutters for all the structures consist of traditional timber formwork.

22. Traditional formwork systems are used for the cut and cover tunnels external walls, which utilise timber facing and backing members, the forms are crane handled.

23. The external wall external forms however utilised a non facing sheet as the steel membrane actually contains the concrete. The external wall internal forms are similar in detail to the above forms except that a facing ply sheet was used. Forms for the central walls are similar to the internal form of the external wall except that the form tie patterns were determined to suit concrete pressures rather than a regular pattern.

24. The scheme principles described above for the cut and cover tunnel are similarly adopted for the tunnel elements except that the internal forms consist of steel panels and the forms are supported on purpose made trolleys to minimise crane usage. The trolley incorporates hydraulic horizontal mounted rams to allow positioning and striking of the forms.

25. Traditional falsework and timber forms are used for the cut and cover roof soffit, while the roof soffit to the tunnel elements is again constructed with steel panels supported by a purpose made trolley. A system of wheels and locking hydraulic rams are utilised to enable travelling and vertical alignment respectively.

Reinforcement and Cast-in Items

26. Approximately 6,500T and 4,100T of reinforcement is to be fixed to the tunnel elements and cut and cover tunnels, respectively.

27. In addition a significant number of temporary work items are required to facilitate element installation and consequently numerous anchorages are to be cast into the tunnel elements. These items of temporary works include bollards, towers, temporary supports, lifting points, ballast tank and bulkheads.

Concrete

28. The design of concrete mixes for the structures have not only to satisfy the normal strength and associated contract requirements but to additionally satisfy the insitu concrete temperature requirements of the contract and for the elements, satisfy a stringent density restraint. Notwithstanding these criteria the concrete mixes have to be produced in large quantities to a very consistent high quality.

29. These requirements were fulfilled by the use of a OPC/PFA cementitious blend with either local dolerite or limestone coarse aggregates with fine aggregate being supplied by two constituents; a crushed rock fine and a natural sand.

30. A ready mix supply is used for the East Cut and Cover tunnel while CTJV batching plants produce concrete on the west side of the river to service the tunnel elements in the Casting Basin, the adjacent West Cut and Cover tunnel and other large volume structures on the west bank. The west side batcher complex comprises three batching plants, two rated at $60m^3/hr$ and a third at $40m^3/hr$. Concrete placement for all structures is by concrete pumps, rated at a nominal $60m^3/hr$.

Concrete Temperature Requirement

31. The contract requires temperature differentials within each tunnel pour and between walls and underlying bases to be limited in magnitude. The selection of the appropriate temperature control measures to be taken were made following an analysis by the British Cement Association and CTJV of the heat generation/dissipation characteristics of the various concrete mixes and tunnel structural components. In summary the temperature control measures comprises the use of a coarsely ground OP cement, maximised proportion of PFA, chilled mixing water and appropriate insulation of the formwork or pours.

Concrete temperatures are monitored in all tunnel sections and the temperature requirements of the contract have been satisfied. The maximum base temperature obtained was 67°C, while the maximum external and central walls and roof temperatures were 52°C, 38°C and 52°C respectively.

Tunnel Elements - Dimensional Control

32. The dimensional control of the elements were identified as being particularly important and cognisance was taken in the design of formwork systems and setting out to ensure that the required displacement and volumetric characteristics of the element could be achieved. In this respect post construction surveys were also carried out to ensure that the required elements dimensional control is achieved. Similarly large concrete cubes are manufactured and insitu coring carried out to determine the elements concrete density. The results of the dimensional and density monitoring will then be combined to confirm that suitable weight and buoyancy characteristics of the elements are obtained.

Fitting Out and Finishings

33. Once the element is structurally complete the fitting out and installation of the immersion temporary works items will commence. These works will be phased in with painting to the internal surfaces of the elements and placing of ballast concrete (if possible).

DREDGING AND RECLAMATION WORKS

Phase 1

34. Phase 1 dredging comprises the dredging and construction of the cofferdam bunds and is described in paragraph 12.

Phase 2

35. Phase 2 dredging and reclamation works comprises the tunnel trench dredging together with the bund removal to the East Cut and Cover tunnel, bund removal to the West Cut and Cover tunnel and Casting Basin bund and dredging the exit channel from the Casting Basin to tunnel trench. Current proposals for these works include the dredging of some 1.3Mm^3 of clays, sands and gravels by use of a cutter suction dredger and land based equipment for parts of the bunds.

36. Subsequent permanent work backfill consists of placing some 0.4Mm^3 of granular material to the elements in the form of locking, loading and amenity fills.

37. Reclamation Area 2 is provided within the contract for unsuitable materials. The area is located on the east bank of the upper estuary, that is upstream of the Causeway. Dredged unsuitable material will be deposited in this area and allowed to settle out of suspension and consolidate to suit later land use.

CONSTRUCTION

Temporary stockpile areas will be used in the lower estuary to store the suitable estuary materials for later use as tunnel backfills.

TUNNEL ELEMENT INSTALLATION
Hydraulic Model Tests

38. As identified above the hydraulic regime of the estuary is complex and is influenced not only by a significant tidal range but also by sedimentation processes and the regular interchanges of saline/fresh water. These factors together with tidal currents of up to 2m/s, significant water density variations (range between 1002 and 1028 Kg/M^3) and a skew tunnel alignment with respect to the current direction (60° from normal attack) have necessitated hydraulic modelling of the tunnel element installation to confirm the suitability of the immersion proposals, equipment and associated temporary works.

39. The hydraulic model tests were carried out by Delft Hydraulics Laboratory and have been extensively reported (internal documents) and summarised by Hamer and De Jong (Ref.1). The model tests consisted of mathematical modelling of the tidal generated currents, using a computer program which determines two dimensional depth averaged currents. The mathematical model was calibrated and verified using measured tidal and current data.

40. A physical model (1:64) of some 23m x 10m in plan dimensions corresponding to 1450m x 710m prototype area was then constructed to model the estuary hydraulic regime and calibrated using the results from the mathematical model. Then using correctly modelled elements, pontoons, mooring lines and the like, the behaviour of the immersion plant and elements were studied and the model forces were measured for the appropriate range of currents (Ebb and Flood), water level, element depth locations and positions. The results of the model tests confirmed that the magnitude of mooring forces acting on the elements would be within the capacity of the immersion equipment and bollards bolted onto the elements (bollards of 50T and 100T capacity) and that no untoward dynamic excitation of the systems should occur.

41. Using the results of the model tests a relationship was obtained for the transverse and longitudinal flow induced forces acting on the mooring lines in the form of:-

$$F = a \times k \times v^2$$

where F = Force, a = factor determined from the model tests whose in magnitude is dependant upon the element position (depth and location), k = (1.1)x factor depending on water density, and ebb/flood water level, v = velocity measured at either the Ebb or Flood reference location.

Tunnel Element Immersion

42. When the elements are structurally complete they will be fitted out with internal ballast tanks and steel bulkheads fitted to each end to render them watertight. A number of items of immersion temporary works and equipment will be connected to the cast-in roof anchorages and they will include bollards (50T and 100T capacity), lifting points (250T capacity), survey tower and Command Tower with access shaft connecting to a shaft cast in to the roof.

43. The ballast tanks will be filled and the casting basin flooded to a level of 1m above the roof of the elements in order that they can be checked for leaks.

44. The tunnel elements will then be transported and immersed as follows:

- a. Two precast reinforced concrete secondary end support slabs will be placed using a floating crane, on a prepared gravel bed within the tunnel trench to suit the element.
- b. Two pontoons will then be warped into position adjacent to the element. When the appropriate water level occurs the pontoons will be warped over the element and on a falling tide will rest on the top of the element roof. Connections will then be made between the pontoons and lifting points and bollards previously installed on the roof. The element will then be deballasted and on the rising tide allowed to float clear of its foundations, supporting the two pontoons which will be connected to it. The element and pontoons will then be warped out of the basin to the Fit-Out Facility which will be located adjacent to the West Cut and Cover bund, but in the Exit Channel.
- c. Whilst at the Fit Out Facility the elements will be prepared for transportation and immersion. These works will include trimming, reballasting, installation of immersion equipment and the like.
- d. On the appropriate tide the element will be warped from the Fitting Out Facility until it is positioned over the intended location. During the passage of the element across the estuary its position will be maintained by transverse mooring lines anchored to anchor piles previously driven in the estuary. The winches for these lines will be located on the pontoons. The longitudinal position of the element will be maintained by fore and aft mooring lines anchored directly to bollards located on the element but controlled from winch stations located on the East and West Cut and Cover tunnel bunds, respectively. During the transportation stage the position of the element will be continuously monitored by surveying the fore and aft survey towers previously connected to the element. The element will then be floating over but some 1-2m west of its final location.

CONSTRUCTION

e. The tunnel element will be ballasted and then sunk by controlled lowering from the pontoons. (Fig.9). Initially the reusable primary end support beam will engage a member connected to the preceding element. Then the secondary support rams will be brought to rest on the previously placed foundation slabs. The element will then be supported on its own temporary support system. The secondary end support consists of a circular solid steel ram incorporated within a cast in steel thrust collar. An hydraulic jacking system with reaction beams and legs will be set on top of the ram to transfer the elements weight into the ram.

f. Hydraulic rams will be activated to draw the element against the preceding element thus providing an initial seal of the Gina Gasket located on the end frame, which will in turn bear against the secondary end of the preceding element. The chamber formed between the two elements will then be dewatered which causes the differential hydrostatic water pressure so generated to further compress the gasket thus providing a water tight joint between the two elements. Following evacuation of the water from the chamber the bulkhead doors may be opened and pedestrian access gained to the element. At this

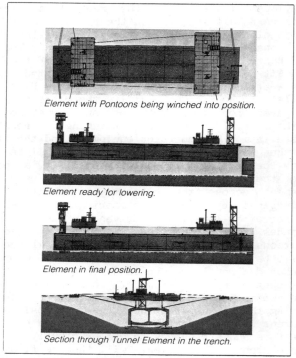

Element with Pontoons being winched into position.

Element ready for lowering.

Element in final position.

Section through Tunnel Element in the trench.

Fig.9 - Tunnel Element - Immersion

stage the element is immersed in its final location and the transportation and immersion sequence is repeated until all six elements are sequentially removed from the Casting Basin and are installed, all basically following the same sequence of operations.

45. Closure panels and longitudinal restraints will be installed to seal the 2m gap between TE6 and the West Cut and Cover Tunnel.

46. The permanent foundation layer beneath the elements will be formed by jetting sand from a specially constructed gantry which travels on top of the elements, sand being fed to the equipment from purpose built pontoons via the Load Out Jetty and equipment stationed alongside the gantry structure. Thereafter the required fills to the sides and top of the element will be placed, using pumped discharge and dumping plant, as appropriate.

47. Finally the closure joint between TE6 and the West Cut and Cover will be constructed in reinforced concrete once the backfilling is complete.

48. Various works have to be undertaken within the tunnel elements and these "finishings" works commence almost immediately the first element is immersed. Similar works to the cut and cover tunnel will commence when the structure is complete.

49. The water ballast is replaced with permanent concrete ballast and the ballast tanks are removed. Internal joints are constructed and the temporary bulkheads are removed. The internal walls and roof are painted and ducts are installed prior to the placing of kerbs and blacktop. On completion of these operations the tunnel will be handed over to the M & E contractor who will install the tunnel lighting and ventilation system.

SITE ORGANISATION

50. The Costain Tarmac Joint Venture operates as an autominous business unit run by a board comprising two directors from each of the parent companies to whom the project manager reports.

51. A staff of 160 comprising 40 administrative, 17 commercial and 103 technical and managerial personnel are based on site and control all construction related activities including procurement, invoice processing, commercial matters, temporary works design, planning and construction management of the works.

52. It is anticipated that the labour force will peak at 700 including sub contractors. Over 100 joiners and a similar number of steelfixers will be required together with some 75-100 plant operatives and 150 general labourers.

53. In view of the specialist nature of many aspects of the construction work, approximately 40% of the works will be sub-contracted. A total of 77 Sub Contracts have been

CONSTRUCTION

let to date with more to follow and multi million pound contracts have been let for work such as Dredging (Zanen UK), Tunnel Float Out and Immersion (HBM), Piling diaphragm walls, "Vibwall", (Bachy Bauer Consortium), Steel Membrane (Pipeline and Energy Contracting) and Bridge Structural Steelworks, (Fairfield Mabey).

REFERENCES

1. DAVIES J.N., WILLIAMS P., and WOODS K.G.
Instrumentation of Temporary Excavations for the A55 Conwy Crossing, Conference on Geotechnical Instrumentation in Civil Engineering Projects, Nottingham University, April 1989.

2. HAMER D.G. and DE JONG R.J. Immersion Eng. and related hydraulic studies for the Conwy Tunnel.
Internal Congress for Tunnel and Water, Madrid June 1988.

28. Fabrication, outfitting and construction techniques proposed for the Storebaelt immersed railway tunnel

A. J. BAST III, PE, BSCE, MEng, Parsons Brinckerhoff Construction Services, Inc., Herndon, Virginia, USA, and P. E. BACH, MSc, Cowiconsult, Virum, Denmark

SYNOPSIS. Tenders were received in June 1988 by a Danish government owned corporation for a tunnel crossing of the Storebaelt in Denmark. This paper reviews the procedures used for surveying fabrication and outfitting sites for the steel immersed tunnel and reports on selected results of the tender process for both immersed options. A review of the proposed construction techniques and a comparison with sea conditions of sites in the United States will be made. It is hoped that the results of the Storebaelt tender process and the sites study will contribute to the transfer of immersed tunnelling technologies among the continents.

INTRODUCTION
1. Two basic approaches to immersed tunnel construction have developed since the first immersed tunnel was placed beneath the Detroit River in 1910. The steel immersed tunnel developed as the standard, first in the United States and then in Japan, circa 1944. Europe saw its first concrete immersed tunnel built in Berlin in 1927. Since that time, the concrete cross section has been the accepted standard in Europe with the highest concentration of this type of tunnel found in the Netherlands. The geographic bifurcation of the technology into concrete and steel thus began early in its development and has continued to this date. An explanation for this has been variously attributed to:

> o   Higher costs of steel construction in Europe than in the United States or Japan.

o Lack of space in the United States and Japan for building basins commonly used for construction of concrete elements.

o A desire for early revenue service in the U.S. where projects were funded by revenue bonds.

o Local preference of the engineering and construction community based on established construction methods or types of available immersion equipment.

2. The Storebaeltforbindelsen (SBF), a government owned corporation established by the Danish government to construct a bridge tunnel crossing of the Storebaelt, was aware of the arguments on each side of this issue. Apparently unconvinced that either technology was superior to the other, they elected to let the construction marketplace re-evaluate the relative merits of the options based on cost. Accordingly, SBF negotiated a contract with a joint venture consisting of Cowiconsult of Virum, Denmark and Parsons Brinckerhoff International of New York City, U.S.A. The joint venture was tasked to develop a tender design for an immersed crossing of the Eastern channel of the Storebaelt approximating 35% of a final design, which would be sufficient to solicit tenders from four construction consortia. A separate tender design was prepared for the concrete and steel immersed tunnels. A similar joint venture was formed to develop a bored tunnel tender design.

3. Preparation of the design required a thorough understanding of the available fabrication and outfitting sites for the steel immersed option and of potential sites for construction basins in the case of the concrete option. Any information gathered was to be provided to the prequalified construction consortia to foster competition. The need for surveying sites for the steel option was considered to be particularly important since this was a new technology to the European construction community. When tenders were finally received in June 1988, the steel immersed option was lower in price. Ultimately, SBF elected to proceed with construction of a bored railway tunnel. They have apparently made the judgement that the bored tunnel will have less of

an environmental impact on the Baltic ecosystems and are willing to accept the arguably higher construction and operational risks which may accompany the bored option. Nevertheless, the results of the Storebaelt immersed tunnel tenders were widely published in Europe and are sure to generate considerable interest in the European steel and shipbuilding industries. These tenders validated the affordability of a steel immersed tunnel in Europe.

DESCRIPTION AND OBJECTIVE OF STOREBAELT CROSSING
4. Storebaeltforbindelsen seeks to design and construct a bridge and tunnel railway crossing of the Storebaelt (see figure 1), an international deep draught shipping channel which lies between the major Danish islands of Saelland and Fyn. This channel connects the North Sea and Baltic Sea. The Storebaelt is divided into the Eastern and Western Channel by the island of Sprogo. The eastern channel is to be traversed by a railway tunnel while the western channel is to be spanned by a low level bridge. Future plans call for a vehicular crossing of these channels. Completion of this fixed rail link will ultimately provide for rail access across the Lillebaelt to Jylland and then on to the European continent. Access to the continent by rail and vehicular traffic is now dependent on ferries operated by the Danish State Railway (DSB). Completion of this crossing will facilitate trade with Denmark and other members of the European Economic Community (EEC) if trade barriers between EEC nations are removed, as now scheduled, in 1992. The Storebaelt Steel Immersed Tunnel was to have 39 tunnel elements each measuring 144 metres long. The channel depth measures up to 60 metres.

SOLICITATION PROCEDURE & COMPETING CONSORTIA
5. A prequalification procedure undertaken by the SBF selected the following four multinational construction consortia:

o GREAT BELT CONTRACTORS (GBC)

- Christiani & Nielsen A/S, Denmark
- Ed. Zublin AG, West Germany
- Spie Batignolles, France
- Armerad Betong Vagforbattringer AB, Sweden
- Cogelfar Construzioni, Italy

CONSTRUCTION

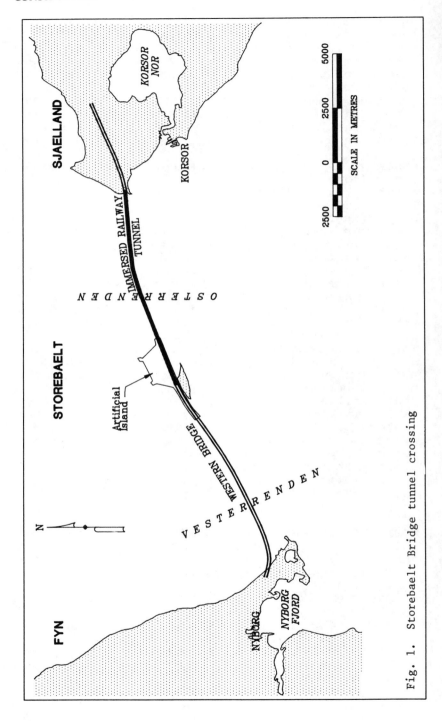

Fig. 1. Storebaelt Bridge tunnel crossing

o MT GROUP (MTG)

- Monberg & Thorsen A/S, Denmark
- Campenon Bernard, France
- Dyckerhoff & Widmann AG, W. Germany
- Kiewit Construction Company, USA
- SOGEA, France

o KAMPSAX TUNNEL JOINT VENTURE (KSX)

- GTM Europe, France
- Hoffman Interlink A/S, Denmark
- Philip Holzmann AG, W. Germany
- Kampsax Enterprise A/S, Denmark
- Volker Stevin Construction Int. bv, Netherlands

o HOJGAARD & SCHULTZ JOINT VENTURE (HSJ)

- Hojgaard & Schultz A/S, Denmark
- Hochtief AG, W. Germany
- Bilfinger & Berger, W. Germany
- Wayss & Freytag AG, W. Germany
- Hollandsche Beton en Waterbouw bv, Netherlands
- G.G. Jensen A/S and Skanska AB, Denmark & Sweden

6. The prequalification procedure required that each team have experience in immersed and bored tunnels. Each consortia was required to submit separate tenders for the concrete, steel and bored options. The tenders were evaluated by the joint venture and by the client to determine if qualifications taken by the consortia were acceptable or would ultimately change the price of the basic tender. These adjustments did not change the low tenderer among the consortia and for ease of presentation will not be addressed in this paper. Costs shown in Table 1 reflect a peninsular ramp approach to the portal from Saelland. The tenderers were asked for alternate tenders to eliminate one half and all of this ramp approach and replace it with tunnel elements. Since prices for eliminating the ramp or portions of the ramp tended to raise the project price, they will not be addressed either. Only the most economical variants proposed by the tenderers are listed.

CONSTRUCTION

Table 1
TENDER PRICES (M DKK)

	MTG	GBC	HSJ	KSX
CONCRETE OPTION	3613	3955	4086	3885
STEEL OPTION	3481	3959	3771	3736

7. With the exception of the GBC tender, the steel variant designs proposed by the contractors were less expensive than the concrete alternatives. MTG was able to outstrip the competition for their steel tender largely due to an alternate joint detail proposed which saved on the materiel quantities required. Their base tender for the joint in the tender documents was actually slightly higher than their concrete alternative. This is only one example where the contractors proposed well engineered cost savings. It represents a benefit of this method of solicitation where the contractors are free to propose variants.

NATURE OF ENVIRONMENTAL CONCERNS

8. The Danish Government, reflecting the concern of its people, desires to minimize the environmental impact of a crossing of the Storebaelt. Their concern is related to impact on the marine environment caused by sediment resulting from dredging or reclamation activities, Also, they are concerned about upsetting ecosystems in the Baltic which are dependent on their survival to the ebb and flow of waters from the North Sea. Any interference with this ebb and flow by bridge piers, for instance, raises serious concern. Recent losses among the Baltic and North Sea seal population have increased awareness of the sensitivity of this ecosystem.

PURPOSE OF FABRICATION AND OUTFITTING SITES STUDY

9. The steel immersed tunnel competitiveness in a European environment was questionable at the time the Cowiconsult/Parsons Brinckerhoff Joint Venture published its Fabrication and Outfitting Sites Study in January 1988. It was apparent that the identification of suitable fabrication and outfitting sites and the transmission of this information to the prequalified contractors was

crucial to the competitiveness of the steel immersed option.

FABRICATING SITES

10. The study conducted by our joint venture first evaluated the availability of fabrication facilities in the Baltic Area. A scan of the Baltic Pilot and of marine charts identified 12 potential sites which are listed in table 2 below and shown at figure 2.

Table 2

Shipyard	Location
Burmeister & Wain	Copenhagen, Denmark
Danyard A/S	Aalborg, Denmark
Odense Lindo	Odense, Denmark
Dannebrog	Aarhus, Denmark
Nakskov	Nakskov, Denmark
Neue Flensburger Shiffbau Gessellschaft	Flensburg, West Germany
Howaldtswerke	Kiel, West Germany
Kockums	Malmo, Sweden
Citgo AB	Goteborg and Landskrona, Sweden
Gotaverken Arendal AB	Goteborg, Sweden
Finnboda AB	Stockholm, Sweden

11. No attempt was made to evaluate the relative ranking of these shipyards. This ranking was ultimately accomplished in the tendering process by the contractors.

12. Five shipyards were identified in Denmark, two in the Federal Republic of Germany and five sites in Sweden. These facilities were further investigated by visiting the site, telephonic interviews and review of literature provided by the shipyards. For each yard, the following data was catalogued and presented in chart form for ease of comparison:

- o   Unique Fabrication Facilities
- o   Access Channel Draught
- o   Probable Method of Subassembly
- o   Probable Method of Final Assembly
- o   Method of Launching
- o   Tow Distances and Time to Tunnel

CONSTRUCTION

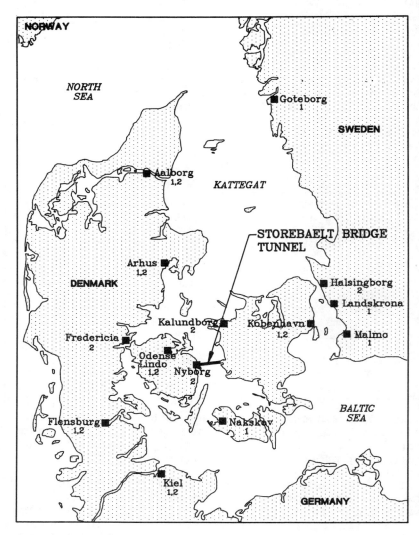

Fig. 2. Potential fabrication and outfitting sites for Storebaelt steel immersed tunnel

o   Other Site Specific Information

13. Information on these fabrication facilities was used during the design process to assess the structural design loadings expected at different stages of construction. Survey results were also forwarded by the Storebaeltforbindelsen to the contractors to assist in formulating their tenders. This fostered competition by encouraging the contractors to contact the shipyards.

14. The survey concluded that seven of the eleven facilities expected to assemble the tunnel elements in graving docks; the remainder expected to end or side launch the elements or to skid the elements to a submersible barge where the keel concrete would be poured. The tender design assumed either a side launching or flotation of the assembled element in a drydock.

15. The following preliminary information was provided to the shipyards in the process of surveying the fabrication facilities:

- o   Weight of steel in each tunnel element = 1025t
- o   Total weight of steel in 39 elements = 40,000t
- o   Draught with keel concrete = 4.5m
- o   Displacement with keel concrete = 8000t
- o   Draught of outfitted tube = 8.9m
- o   Fabrication schedule = Apr 89 - Mar 91
- o   Dimensions of element = 144 x 16 x 10m

OUTFITTING SITES

16. Once fabrication of the steel tunnel elements is completed the elements are normally towed to an outfitting site for completion of the interior liner. A survey of eleven potential outfitting sites was accomplished as part of the COWICONSULT/Parsons Brinckerhoff study and are also shown on the map at figure 2. Eight sites were in Denmark, one in Sweden and two in West Germany. In the process of evaluating potential sites, it became apparent that each of the sites would require some degree of improvement in order to support the seven outfitting berths required. This number of berths was based on the required production rate of one element each three weeks. This production rate was paced by the placement

rate and complemented by the fabrication rate. The construction improvements would require environmental review and a construction permit from the Danish Harbour Authority and other governmental agencies.

17. Since this permitting process had, on other projects, taken up to two years, the project schedule was in jeopardy. It was clear that a site would have to be defined prior to receipt of tenders in order to achieve the April 1993, operational date established by SBF's master project schedule. Selection of the optimal site thus became a priority concern.

18. The outfitting sites were identified and ranked by objectively evaluating the following criteria:

o   Status of other Storebaelt work. The more work anticipated at the site, the lower the score in order to attain equitable work distribution to multiple sites.

o   Ability to concurrently outfit 7 elements including the size of the site and probable labour and materiel availability.

o   Availability of 9.5m water depths alongside the outfitting quay and in the approach channels.

o   Proximity to the tunnel alignment.

o   Pier availability and associated costs for the new construction required.

19. Table 3 summarises the results of the inventory. Points were assigned in the following manner:

o   Significant Advantages   =   2
o   Moderate Advantage   =   1
o   Neutral Condition   =   0
o   Moderate Disadvantage   = -1
o   Significant Disadvantage = -2

The highest in accumulated points in the last column shows the grading of the preferred outfitting site:

Table 3
COMPARATIVE EVALUATION CHART FOR OUTFITTING SITES

SITE	OTHER SBF WORK	SBF SITE	DRAUGHT	PROXIMITY	PIER AVAILABILITY	TOTAL
Aalborg	2	2	2	-1	2	7
Aarhus	2	2	2	0	-1	5
Copenhagen	2	2	-1	0	-2	1
Flensburg	2	1	2	-1	-1	3
Fredericia	2	0	0	1	-2	1
Halsskov	2	2	-1	2	-2	3
Helsingborg	2	1	2	0	-1	4
Kalundborg	2	2	2	1	1	8
Kiel	2	2	2	0	-1	5
Lindo	2	1	-1	1	-2	1
Nyborg	-1	1	2	2	1	5

The outfitting sites study concluded that the most promising outfitting sites were located at Kalundborg. SBF proceeded with the necessary coordination and permitting to acquire that site.

ASSESSMENT OF SEA CONDITIONS IN STOREBAELT

20. The immersion procedures proposed by three of the four tenderers were similar to the procedures used in the United States. As shown in Table 4, the most severe conditions of current to date were experienced on the East River for the 63rd Street Tunnel in New York City. Table 5 shows that the worst wave conditions were experienced during construction of the Chesapeake Bay Bridge Tunnel. A comparison of these sea conditions with those of the Storebaelt are of interest because they indicate the suitability of traditional equipment to the Danish waters.

Table 4
EXTREME ANNUAL CURRENT CONDITIONS

63rd Street Tunnel (NOAA)	2.6 m/s
Chesapeake Bay Bridge Tunnel (NOAA)	1.0 m/s
Storebaelt Tunnel (DHI)	1.6 m/s

Table 5
WAVE CONDITIONS (.00010 EXCEEDANCE FREQUENCY)

Chesapeake Bay Bridge Tunnel (CERC)	H_s = 2.5m T_s = 2 to 4 sec
Storebaelt Tunnel (DHI)	H_s = 1.8m T_s = 4 to 5 sec

The greater significant height of the Chesapeake waves is due to the large fetch caused by exposure to the Atlantic Ocean.

CONCLUSION

21. A number of valuable lessons were learned as a result of the tender process for the Storebaelt Immersed Rail Tunnel. The process itself permitted innovative solutions to be proposed by the construction consortia which is sure to shape the future development of the immersed tunnel technology. Segmental construction schemes were proposed which hold promise for reducing costs of steel immersed tunnels. More structurally efficient steel element joint details were proposed by two of the consortia which significantly reduce the materiel quantities in the end sections. At the same time, the tender appears to have validated the economics of the traditional screeded foundation. The suitability of catamaran immersing equipment to the many European sites was also validated. The Storebaelt, for instance, while an expansive body of water, has less demanding current conditions then experienced on the 63rd Street Tunnel. Similarly, the wave conditions are less severe than those experienced at the mouth of the Chesapeake Bay.

22. It is not unreasonable to forecast that the ailing steel and shipbuilding industry throughout Europe will give serious attention to steel immersed tunnelling options over its concrete alternative. The tender results experienced on the Storebaelt are sure to encourage them in that direction.

23. It is also clear that the parallel concrete and steel immersed tunnel designs commissioned by the Storebaeltforbindelsen resulted in a more economical cross section than has been common on

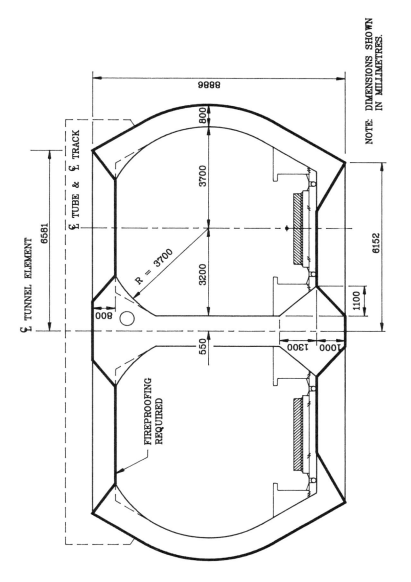

Fig. 3. High efficiency steel shell placement

many other steel immersed tunnels. In fact, the steel tunnel rectangular cross section is notable by its similarity to the concrete tunnel cross sections. The steel plate water barrier becoming popular on concrete immersed tunnels to prevent water leakage is a further indication that these bifurcated technologies are merging.

24. The displacement and height of a rectangular tunnel is less than a tunnel with a circular cross section which has been common in U.S. practice. This results in cost savings due to reduced dredge quantities, approach ramp lengths and ballast quantities. A rectangular cross section, however, carries the hydrostatic backfill and other applied loads in bending transverse to the tunnel axis, rather than in ring compression. Large quantities of reinforcement steel are required to carry the extreme fibre tensile stresses. If the steel shell was located in the tensile areas of the cross section as in Figure 3, reductions in reinforcing steel quantities would be realized. Concurrently, the steel shell would act as a folded plate structure with higher longitudinal and transverse spanning abilities. This would require less integral stiffening and shoring during the fabrication process. Savings from this and from the reduction in reinforcing steel quantities would be partially offset by the additional cost of bending the plate in a brake.

25. The final lesson learned from Storebaelt is that ecological concerns are the Archilles heel of immersed tunnel technology. Its continued development demands close attention to minimizing sedimentation during the dredging and backfilling operations. The intrusion of the tunnel profile above the seabed is another item which raises ecological issues in constricted waterways. The ability of a steel immersed tunnel to produce a dry interior environment and thereby minimize operational costs can be overwhelmed by ecological concerns. Dredging in glacial till may be technically less risky than boring in such conditions. Nevertheless, the higher risk and potentially more expensive bored tunnel may be selected over immersed tunnel options if ecological concerns govern.

29. Eastern Harbour crossing, Hong Kong

Y. MATSUMOTO, Kumagai Gumi Co. Ltd, D. E. OAKERVEE and
A. I. THOMSON, EHC Project Management Co. Ltd, and
D. G. MORTON, Freeman Fox (Far East) Ltd

SYNOPSIS. The Eastern Harbour Crossing, costing about HK$3.4 billion, is the largest single transportation project undertaken by the private sector in Hong Kong. The new tunnel provides a road and rail crossing of Victoria Harbour between Quarry Bay on Hong Kong Island and Cha Kwo Ling in Kowloon Peninsula (Figure 1), and is some 2,300m between portals. The crossing is achieved by means of an immersed tube 1,860m in length across the Harbour, with cut and cover tunnels forming the approaches for the road tunnels and for the rail tunnels on the Kowloon side but bored tunnels on Hong Kong Island. The immersed tube consists of 15 units, each constructed of reinforced concrete and having a deadweight varying between 44 and 46 thousand tonnes. The units house 5 separate conduits, 2 of which accommodate the railway, 2 accommodate the road with the fifth forming a service and ventilation duct.

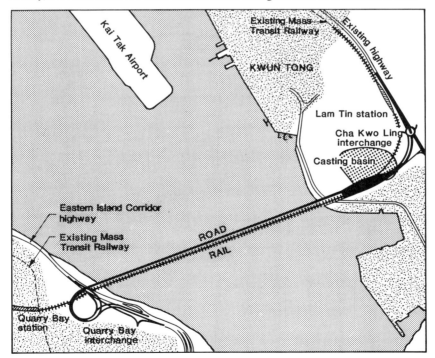

Fig. 1. Location of the Eastern Harbour Crossing, Hong Kong

Immersed tunnel techniques. Thomas Telford, London, 1989.

CONSTRUCTION

HISTORY OF THE PROJECT
1. The Hong Kong Government had, since the late 1970's, been actively considering a second road crossing of Victoria Harbour to overcome ever increasing traffic congestion. However, Government had yet to crystallise their thoughts on this matter when in June 1984 they received a proposal from Kumagai Gumi for a combined Road and Rail crossing. The proposal had been the result of a combined effort by Kumagai Gumi, Marubeni Corporation, Freeman Fox (Far East) Limited and Oakervee Perrett & Partners. After discussion eventually the Government called for open tenders in October 1984 and on the 1st April 1985 nine international bids were submitted. In June of that year a short list of three tenderers was drawn up. At this point, the Kumagai Gumi Group decided to expand its consortium to include the China International Trust and Investment Corporation, Paul Y Construction Company Limited and later, Lilley Construction Limited.

2. After extensive negotiations, the Government announced in December 1985 that the successful bids was that from the consortium led by Kumagai Gumi.

3. The Eastern Harbour Crossing Ordinance, providing the legislation granting the franchise, was passed on 17th July 1986 and construction started on the project on the 17th August of that year.

DESCRIPTION OF THE OVERALL PROJECT
4. The Eastern Harbour Crossing is a complexed project involving major infrastructure comprising of 8.6km of road and a 5.0km extension to the Mass Transit Railway. Thus the Immersed Tube forms only a part of the overall transport facility being provided. When completed, the new tunnel will significantly ease cross-harbour road and rail congestion and make a major contribution to the transportation infrastructure of the territory. It will also open up the eastern areas of Kowloon Peninsula and Hong Kong Island. In addition to the actual harbour crossing, the project includes the total infrastructure needed to integrate the road crossing with the existing highway network and an extension of the existing Mass Transit Railway from its present terminal at Kwun Tong through to a new station at Lam Tin and on to a new terminal station at Quarry Bay.

5. The tunnel crosses Victoria Harbour between Cha Kwo Ling on the Kowloon peninsula and Quarry Bay on Hong Kong Island, and carries both road and rail facilities. Ventilation buildings are situated at each end of the immersed tube, and provide not only a separate ventilation system for the road and rail tunnels but also house the auxiliary electrical and mechanical services equipment.

6. The road tunnel is some 2,300m between the portals of which 1,860m is in an immersed tube. The remaining tunnel forms the approaches and has been constructed using cut and cover techniques. On the Kowloon landfall the road is connected to the Lei Yue Mun Road by means of a grade separated interchange. Likewise, a grade separated interchange of a more complex nature provides the connection with the Island Eastern Corridor on Hong Kong Island. In order to accommodate the Island Eastern Corridor, a new viaduct had to be constructed to accommodate a permanent diversion.

7. The new railway facility forms an extension to the Mass Transit Railway Corporation's Kwun Tong Line from Kwun Tong Station which has been operational since September 1979. This extension, therefore, commences at elevated Kwun Tong

Station and continues eastward on an elevated viaduct until it reached a new Station - Lam Tin which is at ground level for the most part. Upon leaving this station, it enters bored tunnels through rock turning south and continues underground, apart for 180m at grade, through cut and cover tunnels, the immersed tube, and bored tunnels constructed in compressed air until it reaches a new terminal station at Quarry Bay. This Station is sited beneath an existing station on the Island Line, the two having been totally integrated except there is no rail connection.

8. The provision of a new interchange facility at Quarry Bay provides an alternative railway route between Kowloon and the eastern part of the Island, therefore, relieving the existing route to the west which at present is totally saturated beyond its intended capacity.

THE DRY DOCK

9. Whilst the choice of alignment for an eastern crossing was limited, it was not by complete accident that the route taken from the Kowloon landfall was directly through an existing quarry. The quarry, which had been worked since 1974, was nearing its end, and by the removal of 3,000,000 tonnes of rock, the Contractor was able to transform it into a basin suitable for use as a dry dock having an area of 55,000 sq. m. In addition, the quarry provides the aggregate for the entire needs of the Project.

10. In order to conserve both time and money, the quarry bottom was levelled out at its deepest point, which was approximately 4.5m short of that needed to float the units out to the open sea. To overcome this apparent problem the 400m long x 45m wide access channel necessary to connect the quarry to the sea was constructed to the required depth with two floating concrete gates positioned 290m apart. The floating gates enabled the channel to be used as a lock. Therefore, to move the units, sea water was pumped to a level 5m above that of the sea, the units floated into the channel, in pairs, the north gate floated into position and the water level lowered to sea level. Once the south gate had been floated out to its mooring, the units were afforded a free passage to the sea.

THE UNITS

11. Both economics and the shape dictated by the need for several different sized compartments for road and rail led to the early adoption of a rectangular reinforced concrete box containing five tubes: two each for the railway tracks and the road carriageways and one between the road carriageways for ventilation. There are cross adits every 86 metres between road ducts and every 18 metres between rail ducts. For obvious safety reasons there is total separation and no interconnection between the road and rail ducts.

12. There are fifteen units in the immersed tube section of the Eastern Harbour Crossing, there being ten units of 122 metres, four of 128 metres and one of 126.5 metres. Each unit is 35.45 metres wide and 9.75 metres deep. Consideration of the internal spans of walls, roof and base led to a compromise being reached in the choice of member thicknesses to achieve the maximum economy consistent with good floating and sinking characteristics. As a result, the external walls are 1600mm and 1200mm, the internal walls 600mm, the roof 1200mm and the base 1300mm thick (Figure 2).

CONSTRUCTION

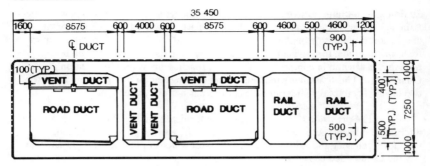

Fig. 2. Cross Section of the Units

13. The units were cast in three batches of five units each. Construction of the first batch of units commenced in November 1986 and the third batch was completed in July 1988 when the last unit was floated out. The caisson gates to the lock were cast with the first batch and the caisson sub-structure for the Cha Kwo Ling ventilation building was cast with the last batch.

The units were constructed in bays of 18 metres length with the end bays of being either 4 metres or 10 metres. Niches, boxouts and adits were located along the tunnel length such that bays were standardized to enable ease of construction.

WATERPROOFING

14. The basic philosophy adopted on waterproofing was to ensure that the primary line of defence against water penetration would be the structural concrete itself. Concrete permeability was minimized by restricting its water/cement ratio, by partial replacement of cement by pulverized full ash and by good site practise and quality control procedures. Cracking of the concrete was controlled by reinforcement which had been designed and detailed in accordance with BS5337 'The structural use of concrete for retaining aqueous liquids'.

15. Laboratory and full scale field tests were conducted on the concrete mix to determine the maximum temperatures and temperature gradients that were likely to occur. Reinforcement detailing took account of these effects and those associated with early tensile strengths in limiting cracks to characteristic widths of 0.2mm.

16. A pragmatic approach was taken to the design of the concrete mix. The best compromise of conflicting requirements on strength, durability, workability and cost was reached. As the casting basin was the site of an existing granite quarry and it was obvious from an economical point of view that all the aggregates would be from that source. Similarly, Ordinary Portland Cement readily available on the market in Hong Kong was specified. To minimize early thermal and drying shrinkage 25% of the cementious material was pulverized fuel ash (PFA) and the maximum permissible water cement ratio was 0.45. Mix proportions which produced all the grade 30/20 concrete for the units are shown in Table 1.

Table 1 : Mix Proportions of Concrete for Immersed Tube Units

Ordinary portland cement	285.0 kg
Pulverized fuel ash	95.0 kg
20mm aggregate	775.0 kg
10mm aggregate	300.0 kg
Crushed rock fines	680.0 kg
Plasticiser	3.8 litre
Water/cement ratio	0.45

17. It was also decided to provide an additional line of defence in the form of an external skin. A comprehensive exercise was undertaken to determine the most economical method of waterproofing which gave the required guarantee of producing a waterproof structure. This took the form of a 6mm thick mild steel plate for the underside of the base of the units. The walls and roof were covered by a nominal 2mm thick epoxy-tar membrane. It is believed that this was the first time that this type of material had been used for immersed tube tunnels and extensive laboratory tests were undertaken to determine its properties. In the event the material was developed into a product specifically providing the requirements for waterproofing concrete structures.

Before application of the waterproofing membrane, every crack in the external surface of the structure was mapped and the crack width measured. Generally, the cracking patterns and widths were in accord with the Designer's assumptions.

JOINTS BETWEEN UNITS

18. The initial seal between units was provided by the well proven Gina-gasket rubber seal. The end of each unit is uniquely cambered to provide the correct vertical alignment between adjacent units. Steel frames were set into the end of each unit into which end plates were welded to an accuracy of $\pm$ 5mm. One end plate carries the Gina-gasket which then seals against the other. The Gina-gasket, therefore, cannot be considered as a permanent seal. An Omega seal is provided as the permanent seal. This can be repaired in-situ by way of a normal vulcanised patch or it can be removed and replaced. The tightness of the joint can be assessed by monitoring pressure and water build-up in the annulus between the Gina and Omega seals. If there is indication that the Gina is leaking, there is a facility for grouting the gap before removing the Omega seal. To date all the Gina joints are completely watertight.

CONSTRUCTION JOINTS

19. A 200mm wide plain web section PVC waterbar was placed centrally in every construction joint. The laitance on each joint was completely removed to expose the coarse aggregate by green cutting with a high pressure water jet.

CASTING BASIN

20. The existing Cha Kwo Ling Quarry on the Kowloon Peninsula provided the opportunity of constructing the casting basin on the line of the proposed alignment. It was particularly suited for this purpose as it had already been excavated to -5.0 PD. However the actual floor level required for floating out was -9.0 PD. In order to start construction of the units at the earliest possible time, it was decided not to excavate the quarry floor further but to excavate the channel between the casting basin and the seafront to -9.0 PD and use that as a lock.

CONSTRUCTION

21. During construction of the first batch of units, the channel or lock connecting the basin to the sea was excavated leaving the existing seawall, supported by a bund in position to protect the basin. Because rockhead gave way to softer material as it approach the original shore line, the lock was completed using contiguous circular diaphragm wall caissons. Sills were formed at each end of the lock into which the inner and outer dockgate caissons were seated during the subsequent casting, flooding and towing out operations.

VENTILATION BUILDINGS

22. The ventilation building sub-structure at the south end of the tunnel at Quarry Bay was constructed on existing reclamation within a 54 metre diameter cofferdam formed by a contiguous diaphragm wall, 1.2 metres thick (Figure 3).

23. During construction, the sub-structure served as access to the immersed tube units for fitting out while the landward tunnels both road and rail were being completed.

24. The diaphragm walls are integrated into the permanent design of the structure. The floors were partially constructed in the first instance to form ring beams to stiffen the structure, whilst leaving sufficient space for top downward construction initially and then for access to the tunnels below.

As Unit One was to be introduced directly to this circular structure, a connecting box section was constructed in situ within a cofferdam on the seaward side of the ventilation building.

25. A channel to the sea was formed within diaphragm walls, running parallel to Unit One and which returned back a sufficient distance to permit removal of the sea wall. These diaphragm walls were supported by a stiffening slab between them and further supported by drag walls.

26. The ventilation structure at the north end of the tunnel at Cha Kwo Ling was built directly over Unit 15. Because of programme requirements it was necessary to construct the sub-structure of this building by a floating caisson method. Initial seal between Unit 15 and the sub-structure was achieved using a Gina section. A permanent rigid joint was formed in concrete immediately after sinking the caisson (Figure 4).

SINKING

27. The units were sunk into position using barge pontoons moored to 4 no 250 tonnes concrete anchors. The units were moored and held 10m clear of the previous unit. Longitudinal movement of the unit was controlled using the brindled control winch wires. Transverse and slow movements were controlled using winch wires at right angles at each corner of the unit. Water ballast was added until the deadweight of the unit was available buoyant weight plus 400 tonnes. Ballasting was monitored by load cells at each lifting point.

28. The unit was then moved to within 5m of the previous unit, with positional monitoring being carried out by traditional surveying techniques. The unit was then lowered 5m and checked for gradient. Lowering continued until the shear connector was about 1 metre above its mate on the previous unit and the unit moved to within 2

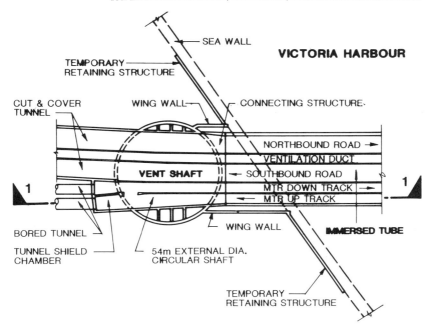

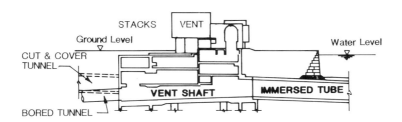

Figure 3   Quarry Bay Ventilation Building

CONSTRUCTION

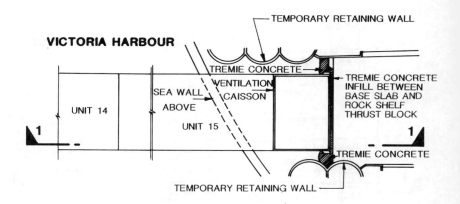

PLAN AT ROOF LEVEL

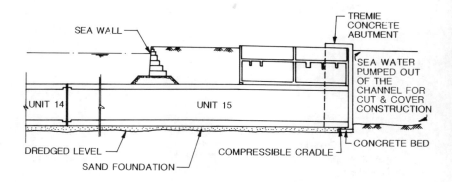

SECTION 1-1

Figure 4   Unit 15 / Cha Kwo Ling Ventilation Building

metres. Divers then entered the water to monitor the final movement. The unit was further lowered until the self aligning brackets on the shear key came into effect and the unit was lowered until half the weight of the inner end of the unit was supported on the inner pontoon and half on the shear key.

29. Pulling jacks were then installed by divers and the unit pulled with a force of 150 tonnes on to the previous unit. At the same time hydraulic hoses were connected to the 2 bag jacks at the outer end of the unit. These were inflated until again the outer pontoon was taking only half the weight of the unit.

30. At this stage personnel entered the previous unit and bled the trapped water from between the bulkheads and full hydrostatic pressure caused full compression of the Gina.

31. Fine tuning of the level of the free end of the unit was then carried out using the bag jacks and the required negative buoyancy added.

SAND PLACING

32. The sand placing method adopted by the Contractor was a variation on the well established sand jetting method. Sand flow methods, involving the casting of pipes into the structure except, at Unit 15, for reasons set out below, were not favoured.

33. Generally sand placing systems tend to be adopted to suit the conditions encountered at the site. In Hong Kong, the Contractor was required to move all floating plant to refuge in typhoon shelters as soon as No. 3 typhoon signal was hoisted. Therefore sand placing operations had to be such that they could be abandoned quickly with minimum demobilising and remobilising time.

34. The method developed involved the pumping of a sand water mixture through nozzles which moved transversly along the underside of the unit as the sand, jetted horizontally, packed progressively into the void between the unit and the trench bottom. Once a strip had been placed successfully across the full width then the system was moved forward along the axis of the tunnel unit. Monitoring of the progress was carried out by echo sounder. Two delivery systems working in parallel were employed, one on each side of the unit. Settlements of the order of 50 - 150 mm were predicted and obtained in practice.

35. Water bags jacks were employed as temporary supports to hold the free end of units and not released until more than 100m of the unit had been supported by sand foundation.

36. Because Unit 15 is located within the lock, it was impossible to use the equipment for the sand jetting method so sand flow methods were employed but with the pipes exiting at side of the unit and not penetrating it.

30. The immersed part of the 'Spoortunnel': a railway Tunnel in Rotterdam

Ir E. H. M. GROOT, Dirk Verstoep bv, and Ir G. M. WOLSINK,
Ministry of Public Works, Department of Locks and Weirs,
The Netherlands

SYNOPSIS. The construction of the railway tunnel under the river Nieuwe Maas in Rotterdam is intended to remove a considerable bottleneck in the railway network in the Netherlands. As this gigantic project is being carried out in the very centre of a lively city, designers and contractor are faced with various new technical problems. In addition to giving a general description of the project, the present paper also discusses in more detail a few of these problem areas.

INTRODUCTION
1. The construction of the railway tunnel in Rotterdam was started at the end of april 1987. In the present railway connection between the cities of Rotterdam and Dordrecht the river Nieuwe Maas is crossed by means of a fixed bridge and a lifting bridge, which places considerable constraints on the railway service schedules. Because of their advanced age, the existing viaduct and bridge structures need to be replaced from a technical point of view as well.
2. Considerations connected with the infrastructure ultimately led to the choice of a tunnel rather than of a bridge connection. Less noise, new possibilities for urban development and the removal of a considerable barrier to the east-west traffic are a few of the advantages for the city of Rotterdam.
3. The engineering of full project is taken care of by three consultants, viz. N.V. Nederlandse Spoorwegen (Netherlands Railways) - infrastructure department, the Rotterdam public works department and "Rijkswaterstaat" (ministry of public works) - department of locks and weirs. The implementation of the project is in the hands of a combination of contractors called "Kombinatie Willemsspoortunnel" (KWT), made up of Ballast Nedam Beton en Waterbouw bv, Van Hattum en Blankevoort bv, Hollandsche Beton - en Waterbouw bv, bv Strukton Bouw and Dirk Verstoep bv.

Immersed tunnel techniques. Thomas Telford, London, 1989.

CONSTRUCTION

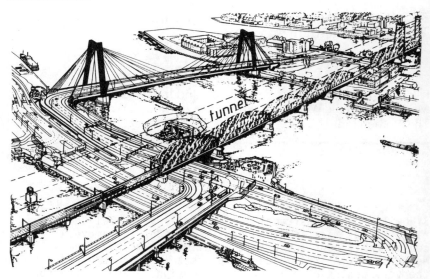

Fig. 1. bird's eye view of the lay out of the immersed part of the railway tunnel under the river Nieuwe Maas in Rotterdam

SIZE OF THE PROJECT
4. The project is very large, both in terms of time and money. The railway tunnel has an overall length of more than 3 km, along which distance conditions vary rather widely.

North bank of the river Nieuwe Maas
5. On the north bank route runs from the central station of the waterfront of the river, the "Maasboulevard". This part will be constructed in situ under difficult conditions. The engineering for this part of the tunnel is done by the public works deparment of the municipality of Rotterdam.

The immersed part
6. The immersed part of the tunnel extends from the "Maasboulevard" to a point 250 m from the water on the south bank. It passes under the river Nieuwe Maas along about 385 m, then under the "Noordereiland" (north island) along 230 m, then passes under the "Koningshaven" (king's harbour) along about 140 m, and finally projects into the south bank along about 250 m. The consulting engineer for the immersed part is the Ministry of public works, department of locks and weirs.
7. The immersed part consists of eight sections, with an overall length of 1015 m.
8. For a view of the route through the city and a longitudinal section of the immersed tunnel, reference is made to figs. 1,2 and 3.

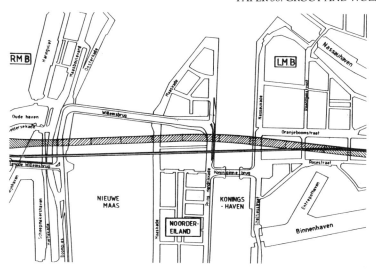

Fig. 2. immersed part of the Spoortunnel Rotterdam

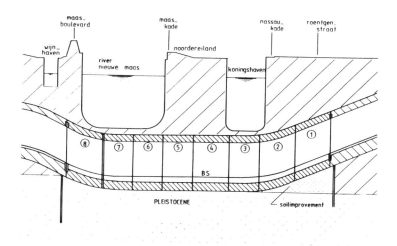

Fig. 3. schematic longitudinal section

South bank of the river Nieuwe Maas

9. The south part of the tunnel will be constructed in situ and has a length of about 850 m. The engineering consultant for this part of the tunnel is the infrastructure department of the Netherlands Railways.

10. The present paper will further be restricted mainly to the construction of the immersed part.

11. Some further characteristics of the project are:
- the total building costs (including engineering etc.) are about Nfl 720 million (about $ 350 million);
- the project is financed by the national government;
- after seven years' construction work two tracks of the tunnel will be put into service;
- after nine years the railway connection will be available with four tracks.

MAIN FEATURES OF THE DESIGN

12. The use of an immersed tunnel to cross two waterways is a fairly obvious choice. In view of the situation prevailing in the Netherlands with regard to soil conditions and the geometry of the waterways, much experience has been gained with the construction of tunnel sections in dry docks and the transport of the tunnel sections to the site by water. In the dry dock, construction can proceed under nearly ideal conditions. As a result, a satisfactory, crack-free concrete can be cast, which can be ensured even more by cooling of the side walls. This is an important point in connection with the required water-tightness of tunnels. In Holland tunnels have, in recent years, no longer been provided with external water-tight coatings. A further considerable advantage of the immersed method is that interference with shipping is reduced to a minimum. Moreover, the method has been found to be attractive financially.

13. Financial reasons also decided the use of this construction method for the sections under the Noordereiland and on a considerable stretch on the south bank: in these places it is not allowed to lower the groundwater level, so that with the considerable depths involved in-situ methods would be considerably more expensive.

14. In general, the limits between which it is technically feasible to build tunnels by the immersed method are dependent on the points to which it is possible temporarily to move river dykes. Thus, on the north bank the immersed limit is determined by the existing dyke, the Maasboulevard, which cannot be moved at all. On the south bank, on the other hand, the tunnel runs entirely outside the dyke, which makes it technically possible to float the tunnel sections along a considerable distance into the south bank and immerse them into their position.

15. The locations of the section joints is governed by soil mechanical and drainage factors. Under the immersed section in the south bank the poor quality of the subsoil will require an extra portion of soil to be removed and to be replaced with material of higher quality.

THE IMMERSION TRENCH

16. In view of the widely varying conditions along the tunnel route, the technical solutions used for the formation of the immersion trench cannot be the same everywhere. With the tunnels built in the Netherlands sofar it was practically always possible to dredge sloping banks on either side of the trench, with inclinations ranging from 1:5 to 1:6. The very limited space between the present railroad bridge and the vehicular bridge does not allow this. Dredging a trench with sloping banks is only possible along a stretch in the river Nieuwe Maas.

17. In the river Nieuwe Maas a bank can be dredged only on the side of the traffic bridge (the "Willemsbrug"). Along certain stretches on the other side it is impossible because of the presence ot the supports of the existing railroad bridge very near the projected trench (see fig. 4).

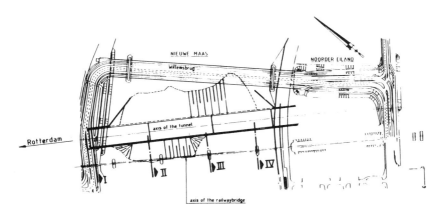

Fig. 4. position of the tunnel in relation to the supports of the existing railway bridge

18. In figure 4 it can be seen that the bridge supports III and IV present the worst problems, because they have been founded not very deeply on wooden piles. At these supports it is therefore impossible to dredge away any soil. Hence, the trench will be formed here by a non-anchored sheetpiling that is driven into the ground underwater. The sheetpiling will consist of tube piles with locks welded to them and separated from one another by two conventional Larssen sheetpile profiles. This type of sheetpiling, which will be further referred to as "combi-wall", will also be used in several other places along the tunnel route.

CONSTRUCTION

19. At bridge support II the situation is more favourable, because this support is placed on a caisson with a fairly deep foundation level. Hence, soil can be dredged away to some extent here, so that a sloping bank can be formed.

20. From a hydraulic point of view a trench with sloping banks is more satisfactory than one with vertical sheetpiling. With sloping banks the water can easily pass under the tunnel section during the immersing operation, as a result of which hydraulic forces exerted on the tunnel section will be smaller. By a happy coincidence, the section that is to be immersed last (no. 7), for which conditions are always the most critical, is envisaged to be placed in the banked part of the immersion trench.

21. At the connection with the tunnel part on the north bank, section no. 8 is envisaged to lie under the existing access to the traffic bridge along a distance of about 30 m. During the construction of the access account was taken of the future construction of the tunnel: the access is locally constructed as a bridging slab supported on two sheetpilings, which will constitute the walls of the future immersion trench (see fig. 5).

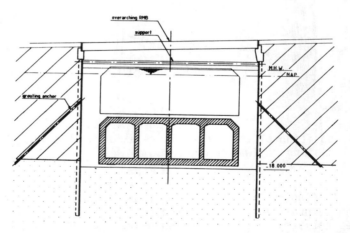

Fig. 5. tunnel section 8 positioned in the trench under the access of the existing traffic bridge

However, the dimensions of these existing walls are such that it is not sufficient just to provide a single frame of struts at the top of the walls. The safety of the walls is only acceptable if a second series of struts is provided at a lower level. By leaving a ground dam it is possible, after dry digging, to strut the walls at the lower level by means of injected cement anchors.

At this stage of the construction it will also be necessary to drain the deep pleistocene sand, to avoid cracking of a clay layer that will then still cover the sand. This drainage will additionally have a favourable effect on the moments exerted in the sheetpiling. After removal of the ground dam and deepening the trench, the tunnel section can be floated into the roofed trench.

THE "NOORDEREILAND" (NORTH ISLAND)

22. In the Noordereiland there is so little space available that the immersion trench has to be formed with sheetpiles. To provide free access for the tunnel sections, the walls will only be supported at the top by means of struts, with their centre line at 2.25 m + NAP (new Amsterdam level). A cross-section of the immersion trench can be seen in fig. 6.

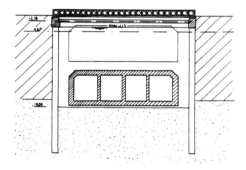

Fig. 6. tunnel section in the strutted immersion trench in the Noordereiland

23. The effective height of the sheetpiles, which here again will be a "combi-wall", is 22 m. The length of the tubes is 32 m, so that they will be driven into the pleistocene sand for about 10 m. The diameter of the tubes is 1420 mm, and their wall thickness 15.7 or 18.7 mm. For a cross-sectional view of the "combi-wall" system, reference is made to fig. 7.

THE "COMBI-WALL"

24. The "combi-wall", which is used at the suggestion of the Kombinatie Willemsspoortunnel, has been found to have a

CONSTRUCTION

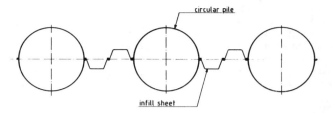

Fig. 7. the "Combi-wall" system in section

number of advantages in connection with the moments exerted. In the first place, the tubes are commercially available in fairly high steel qualities. For the most heavily loaded parts use is made of steel with a guaranteed yield stress of 480 N/mm2. Further, the diameter of the steel tubes far exceeds the maximum dimensions of the conventional sheet-piling systems. In spite of the fact that with circular cross-sections much of the material lies in the neutral area, the considerable length of the tubes still permits an economic use of the material. Also, as their stiffness is homogeneous in every direction, the tubes withstand driving, the tubes with their circular cross-section will, unlike the box-shaped piles, be deformed only little. During pile driving on the Noordereiland, plug formation was indeed observed, as a result of which driving of the intermediate sheet piles had to be preceeded by drilling.

25. Apart from a good drivability of the tubes, consideration should also be given to the possibility of uckling under bending loads if wall thicknesses are small. When wall thicknesses are so small that the buckling mechanism has a strong negative effect on the wall's capacity for taking up strain before collapsing, the safety coefficient required will be higher than usual. With a tube having a diameter of 1420 mm and a wall thickness of 15.7 mm dimensioning is governed by the buckling mechanism, whilst buckling is no longer decisive at a wall thickness of 18.7 mm.

26. In the longitudinal direction of the sheetpiling walls, loads exerted on the wall are not homogeneous. As the steel tubes are much stiffer than the intermediate sheets, pressure curves will be formed in the soil, extending from tube to tube. The two intermediate sheet piles will have to absorb hardly any ground pressure. On the other hand, they will have to absorb the water pressure. Measurements on the Noordereiland have revealed that the water table in the sand follows the varying level of the river water rather well. The water level difference across the "combi-wall" is limited here to half a metre.

THE KONINGSHAVEN
27. At the Koningshaven the minimum distance between the sides of the immersion trench and the foundation of the lifing bridge towers is as little as 18 m (see fig. 8).

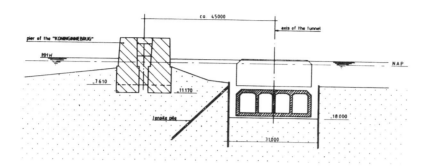

Fig. 8. the tunnel position in relation to the tower foundations of the lifting bridge in the Koningshaven

For a continued proper functioning of the lifting bridge, it is essential that deformations be kept small. Calculations be the finite-element method have indicated that this can only be realized by anchoring the "combi-wall" at river bottom level by a tension beam, which makes the construction so stiff that the lifting bridge towers can be expected not to exceed the maximum deformation limits. The calculations have taken into account any plastic behaviour of the soil. As deformations of constructions in a soil can only be calculated with limited accuracy, the immersion trench will be dredged out layer after layer, while the sheet pile construction and the lifting towers will simultaneously be monitored (with the aid of slope meters).
This will enable us to compare at any stage the deformations measured with those calculated. As everything is to be constructed under water, use is not made of girders, but every tube pile is individually anchored by a separate tension beam rammed into the ground. The tension beam is passed through an opening in the tube, the connection being ultimately formed by a lump of concrete to be cast under water. For ensuring a thorough connection, pins have been welded on to the tension beam.

CONSTRUCTION

THE SOUTH BANK

28. On the south bank the soil in the immersion trench mainly consists of clay and peat. As would be expected, water tension measurements have shown that the water table levels in these poorly permeable layers are not affected by the level of the river water. As a result, the water level difference across the sheetpiling is considerably larger here than on the sandy Noordereiland (about 3.5 m water column). Hence, the intermediate sheet piles are rather heavily loaded.

29. The sheetpilings on either side of the trench are only supported by struts at the top, because the tunnel sections must be able to pass under them. Owing to the low stiffness of the intermediate sheet piles compared with that of the tube piles, loads will be transmitted in transverse direction on either side to the tubes. In this direction, however, the stiffness of the intermediate sheets is several orders of magnitude lower than in the longitudinal direction of the sheet piles. On the other hand, provided that the wall sheet piles remain interlocked, a membrane tension can be formed in the event of very large deformations. The collapsing strength is extremely high then. It should be added that a prerequisite is that the design criterion is not the yield stress of the steel, but the resistance to deformation of the intermediate sheets.

This is of particular importance on the side where close to the immersion trench a temporary auxiliary viaduct on steel piles is envisaged. The maximum allowed deformation here is about 150 mm. The main system, consisting of the large-diameter tube piles, is sufficiently stiff to satisfy the requirements.

30. The question was, however, what would be the additional deformation of the intermediate sheet piles. In this respect factors such as the stresses due to pile driving inaccuracies and the behaviour of the locks (the middle lock between the two intermediate sheets must be pinched) were incertain points. To find the answers to them, Delft Technical University conducted a number of relatively simple experiments. The relationship between a uniform load such as a water load and the resulting deformations of the intermediate sheets was determined by inflating a large synthetic-material bag enclosed in a space formed by 2 x 2 intermediate sheets. The experiments revealed that at a water pressure of 3.5 m water column, the intermediate sheets will be deformed not more than 30 mm; this deformation, together with the deformations of the main system, is considered acceptable.

31. A favourable condition on the south bank is also that it was possible to eliminate an originally critical construction phase. As appears from the depth of the tunnel position relative to the pleistocene sand (fig. 3), a 3.5 m layer of poor-quality soil has to be replaced with dense sand along a rather considerable length. Now, if digging the immersion trench to the level of the sand coincides with an extremely low level of the river water., this is a critical condition for the "combi-wall". However, it is possible to leave a ground dam on the Koningshaven side, so that it is no longer necessary to take into account an extremely low level of the river water, which makes it possible to save on materials.

THE TUNNEL SECTIONS

32. The tunnel sections are built in the available dry dock at Barendrecht on the river Oude Maas near the Heinenoord tunnel. Unlike other docks that are situated nearer by, this particular dock is sufficiently large to accommodate all the sections in one docking. Four of the tunnel sections each have a length of about 115 m, the other four sections each being 138 m long. A typical cross-section through the tunnel is shown in figure 9. The section lengths were chosen on the basis of orienting calculations of the "barrier factors" of the river Nieuwe Maas and the Koningshaven and of flow resistances during transport.
It was found that longer sections would be exposed to too high retaining forces.

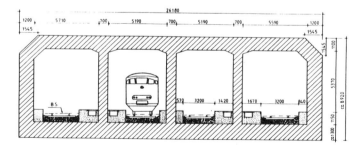

Fig. 9. typical cross-section of the tunnel

CONSTRUCTION

33. The sections are composed of 5 (L = 115 m) or 6 (L = 138 m) casting pieces, each with a length of 23 m. These pieces are cast in the order: first the floor, then the three intermediate walls and finally the two outside walls together with the roof. To prevent the hydration process from causing too high tensile stresses in the outside walls, which could lead to crack formation, these walls are cooled during hardening of the concrete. For the same reason the cement content is restricted to about 30 kg furnace cement per m3 concrete.

34. The pieces are separated from one another by dilatation joints. These joints are toothless joints provided with injectable rubber-metal joint strips. In this way the tunnel is given some flexibility. This is necessary in view of the variations in setting to be expected because of future building activities on or near the tunnel.

35. The sections are so dimensioned that the section joints will lie at the transitions from water to land. Large differences in setting due to considerable variations in the load on the roof can be followed well by the rubber GINA sealing profiles. The construction of section joints and the final joint is shown in figure 10.

36. The dimensions of the concrete are governed by the fact that on the one hand the tunnel must just float during transport and must be kept in its final position on the river bottom with the aid of ballast on the other. As a result, the dimensions in relation to the width spanned are far more compact than usual for traffic tunnels. Accordingly, the amount of reinforcement is not very large.

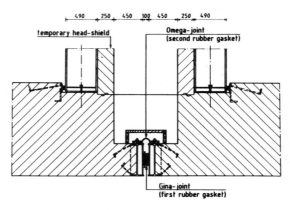

Fig. 10a. the principle of section joints

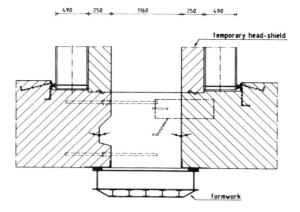

Fig. 10b. the principle of the final joint

37. With the sections lying on land the load on the roof is much heavier than with the underwater sections. Consequently, the roof thickness of the land sections is 200 mm more than the underwater sections (1100 and 900 mm, resp.). With the underwater sections the 200 mm difference is used for a non-structural protective covering against falling and dragging anchors. The land sections can simply be weighted down in the immersed operation by covering them with a few meters of backfill. With the underwater sections this will be effected by the concrete in the anti-derailment provisions, the foot paths and the gravel bed under the rails. The basic principle is that from the moment the temporary bulkheads are removed, the pressure of the sections of the river bottom is always at least 6 kN/m2. In this connection account should be taken of the possible presence of a salt-water bottom layer.

38. The tunnel sections are prestressed in the longitudinal direction, in order to obtain an integrated and homogeneous behaviour of the separate pieces during transport and sinking. When the tunnel has been immersed the cables are cut through at every dilatation joint, so that these joints regain some flexibility. Within the casting pieces between the joints, the concrete remains permanently prestressed owing to the adhesion between concrete and steel. This helps to prevent cracking due to possible setting and temperature variations.

TRANSPORT AND IMMERSING

39. The sections have to travel a distance of about 30 km form the dry dock to the immersing location, and will have to pass a few bottlenecks, which are indicated in fig. 11.

CONSTRUCTION

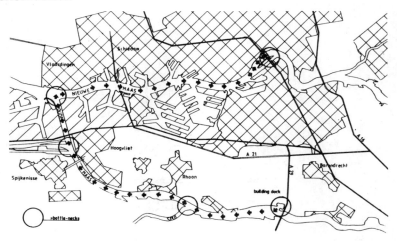

Fig. 11. route followed by the tunnel sections from the dry dock to the site

The entire distance will take about 12 hours per section. After transport, the sections will be parked on the river Nieuwe Maas. From this point tugs will turn the sections during the turning of the tide and then tow them into the immersion trench on the Noordereiland. On the Noordereiland and also on the south bank a warping system will pull the sections under the struts mounted at the top of the trench, after which the sections will be immersed with the aid of construction girders supported on the sheet piles. This immersing procedure is also used in the Koningshaven, where the construction girders will be placed on tubes of the "combi-wall" that project above the water near the bank. In the river Nieuwe Maas, the tunnel sections will be positioned with the aid of tugs and warps, and by means of pontoons.
    40. As immersing of sections between sheet piles is not the usual practice, model experiments on a scale of 1:30 have been made, to assess the strength of the various currents in the several phases of immersing.
    41. Before immersing and the sand flow procedure it is essential that the trench be thoroughly cleaned up, to remove any remaining mud layers. It is known from experience with previous tunnels that enclosed mud layers can lead to very considerable settings.
This is important, in particular, with the land sections, where loads on the roof are high. For the sake of brevity we will not further discuss the sand flow procedure here; the method used is not principally different from those followed with previous tunnels.

42. On the Noordereiland and on the south bank the trench is filled with sand to the ground level. With a view to ensuring stability of the jetted sand layer, the sheet piles must not be pulled out. Instead they are cut off 2.5 m below ground level. When the tunnel has been fitted out, it is ready to receive the railway traffic.

FINAL REMARKS

43. Fitting out of the dry dock was started in the second half of 1987. Work on the immersion trench was begun in the second half of 1988. After the immersing operations, it will be possible to begin with the fitting out of the tunnel sections at the end of 1990, after which the tunnel can be put into service on two tracks at the end of 1993. It is expected that after a construction period of about 9 years, the full project will be finished at the beginning of 1996.

44. In 1996 the Netherlands Railways will have a river crossing which can face the challenges of the future. After long, long years the worst bottleneck in the railway network will at last have been removed.

Discussion

MR N.S. RASMUSSEN, Christiani & Nielsen A/S, Herlev, Denmark
We have already learned - and it is also my personal experience - that the basic design concepts and construction methods were established as early as the mid 1950s and that only relatively small changes, or developments, have taken place since then. There are good reasons for this.

(a) The market for immersed tunnels has remained rather small and immersed tunnelling has become a speciality, for designers as well as for contractors. This has permitted conservatism in design and some mystery-mongering on the construction side.
(b) From the public clients' side, protection of national firms has been widespread and this, coupled with prudence, has not urged development.

In the meantime, we have - during the latest decades in particular - seen a remarkable development of the bored tunnelling technology and thereby an increasing competition in places where a bored tunnel could be a technically acceptable alternative to an immersed tunnel.
Development in design and construction will therefore be necessary in the immersed tunnelling industry - ultimately to the benefit of our societies. In this context, private customers undertaking build-operate and transfer tunnel projects can be expected to inspire the immersed tunnelling industry; interesting examples of this have already been seen in Hong Kong and Australia.
Also, it is recognized that the number of design and contracting firms in the immersed tunnelling club is increasing. In this context, it is interesting to note the increasing British involvement during the last decades.
Immersed tunnelling has been governed by a considerable conservatism. Two main subjects of concern during execution remain

(i) the watertightness of the tunnel in the permanent stage
(ii) the provision of a clean tunnel trench of appropriate depth in the temporary stage - and thereby a

CONSTRUCTION

satisfactory vertical alignment and permanent foundation for the tunnel.

MR M.N. KELLEY, Kiewit Engineering Company, Omaha, Nebraska
In the United States the screeded bed method of foundation preparation seems to be used exclusively with steel tubes, while in Europe the sand-jetted bed seems to be used exclusively with concrete tubes. The question has been raised as to whether this has to do with the characteristics of the tube or with the relative experiences on the two continents. The sand-jetted method of foundation preparation has been used in four tube projects in North America. On the Mobile, Alabama, Tube Project, the sand-jetted bed method was used for a steel shelled tube. In the case of the Fort McHenry Tube and Tunnel Project, the contractor was given the option of bidding either the pump sand foundation method or the screeded bed method. Both bids received on the project were based on using the screeded bed method, and our analysis at bid time showed a saving of a minimum of $5-6 million by using the screeded bed method. In my opinion, either the screeded bed or the sand-jetted method will give equally satisfactory foundations but, as such site conditions as currents, depth of water, number of tubes to be placed and time allowed for tube placing, all enter into the economic analysis, the choice should be left to the contractor.

It has been suggested that with the steel tube method of construction, the first steel fabricated element can be delivered in nine months. In our experience, it usually takes closer to 12 months for the tube fabricator to complete, launch and deliver the first element.

MR F.J. HANSEN, Paper 24
Immersed tunnelling is a clever way of dealing with the presence of water and of constructing a tunnel under water, and the account by Mr Kelley (Session 1) of the strong currents that his firm had dealt with successfully, confirms further that the established techniques can be applied over a wide range of physical conditions. However, strong currents and severe wave action make the various operations more difficult and call for more costly and elaborate precautions, which at some stage probably could make the technique non-competitive.

The placing of a tunnel element by the conventional method on a screeded gravel bed and joining it successfully to the previous element depend on two factors

(a) the accuracy of the screeded mattress
(b) the accuracy of the placing technique, i.e. the position control.

With stronger currents, larger stones are required to keep

DISCUSSION

the bed stable, which in itself makes the screeding less accurate. This, combined with the increased difficulty of holding the screeding apparatus still, will ultimately make the bed unacceptable.

The conventional technique of lowering the tunnel element from the surface becomes less accurate with increased water depth and, at some stage, with the added risk of wave action, possibly so inaccurate that joining to the previous element becomes impracticable.

With positive sinking, as outlined in our Paper 24, water depth and wave action are not the obvious limiting factors. To keep the unit still at the correct position to join it to the previous element is naturally a question of being able to exert the required forces and to apply them rigidly enough. These forces are governed by the current velocities, but they can be ascertained, and applied, by means of quite ordinary hydraulic jacking equipment.

The problems associated with sand placing after the tunnel element is in position are an entirely different proposition from those of placing gravel beforehand. If there is a buoyancy box underneath the element, as indicated in Fig.6 of our Paper, it would be natural to attach some curtain at the ends to divert unwanted currents away from the zone being filled with sand. On the other hand, if siltation is the problem, it would be equally natural to incorporate a suction or jetting system in the buoyancy box, so that accumulated silt or mud could be removed immediately before the sand was to be placed.

Immersed tunnelling has converted conventional tunnelling from a tough fight against water and uncertain soil conditions to a generally well-defined task of prefabrication under workshop conditions, and it is without a shadow of doubt a competitive construction method whether the tunnel is called steel or concrete. It is the approach to the marine works aspects of the technique that will determine how widely it is used.

MR D. KERR, Sir Robert McAlpine and Sons Ltd
Would the Authors comment on tidal conditions during foundation preparation (screeded bed or jetted sand) and placement of units. What do they consider to be the practical limits of tidal current for these operations? Could they give examples of working in high currents?

MR T.J. LLEWELLYN, Taylor Woodrow Construction Ltd
Why was the 6 mm waterproofing steel membrane on the side of the Conwy Tunnel taken to the full height of the walls? Was it because of concern about aggregate quality or chloride ion attack, bearing in mind that steel membranes on most concrete submerged tubes are confined to the base only, plus perhaps

CONSTRUCTION

up to the kicker level of the walls?

By way of comparison, could I ask Mr Bast if full height steel waterproof membrane had been considered for the Storebaelt Eastern Tunnel solution?

MR J. LINNEKAMP, Hollandsche Beton- en Waterbouw, Gouda, The Netherlands

With reference to Paper 29, in the case of the Hong Kong Tunnel the secondary supports were replaced by inflatable bags. In the case of the Sydney Harbour Tunnel the secondary supports returned again. As both tunnels had the same designer and the same contractor, I should like to know why this was.

MR M.W. MORRIS, Acer Consultants (Far East) Ltd

With reference to the question regarding waterproofing of the Conwy Tunnel, it is interesting to compare the waterproofing systems used on the Eastern Harbour Crossing (EHC), Hong Kong, with that used on the Sydney Harbour Tunnel (SHT).

The EHC has a 6 mm steel plate on the bottom. Use of the steel plate on the side walls was considered but we were concerned about the effects of welding distortion. However, we understand from discussion with RTM that they have successfully overcome this problem. In the case of the EHC, it was intended to use a bitumen impregnated fabric on the walls and roof, but the availability of spray-on epoxy polyurethane materials offered the possibility of better adhesion to the substrate and greater speed.

For the SHT, we were unable to convince the authorities that a 6 mm plate offered sufficient long-term corrosion resistance. Consequently, a ribbed PVC membrane, originally developed for lining sewer pipes, has been used across the base of the unit. Clearly, this is a less robust membrane both in the long-term and as a working surface for steel fixing etc., but has been fairly successful so far. An epoxy-polyurethane membrane, similar to that of the EHC, has been used on the walls and roof.

An earlier questioner asked why the use of water bag jacks for support of the free end of the unit on the EHC had not been extended to the SHT. The reason relates mainly to the sequence of sinking and sandplacing of the foundation.

For the EHC, each unit has its foundation placed as soon as it has been sunk, thus releasing the bag jacks for use on the next unit. The bags, while very successful, are also very expensive. For the SHT, restrictions on the use of floating plant mean that it is likely that up to three units will first be sunk and the foundation for all three placed in one operation. In this case, it was considered that the use of conventional hydraulic jacks would be more economic.

DISCUSSION

MR A.R. UMNEY, G. Maunsell & Partners
I wish to comment on two aspects of the MRT immersed tunnel in Hong Kong that have not been discussed so far.

The first relates to connection of bored tunnels driven under compressed air to both ends of the immersed tube. Provisions were made for special backfill around the ends of the tubes to provide protection for the tunnels from the harbour waters. The tunnel centres had to be reduced to a minimum to avoid any unnecessary flawing at the ends of the immersed tubes. Finally, a double bulkhead, referred to by Per Hall in Paper 13, was constructed to allow the bored tunnels to be sealed into the end of the tube before the second temporary bulkhead was removed. I am not aware of any other immersed tube project that has had to accommodate a bored tunnel connection. Perhaps there is someone who can confirm or advise us otherwise.

The second aspect relates to programme and the order of placing the units. The MRT tunnel in Hong Kong was constructed from the south to the north shores. To have waited for the final unit at the north end to be placed before the bored tunnel connection was made would have put that section of the project on the critical path. The end unit was therefore placed earlier in the programme and a special closing unit was used.

The only reference we have had has been on the Spoortunnel Railway Tunnel described in Paper 30. Could Mr Groot explain how he proposes to complete the final joint on that project?

MR J.M. TURNER, Hong Kong Government Highways Department
The Eastern Harbour Crossing (EHC) project was an important link in the Hong Kong transport infrastructure programme. The EHC is a privatized project and is being constructed and operated under a 30 year franchise agreement. The EHC is probably the largest immersed tube tunnel project under construction at the moment, and in terms of volume is arguably the world's largest immersed tube tunnel.

With regard to privatized tunnel projects, Paper 15 should be regarded as essential reading for those involved or likely to become involved in such projects. In particular, I would highlight the complexity of the organizational structure within a privatized project, and the large number of legal, contractual and statutory documents involved. Fig.2 of Paper 27 clearly illustrates this factor.

In considering the view of the Hong Kong Government towards privatization, I would recall a comment made by Hong Kong's previous Financial Secretary, Sir John Bembridge, who stated that, in his view, Government should not be involved in activities that could be done better in the private sector – but there could be no privatization without the prospects of reasonable profits.

While the boundary between public expenditure and privatization was often arguable, the benefits to Government or public authorities of a project well suited for privatization clearly favoured the privatization process for the following reasons: it allows a more flexible approach and hence, by inference, a better product to the public; it saves the public purse; it creates competition; it promotes efficiency; it makes use of the expertise and resources of the private sector.

Although privatization of infrastructure projects does have disadvantages, the advantages under the right circumstances clearly far outweigh any disadvantages. The award of a franchise does, however, create a monopoly situation, making it more difficult for the authorities to protect their and the public interests. It is therefore important that the rules regulating Government control and quality assurance are clearly defined within a carefully documented legal and statutory framework.

MR PHILLIPS, Paper 27
The 6 mm steel membrane did suffer some weld distortion; however, selection of the most appropriate form of steel plate and welding techniques limited the extent of the problems.

The external wall steel membrane was held firmly in position by the specially designed formwork to facilitate the panel welds connection carried out in situ. In this respect, careful planning and formwork detailing proved to be highly effective, as discussed in our Paper.

In certain areas, the steel membrane was also detailed to envelope the top surface of the roof slab. Concreting of the roof slab in these areas was successfully carried out by incorporation of poor windows in the steel membrane.

MR LUNNISS, Paper 22
In reply to Mr Llewellyn, the extent of the steel waterproofing membrane at Conwy was not related to concern over aggregate quality or chloride ion attack. The membrane is an additional waterproofing measure which, combined with measures to reduce cracking and to reduce the permeability of the concrete, gives the best long-term assurance against leakage.

The difficulties experienced in previous tunnels with bituminous membranes on the walls led to the conclusion that the steel membrane would be a 'better buy'. Its greater capital cost is to some extent offset by savings in the formwork, and it provides a more robust and durable membrane to the walls.

MR BAST, Paper 28
In their discussion, Messrs Rasmussen and Hansen raise the issue of the competitiveness of immersed tunnels relative to

bored tunnels. I agree that innovations in immersed tunnelling have lagged behind those in the bored tunnelling sector, and that the continued competitiveness of the immersed technology is largely dependent on the marine works, including placing and dredging schemes. Various proprietary approaches which were proposed as part of the tender process for the English Channel crossing and the Storebaelt crossing are sure to surface in upcoming immersed projects. Perhaps the upcoming projects announced by Mr Turner in Hong Kong will be the platform for advancing the state of the art.

Mr Kerr asks about the practical limit of tidal current and examples of working in tidal currents. At least two issues arise when discussing tidal currents: the first pertains to the erosion of the foundation materials; the second pertains to the ability accurately to place the tunnel element.

In the case of foundation erosion, the solution depends on the nature of the foundation. If a sand-jetted or sand-flow foundation is used, skirts mounted on the bottom sides of the tunnel elements have been suggested to prevent currents beneath the elements from eroding the foundation. In the case of a screeded foundation, the size of the gravel foundation can be selected to accommodate large currents. In any event, the actual placement is best accomplished during periods of slack water (< 0.5 knots).

The worst case of high current of which I am aware was experienced during construction of the 63rd Street Tunnel in New York, where maximum currents are 2.6 m/s. Slack water was present for less than 30 minutes and the contractor experienced difficulty with erosion of the foundation materials.

In response to Mr Llewellyn's question, a 6 mm thick steel waterproof membrane surrounded the Storebaelt concrete alternate tunnel section to achieve a watertight condition in water depths of up to 60 m.

Mr Umney states that he was not aware of other immersed tunnels which required a bored tunnel connection. The 63rd Street Immersed Tunnel had a mined approach on both sides of the East River. Similarly, the BART tunnel ventilation building in San Francisco was connected to a bored approach on one side and an immersed tunnel on the other.

MR MORTON, Paper 29
In reply to Mr Linnekamp, although there are specific programme reasons given for the use of secondary support at Sydney, a large part of the different approach between Hong Kong and Sydney lies at the root of the design and construction of immersed tunnels and that is the confidence that the constructor has in a particular technique.

The above has been alluded to by Mr Kelley. In this case, one project manager felt happier than the other with the new technique of bag jacks. Designers of immersed tube have to

CONSTRUCTION

recognize that, ultimately, the constructor must choose and be confident in the method of sinking, bedding and jointing, and whatever the designer thinks or feels, the design must reflect the chosen method of construction.